Phonons in nanostructures

This book focuses on the theory of phonon interactions in nanoscale structures with particular emphasis on modern electronic and optoelectronic devices.

The continuing progress in the fabrication of semiconductor nanostructures with lower dimensional features has led to devices with enhanced functionality and even to novel devices with new operating principles. The critical role of phonon effects in such semiconductor devices is well known. There is therefore a pressing need for a greater awareness and understanding of confined phonon effects. A key goal of this book is to describe tractable models of confined phonons and how these are applied to calculations of basic properties and phenomena of semiconductor heterostructures.

The level of presentation is appropriate for undergraduate and graduate students in physics and engineering with some background in quantum mechanics and solid state physics or devices. A basic understanding of electromagnetism and classical acoustics is assumed.

DR MICHAEL A. STROSCIO earned a Ph.D. in physics from Yale University and held research positions at the Los Alamos Scientific Laboratory and the Johns Hopkins University Applied Physics Laboratory, before moving into the management of federal research and development at a variety of US government agencies. Dr Stroscio has served as a policy analyst for the White House Office of Science and Technology Policy and as Vice Chairman of the White House Panel on Scientific Communication. He has taught physics and electrical engineering at several universities including Duke University, the North Carolina State University and the University of California at Los Angeles. Dr Stroscio is currently the Senior Scientist in the Office of the Director at the US Army Research Office (ARO) as well as an Adjunct Professor at both Duke University and the North Carolina State University. He has authored about 500 publications, presentations and patents covering a wide variety of topics in the physical sciences and electronics. He is the author of *Quantum Heterostructures: Microelectronics and Optoelectronics* and the joint editor of two World Scientific books entitled *Quantum-based Electronic Devices and Systems* and *Advances in Semiconductor Lasers and Applications to Optoelectronics*. He is a Fellow of both the Institute of Electrical and Electronics Engineers (IEEE) and the American Association for the Advancement of Science and he was the 1998 recipient of the IEEE Harry Diamond Award.

DR DUTTA earned a Ph.D. in physics from the University of Cincinnati; she was a research associate at Purdue University and at City College, New York, as well as a visiting scientist at Brookhaven National Laboratory before assuming a variety of government posts in research and development. Dr Dutta was the Director of the Physics Division at the US Army's Electronics Technology and Devices Laboratory as well as at the Army Research Laboratory prior to her appointment as the Associate Director for Electronics in the Army Research Office's Engineering Sciences Directorate. Dr Dutta recently assumed a senior executive position as ARO's Director of Research and Technology Integration. She has over 160 publications, 170 conference presentations, 10 book chapters, and has had 24 US patents issued. She is the joint editor of two World Scientific books entitled *Quantum-Based Electronic Devices and Systems* and *Advances in Semiconductor Lasers and Applications to Optoelectronics*. She is an Adjunct Professor of the Electrical and Computer Engineering and Physics departments of North Carolina State University and has had adjunct appointments at the Electrical Engineering departments of Rutgers University and the University of Maryland. Dr Dutta is a Fellow of both the Institute of Electrical and Electronics Engineers (IEEE) and the Optical Society of America, and she was the recipient in the year 2000 of the IEEE Harry Diamond Award.

Phonons in Nanostructures

Michael A. Stroscio and Mitra Dutta
US Army Research Office, US Army Research Laboratory

CAMBRIDGE
UNIVERSITY PRESS

CAMBRIDGE UNIVERSITY PRESS
Cambridge, New York, Melbourne, Madrid, Cape Town, Singapore, São Paulo

Cambridge University Press
The Edinburgh Building, Cambridge CB2 2RU, UK

Published in the United States of America by Cambridge University Press, New York

www.cambridge.org
Information on this title: www.cambridge.org/9780521792790

First published 2001
This digitally printed first paperback version 2005

A catalogue record for this publication is available from the British Library

Library of Congress Cataloguing in Publication data

Stroscio, Michael A., 1949–
    Phonons in nanostructures / Michael A. Stroscio and Mitra Dutta.
        p.   cm.
    Includes bibliographic references and index.
    ISBN 0 521 79279 7
        1. Nanostructures.   2. Phonons.   I. Dutta, Mitra.   II. Title.

QC176.8.N35 S77 2001
530.4′16–dc21   00-54669

ISBN-13 978-0-521-79279-0 hardback
ISBN-10 0-521-79279-7 hardback

ISBN-13 978-0-521-01805-0 paperback
ISBN-10 0-521-01805-6 paperback

Mitra Dutta dedicates this book to her parents
Dhiren N. and Aruna Dutta

and

Michael Stroscio dedicates this book to
his friend and mentor Morris Moskow and
his friend and colleague Ki Wook Kim

Contents

Preface xi

Chapter 1 **Phonons in nanostructures** 1

1.1 Phonon effects: fundamental limits on carrier mobilities and dynamical
      processes 1
1.2 Tailoring phonon interactions in devices with nanostructure components 3

Chapter 2 **Phonons in bulk cubic crystals** 6

2.1 Cubic structure 6
2.2 Ionic bonding – polar semiconductors 6
2.3 Linear-chain model and macroscopic models 7
2.3.1 *Dispersion relations for high-frequency and low-frequency modes* 8
2.3.2 *Displacement patterns for phonons* 10
2.3.3 *Polaritons* 11
2.3.4 *Macroscopic theory of polar modes in cubic crystals* 14

Chapter 3 **Phonons in bulk würtzite crystals** 16

3.1 Basic properties of phonons in würtzite structure 16
3.2 Loudon model of uniaxial crystals 18
3.3 Application of Loudon model to III-V nitrides 23

Chapter 4 **Raman properties of bulk phonons** 26

4.1 Measurements of dispersion relations for bulk samples 26
4.2 Raman scattering for bulk zincblende and würtzite structures 26

4.2.1   *Zincblende structures* 28
4.2.2   *Würtzite structures* 29
4.3   Lifetimes in zincblende and würtzite crystals 30
4.4   Ternary alloys 32
4.5   Coupled plasmon–phonon modes 33

Chapter 5 **Occupation number representation** 35

5.1   Phonon mode amplitudes and occupation numbers 35
5.2   Polar-optical phonons: Fröhlich interaction 40
5.3   Acoustic phonons and deformation-potential interaction 43
5.4   Piezoelectric interaction 43

Chapter 6 **Anharmonic coupling of phonons** 45

6.1   Non-parabolic terms in the crystal potential for ionically bonded atoms 45
6.2   Klemens' channel for the decay process LO → LA(1) + LA(2) 46
6.3   LO phonon lifetime in bulk cubic materials 47
6.4   Phonon lifetime effects in carrier relaxation 48
6.5   Anharmonic effects in würtzite structures: the Ridley channel 50

Chapter 7 **Continuum models for phonons** 52

7.1   Dielectric continuum model of phonons 52
7.2   Elastic continuum model of phonons 56
7.3   Optical modes in dimensionally confined structures 60
7.3.1   *Dielectric continuum model for slab modes: normalization of interface modes* 61
7.3.2   *Electron–phonon interaction for slab modes* 66
7.3.3   *Slab modes in confined würtzite structures* 71
7.3.4   *Transfer matrix model for multi-heterointerface structures* 79
7.4   Comparison of continuum and microscopic models for phonons 90
7.5   Comparison of dielectric continuum model predictions with Raman measurements 93
7.6   Continuum model for acoustic modes in dimensionally confined structures 97
7.6.1   *Acoustic phonons in a free-standing and unconstrained layer* 97
7.6.2   *Acoustic phonons in double-interface heterostructures* 100
7.6.3   *Acoustic phonons in rectangular quantum wires* 105
7.6.4   *Acoustic phonons in cylindrical structures* 111
7.6.5   *Acoustic phonons in quantum dots* 124

Chapter 8   **Carrier–LO-phonon scattering** 131

8.1   Fröhlich potential for LO phonons in bulk zincblende and würtzite
            structures 131
8.1.1   *Scattering rates in bulk zincblende semiconductors* 131
8.1.2   *Scattering rates in bulk würtzite semiconductors* 136
8.2   Fröhlich potential in quantum wells 140
8.2.1   *Scattering rates in zincblende quantum-well structures* 141
8.2.2   *Scattering rates in würtzite quantum wells* 146
8.3   Scattering of carriers by LO phonons in quantum wires 146
8.3.1   *Scattering rate for bulk LO phonon modes in quantum wires* 146
8.3.2   *Scattering rate for confined LO phonon modes in quantum wires* 150
8.3.3   *Scattering rate for interface-LO phonon modes* 154
8.3.4   *Collective effects and non-equilibrium phonons in polar quantum wires* 162
8.3.5   *Reduction of interface–phonon scattering rates in metal–semiconductor
            structures* 165
8.4   Scattering of carriers and LO phonons in quantum dots 167

Chapter 9   **Carrier–acoustic-phonon scattering** 172

9.1   Carrier–acoustic-phonon scattering in bulk zincblende structures 172
9.1.1   *Deformation-potential scattering in bulk zincblende structures* 172
9.1.2   *Piezoelectric scattering in bulk semiconductor structures* 173
9.2   Carrier–acoustic-phonon scattering in two-dimensional structures 174
9.3   Carrier–acoustic-phonon scattering in quantum wires 175
9.3.1   *Cylindrical wires* 175
9.3.2   *Rectangular wires* 181

Chapter 10 **Recent developments** 186

10.1   Phonon effects in intersubband lasers 186
10.2   Effect of confined phonons on gain of intersubband lasers 195
10.3   Phonon contribution to valley current in double-barrier structures 202
10.4   Phonon-enhanced population inversion in asymmetric double-barrier quantum-well
            lasers 205
10.5   Confined-phonon effects in thin film superconductors 208
10.6   Generation of acoustic phonons in quantum-well structures 212

Chapter 11 **Concluding considerations** 218

11.1   Pervasive role of phonons in modern solid-state devices 218
11.2   Future trends: phonon effects in nanostructures and phonon engineering 219

Appendices 221

Appendix A: Huang–Born theory   221

Appendix B: Wendler's theory   222

Appendix C: Optical phonon modes in double-heterointerface structures   225

Appendix D: Optical phonon modes in single- and double-heterointerface würtzite
   structures   236

Appendix E: Fermi golden rule   250

Appendix F: Screening effects in a two-dimensional electron gas   252

References   257

Index   271

Preface

This book describes a major aspect of the effort to understand nanostructures, namely the study of phonons and phonon-mediated effects in structures with nanoscale dimensional confinement in one or more spatial dimensions. The necessity for and the timing of this book stem from the enormous advances made in the field of nanoscience during the last few decades.

Indeed, nanoscience continues to advance at a dramatic pace and is making revolutionary contributions in diverse fields, including electronics, optoelectronics, quantum electronics, materials science, chemistry, and biology. The technologies needed to fabricate nanoscale structures and devices are advancing rapidly. These technologies have made possible the design and study of a vast array of novel devices, structures and systems confined dimensionally on the scale of 10 nanometers or less in one or more dimensions. Moreover, nanotechnology is continuing to mature rapidly and will, no doubt, lead to further revolutionary breakthroughs like those exemplified by quantum-dot semiconductor lasers operating at room temperature, intersubband multiple quantum-well semiconductor lasers, quantum-wire semiconductor lasers, double-barrier quantum-well diodes operating in the terahertz frequency range, single-electron transistors, single-electron metal-oxide–semiconductor memories operating at room temperature, transistors based on carbon nanotubes, and semiconductor nanocrystals used for fluorescent biological labels, just to name a few!

The seminal works of Esaki and Tsu (1970) and others on the semiconductor superlattice stimulated a vast international research effort to understand the fabrication and electronic properties of superlattices, quantum wells, quantum wires, and quantum dots. This early work led to truly revolutionary advances in nanofabrication technology and made it possible to realize band-engineering and atomic-level structural tailoring not envisioned previously except through the molecular and atomic systems found in nature. Furthermore, the continuing reduction of dimensional features in electronic and optoelectronic devices coupled with revolutionary advances in semiconductor growth and processing technologies have opened many avenues for increasing the performance levels and functionalities of electronic and optoelectronic devices. Likewise, the discovery of the buckyball by Kroto *et al.* (1985) and the carbon nanotube by Iijima (1991) led to an intense worldwide program to understand the properties of these nanostructures.

During the last decade there has been a steady effort to understand the optical and acoustic phonons in nanostructures such as the semiconductor superlattice, quantum wires, and carbon nanotubes. The central theme of this book is the description of the optical and acoustic phonons in these nanostructures. It deals with the properties of phonons in isotropic, cubic, and hexagonal crystal structures and places particular emphasis on the two dominant structures underlying modern semiconductor electronics and optoelectronics – zincblende and würtzite. In view of the successes of continuum models in describing optical phonons (Fasol *et al.*, 1988) and acoustic phonons (Seyler and Wybourne, 1992) in dimensionally confined structures, the principal theoretical descriptions presented in this book are based on the so-called dielectric continuum model of optical phonons and the elastic continuum model of acoustic phonons. Many of the derivations are given for the case of optical phonons in würtzite crystals, since the less complicated case for zincblende crystals may then be recovered by taking the dielectric constants along the c-axis and perpendicular to the c-axis to be equal.

As a preliminary to describing the dispersion relations and mode structures for optical and acoustic phonons in nanostructures, phonon amplitudes are quantized in terms of the harmonic oscillator approximation, and anharmonic effects leading to phonon decay are described in terms of the dominant phonon decay channels. These dielectric and elastic continuum models are applied to describe the deformation-potential, Fröhlich, and piezoelectric interactions in a variety of nanostructures including quantum wells, quantum wires and quantum dots. Finally, this book describes how the dimensional confinement of phonons in nanostructures leads to modifications in the electronic, optical, acoustic, and superconducting properties of selected devices and structures including intersubband quantum-well semiconductor lasers, double-barrier quantum-well diodes, thin-film superconductors, and the thin-walled cylindrical structures found in biological structures known as microtubulin.

The authors wish to acknowledge professional colleagues, friends and family members without whose contributions and sacrifices this work would not have been undertaken or completed. The authors are indebted to Dr C.I. (Jim) Chang, who is both the Director of the US Army Research Office (ARO) and the Deputy Director of the US Army Research Laboratory for Basic Science, and to Dr Robert W. Whalin and Dr John Lyons, the current director and most recent past director of the US Army

Research Laboratory; these leaders have placed a high priority on maintaining an environment at the US Army Research Office such that it is possible for scientists at ARO to continue to participate personally in forefront research as a way of maintaining a broad and current knowledge of selected fields of modern science.

Michael Stroscio acknowledges the important roles that several professional colleagues and friends played in the events leading to his contributions to this book. These people include: Professor S. Das Sarma of the University of Maryland; Professor M. Shur of the Rensselaer Polytechnic Institute; Professor Gerald J. Iafrate of Notre Dame University; Professors M.A. Littlejohn, K.W. Kim, R.M. Kolbas, and N. Masnari of the North Carolina State University (NCSU); Dr Larry Cooper of the Office of Naval Research; Professor Vladimir Mitin of the Wayne State University; Professors H. Craig Casey Jr, and Steven Teitsworth of Duke University; Professor S. Bandyopadhyay of the University of Nebraska; Professors G. Belenky, Vera B. Gorfinkel, M. Kisin, and S. Luryi of the State University of New York at Stony Brook; Professors George I. Haddad, Pallab K. Bhattacharya, and Jasprit Singh and Dr J.-P. Sun of the University of Michigan; Professors Karl Hess and J.-P. Leburton at the University of Illinois; Professor L.F. Register of the University of Texas at Austin; Professor Viatcheslav A. Kochelap of the National Academy of Sciences of the Ukraine; Dr Larry Cooper of the Office of Naval Research; and Professor Paul Klemens of the University of Connecticut. Former graduate students, postdoctoral researchers, and visitors to the North Carolina State University who contributed substantially to the understanding of phonons in nanostructures as reported in this book include Drs Amit Bhatt, Ulvi Erdogan, Daniel Kahn, Sergei M. Komirenko, Byong Chan Lee, Yuri M. Sirenko, and SeGi Yu. The fruitful collaboration of Dr Rosa de la Cruz of the Universidad Carlos III de Madrid during her tenure as a visiting professor at Duke University is acknowledged gratefully. The authors also acknowledge gratefully the professionalism and dedication of Mrs Jayne Aldhouse and Drs Simon Capelin and Eoin O'Sullivan, of Cambridge University Press, and Dr Susan Parkinson.

Michael Stroscio thanks family members who have been attentive during the periods when his contributions to the book were being written. These include: Anthony and Norma Stroscio, Mitra Dutta, as well as Gautam, Marshall, and Elizabeth Stroscio. Moreover, eight-year-old Gautam Stroscio is acknowledged gratefully for his extensive assistance in searching for journal articles at the North Carolina State University.

Mitra Dutta acknowledges the interactions, discussions and work of many colleagues and friends who have had an impact on the work leading to this book. These colleagues include Drs Doran Smith, K.K. Choi, and Paul Shen of the Army Research Laboratory, Professor Athos Petrou of the State University of New York at Buffalo, and Professors K.W. Kim, M.A. Littlejohn, R.J. Nemanich, Dr Leah Bergman and Dimitri Alexson of the North Carolina State University, as well as Professors Herman Cummins, City College, New York, A.K. Ramdas, Purdue

University and Howard Jackson, University of Cincinnati, her mentors in various facets of phonon physics. Mitra Dutta would also like to thank Dhiren Dutta, without whose encouragement she would never have embarked on a career in science, as well as Michael and Gautam Stroscio who everyday add meaning to everything.

Michael Stroscio and Mitra Dutta

Chapter 1

Phonons in nanostructures

There are no such things as applied sciences, only applications of sciences.

Louis Pasteur, 1872

1.1 Phonon effects: fundamental limits on carrier mobilities and dynamical processes

The importance of phonons and their interactions in bulk materials is well known to those working in the fields of solid-state physics, solid-state electronics, optoelectronics, heat transport, quantum electronics, and superconductivity.

As an example, carrier mobilities and dynamical processes in polar semiconductors, such as gallium arsenide, are in many cases determined by the interaction of longitudinal optical (LO) phonons with charge carriers. Consider carrier transport in gallium arsenide. For gallium arsenide crystals with low densities of impurities and defects, steady state electron velocities in the presence of an external electric field are determined predominantly by the rate at which the electrons emit LO phonons. More specifically, an electron in such a polar semiconductor will accelerate in response to the external electric field until the electron's energy is large enough for the electron to emit an LO phonon. When the electron's energy reaches the threshold for LO phonon emission – 36 meV in the case of gallium arsenide – there is a significant probability that it will emit an LO phonon as a result of its interaction with LO phonons. Of course, the electron will continue to gain energy from the electric field.

In the steady state, the processes of electron energy loss by LO phonon emission and electron energy gain from the electric field will come into balance and the electron will propagate through the semiconductor with a velocity known as the saturation velocity. As is well known, experimental values for this saturated drift velocity generally fall in the range 10^7 cm s^{-1} to 10^8 cm s^{-1}. For gallium arsenide this velocity is about 2×10^7 cm s^{-1} and for indium antimonide 6×10^7 cm s^{-1}.

For both these polar semiconductors, the process of LO phonon emission plays a major role in determining the value of the saturation velocity. In non-polar materials such as Si, which has a saturation velocity of about 10^7 cm s^{-1}, the deformation-potential interaction results in electron energy loss through the emission of phonons. (In Chapter 5 both the interaction between polar-optical-phonons and electrons – known as the Fröhlich interaction – and the deformation-potential interaction will be defined mathematically.)

Clearly, in all these cases, the electron mobility will be influenced strongly by the interaction of the electrons with phonons. The saturation velocity of the carriers in a semiconductor provides a measure of how fast a microelectronic device fabricated from this semiconductor will operate. Indeed, the minimum time for the carriers to travel through the active region of the device is given approximately by the length of the device – that is, the length of the so-called gate – divided by the saturation velocity. Evidently, the practical switching time of such a microelectronic device will be limited by the saturation velocity and it is clear, therefore, that phonons play a major role in the fundamental and practical limits of such microelectronic devices. For modern integrated circuits, a factor of two reduction in the gate length can be achieved in many cases only through building a new fabrication facility. In some cases, such a building project might cost a billion dollars or more. The importance of phonons in microelectronics is clear!

A second example of the importance of carrier–phonon interactions in modern semiconductor devices is given by the dynamics of carrier capture in the active quantum-well region of a polar semiconductor quantum-well laser. Consider the case where a current of electrons is injected over a barrier into the quantum-well region of such a laser. For the laser to operate, an electron must lose enough energy to be 'captured' by the quasi-bound state which it must occupy to participate in the lasing process. For many quantum-well semiconductor lasers this means that the electron must lose an energy of the order of a 100 meV or more. The energy loss rate of a carrier – also known as the thermalization rate of the carrier – in a polar-semiconductor quantum well is determined by both the rate at which the carrier's energy is lost by optical-phonon emission and the rate at which the carrier gains energy from optical-phonon absorption. This latter rate can be significant in quantum wells since the phonons emitted by energetic carriers can accumulate in these structures. Since the phonon densities in many dimensionally confined semiconductor devices are typically well above those of the equilibrium phonon population, there is an appreciable probability that these non-equilibrium – or 'hot' – phonons will be reabsorbed. Clearly, the net loss of energy by an electron in such a situation depends on the rates for both phonon absorption and phonon emission. Moreover, the lifetimes of the optical phonons are also important in determining the total energy loss rate for such carriers. Indeed, as will be discussed in Chapter 6, the longitudinal optical (LO) phonons in GaAs and many other polar materials decay into acoustic phonons through the Klemens' channel. Furthermore, over a wide range of temperatures and phonon wavevectors, the lifetimes of longitudinal optical phonons in GaAs vary from a few picoseconds to about 10 ps (Bhatt *et al.*, 1994). (Typical lifetimes for other polar semiconductors are also of this magnitude.) As a result of the Klemens' channel, the 'hot' phonons decay into acoustic phonons in times of the order of 10 ps. The LO phonons undergoing decay into acoustic phonons are not available for absorption by the electrons and as a result of the Klemens' channel the electron thermalization is more rapid than it would be otherwise; this phenomenon is referred to as the 'hot-phonon-bottleneck effect'.

The electron thermalization time is an important parameter for semiconductor quantum-well lasers because it determines the minimum time needed to switch the laser from an 'on' state to an 'off' state; this occurs as a result of modulating the electron current that leads to lasing. Since the hot-phonon population frequently decays on a time scale roughly given by the LO phonon decay rate (Das Sarma *et al.*, 1992), a rough estimate of the electron thermalization time – and therefore the minimum time needed to switch the laser from an 'on' state to an 'off' state – is of the order of about 10 ps. In fact, typical modulation frequencies for gallium arsenide quantum-well lasers are about 30 GHz. The modulation of the laser at significantly higher frequencies will be limited by the carrier thermalization time and ultimately by the lifetime of the LO phonon. The importance of the phonon in modern optoelectronics is clear.

The importance of phonons in superconductors is well known. Indeed, the Bardeen–Cooper–Schrieffer (BCS) theory of superconductivity is based on the formation of bosons from pairs of electrons – known as Cooper pairs – bound through the mediating interaction produced by phonons. Many of the theories describing the so-called high-critical-temperature superconductors are not based on phonon-mediated Cooper pairs, but the importance of phonons in many supercon-ductors is of little doubt. Likewise, it is generally recognized that acoustic phonon interactions determine the thermal properties of materials.

These examples illustrate the pervasive role of phonons in bulk materials. Nanotechnology is providing an ever increasing number of devices and structures having one, or more than one, dimension less than or equal to about 100 ångstroms. The question naturally arises as to the effect of dimensional confinement on the properties on the phonons in such nanostructures as well as the properties of the phonon interactions in nanostructures. The central theme of this book is the descrip-tion of the optical and acoustic phonons, and their interactions, in nanostructures.

1.2   Tailoring phonon interactions in devices with nanostructure components

Phonon interactions are altered unavoidably by the effects of dimensional confine-ment on the phonon modes in nanostructures. These effects exhibit some similarities to those for an electron confined in a quantum well. Consider the well-known wavefunction of an electron in a infinitely deep quantum well, of width L_z in the z-direction. The energy eigenstates $\Psi_n(z)$ may be taken as plane-wave states in the directions parallel to the heterointerfaces and as bound states in an infinitely deep quantum well in the z-direction:

$$\Psi_n(z) = \frac{e^{i\mathbf{k}_\parallel \cdot \mathbf{r}_\parallel}}{\sqrt{A}} \sqrt{\frac{2}{L_z}} \sin k_z z, \tag{1.1}$$

where $\mathbf{r}_\parallel$ and $\mathbf{k}_\parallel$ are the position vector and wavevector components in a plane parallel to the interfaces, $k_z = n\pi/L_z$, and $n = 1, 2, 3, \ldots$ labels the energy eigenstates, whose energies are

$$E_n(\mathbf{k}_\parallel) = \frac{\hbar^2 (\mathbf{k}_\parallel)^2}{2m} + \frac{\hbar^2 \pi^2 n^2}{2m L_z^2}. \tag{1.2}$$

A is the area of the heterointerface over which the electron wavefunction is normalized. Clearly, a major effect of dimensional confinement in the z-direction is that the z-component of the bulk continuum wavevector is restricted to integral multiples of π/L_z. Stated in another way, the phase space is restricted.

As will be explained in detail in Chapter 7, the dimensional confinement of phonons results in similar restrictions in the phase space of the phonon wavevector q. Indeed, we shall show that the wavevectors of the optical phonons in a dielectric layer of thickness L_z are given by $q_z = n\pi/L_z$ (Fuchs and Kliewer, 1965) in analogy to the case of an electron in an infinitely deep quantum well. In fact, Fasol $et\ al.$ (1988) used Raman scattering techniques to show that the wavevectors $q_z = n\pi/L_z$ of optical phonons confined in a ten-monolayer-thick AlAs/GaAs/AlAs quantum well are so sensitive to changes in L_z that a one-monolayer change in the thickness of the quantum well is readily detectable as a change in q_z! These early experimental studies of Fasol $et\ al.$ (1988) demonstrated not only that phonons are confined in nanostructures but also that the measured phonon wavevectors are well described by relatively simple continuum models of phonon confinement.

Since dimensional confinement of phonons restricts the phase space of the phonons, it is certain that carrier–phonon interactions in nanostructures will be modified by phonon confinement. As we shall see in Chapter 7, the so-called dielectric and elastic continuum models of phonons in nanostructures may be applied to describe the deformation-potential, Fröhlich, and piezoelectric interactions in a variety of nanostructures including quantum wells, quantum wires, and quantum dots. These interactions play a dominant role in determining the electronic, optical and acoustic properties of materials (Mitin $et\ al.$, 1999; Dutta and Stroscio, 1998b; Dutta and Stroscio, 2000); it is clearly desirable for models of the properties of nanostructures to be based on an understanding of how the above-mentioned interactions change as a result of dimensional confinement. To this end, Chapters

8, 9 and 10 of this book describe how the dimensional confinement of phonons in nanostructures leads to modifications in the electronic, optical, acoustic, and superconducting properties of selected devices and structures, including intersubband quantum-well semiconductor lasers, double-barrier quantum-well diodes, thin-film superconductors, and the thin-walled cylindrical structures found in the biological structures known as microtubulin. Chapters 8, 9, and 10 also provide analyses of the role of collective effects and non-equilibrium phonons in determining hot-carrier energy loss in polar quantum wires as well as the use of metal–semiconductor structures to tailor carrier–phonon interactions in nanostructures. Moreover, Chapter 10 describes how confined phonons play a critical role in determining the properties of electronic, optical, and superconducting devices containing nanostructures as essential elements. Examples of such phonon effects in nanoscale devices include: phonon effects in intersubband lasers; the effect of confined phonons on the gain of intersubband lasers; the contribution of confined phonons to the valley current in double-barrier quantum-well structures; phonon-enhanced population inversion in asymmetric double-barrier quantum-well lasers; and confined phonon effects in thin film superconductors.

Chapter 2

Phonons in bulk cubic crystals

The Creator, if He exists, has a special preference for beetles.
J.B.S. *Haldane, 1951*

2.1 Cubic structure

Crystals with cubic structure are of major importance in the fields of electronics and optoelectronics. Indeed, zincblende crystals such as silicon, germanium, and gallium arsenide may be regarded as two face-centered cubic (fcc) lattices displaced relative to each other by a vector $(a/4, a/4, a/4)$, where a is the size of the smallest unit of the fcc structure. Figure 2.1 shows a lattice with the zincblende structure.

A major portion of this book will deal with phonons in cubic crystals. In addition, we will describe the phonons in so-called isotropic media, which are related mathematically to cubic media as explained in detail in Section 7.2. The remaining portions of this book will deal with crystals of würtzite structure, defined in Chapter 3. More specifically, the primary focus of this book concerns phonons in crystalline structures that are dimensionally confined in one, two, or three dimensions. Such one-, two-, and three-dimensional confinement is realized in quantum wells, quantum wires, and quantum dots, respectively. As a preliminary to considering phonons in dimensionally confined structures, the foundational case of phonons in bulk structures will be treated. The reader desiring to supplement this chapter with additional information on the basic properties of phonons in bulk cubic materials will find excellent extended treatments in a number of texts including Blakemore (1985), Ferry (1991), Hess (1999), Kittel (1976), Omar (1975), and Singh (1993).

2.2 Ionic bonding – polar semiconductors

As is well known, the crystal structure of silicon is the zincblende structure shown in Figure 2.1. The covalent bonding in silicon does not result in any net transfer of charge between silicon atoms. More specifically, the atoms on the two displaced face-centered cubic (fcc) lattices depicted in Figure 2.1 have no excess or deficit of charge relative to the neutral situation. This changes dramatically for polar semiconductors like gallium arsenide, since here the ionic bonding results in charge transfer from the Group V arsenic atoms to the Group III gallium atoms: Since Group V atoms have five electrons in the outer shell and Group III atoms have three electrons in the outer shell, it is not surprising that the gallium sites acquire a net negative charge and the arsenic sites a net positive charge. In binary polar semiconductors, the two atoms participating in the ionic bonding carry opposite charges, e^* and $-e^*$, respectively, as a result of the redistribution of the charge associated with polar bonding. In polar materials such ionic bonding is characterized by values of e^* within an order of magnitude of unity. In the remaining sections of this chapter, it will become clear that e^* is related to the readily measurable or known ionic masses, phonon optical frequencies, and high-frequency dielectric constant of the polar semiconductor.

2.3   **Linear-chain model and macroscopic models**

The linear-chain model of a one-dimensional diatomic crystal is based upon a system of two atoms with masses, m and M, placed along a one-dimensional chain as depicted in Figure 2.2. As for a diatomic lattice, the masses are situated alternately along the chain and their separation is a. On such a chain the displacement of one atom from its equilibrium position will perturb the positions of its neighboring atoms.

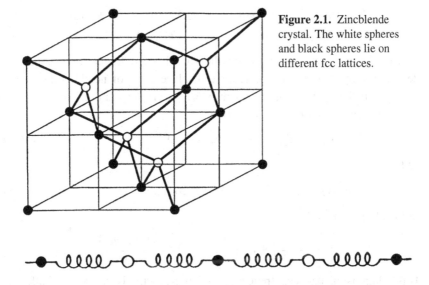

Figure 2.1. Zincblende crystal. The white spheres and black spheres lie on different fcc lattices.

Figure 2.2. One-dimensional linear-chain representation of a diatomic lattice.

In the simple linear-chain model considered in this section, it is assumed that only nearest neighbors are coupled and that the interaction between these atoms is described by Hooke's law; the spring constant α is taken to be that of a harmonic oscillator. This model describes many of the basic properties of a diatomic lattice. However, as will become clear in Chapter 6, it is essential to supplement the so-called 'harmonic' interactions with anharmonic interactions in order to describe the important process of phonon decay.

2.3.1   Dispersion relations for high-frequency and low-frequency modes

To model the normal modes of this system of masses, the atomic displacements along the direction of the chain – the so-called longitudinal displacements of each of the two types of atoms – are taken to be

$$u_{2r} = A_1 e^{i(2rqa-\omega t)} \tag{2.1}$$

and

$$u_{2r+1} = A_2 e^{i[(2r+1)qa-\omega t]} \tag{2.2}$$

where q is the phonon wavevector and ω is its frequency. In the nearest-neighbor approximation, these longitudinal displacements satisfy

$$m(d^2 u_{2r}/dt^2) = -\alpha(u_{2r} - u_{2r-1}) - \alpha(u_{2r} - u_{2r+1})$$
$$= \alpha(u_{2r+1} + u_{2r-1} - 2u_{2r}) \tag{2.3}$$

and

$$M(d^2 u_{2r+1}/dt^2) = -\alpha(u_{2r+1} - u_{2r}) - \alpha(u_{2r+1} - u_{2r+2})$$
$$= \alpha(u_{2r+2} + u_{2r} - 2u_{2r+1}). \tag{2.4}$$

The signs in the four terms on the right-hand sides of these equations are determined by considering the relative displacements of neighboring atoms. For example, if the positive displacement of u_{2r} is greater than that of u_{2r-1} there is a restoring force $-\alpha(u_{2r+1} - u_{2r})$. Hence

$$-m\omega^2 A_1 = \alpha A_2(e^{iqa} + e^{-iqa}) - 2\alpha A_1 \tag{2.5}$$

and

$$-M\omega^2 A_2 = \alpha A_1(e^{iqa} + e^{-iqa}) - 2\alpha A_2. \tag{2.6}$$

Eliminating A_1 and A_2,

$$\omega^2 = \alpha\left(\frac{1}{m} + \frac{1}{M}\right) \pm \alpha\left[\left(\frac{1}{m} + \frac{1}{M}\right)^2 - \frac{4\sin^2 qa}{mM}\right]^{1/2}. \tag{2.7}$$

This relationship between frequency and wavevector is commonly called a dispersion relation. The higher-frequency solution is known as the optical mode since, for many semiconductors, its frequency is in the terahertz range, which happens to coincide with the infrared portion of the electromagnetic spectrum. The lower-frequency solution is known as the acoustic mode. More precisely, since only longitudinal displacements have been modeled, these two solutions correspond to the longitudinal optical (LO) and longitudinal acoustic (LA) modes of the linear-chain lattice. Clearly, the displacements along this chain can be described in terms of wavevectors q in the range from $-\pi/2a$ to $\pi/2a$. From the solution for ω, it is evident that over this Brillouin zone the LO modes have a maximum frequency $[2\alpha(1/m + 1/M)]^{1/2}$ at the center of the Brillouin zone and a minimum frequency $(2\alpha/m)^{1/2}$ at the edge of the Brillouin zone. Likewise, the LA modes have a maximum frequency $(2\alpha/M)^{1/2}$ at the edge of the Brillouin zone and a minimum frequency equal to zero at the center of the Brillouin zone.

In polar semiconductors, the masses m and M carry opposite charges, e^* and $-e^*$, respectively, as a result of the redistribution of the charge associated with polar bonding. In polar materials such ionic bonding is characterized by values of e^* equal to 1, to an order-of-magnitude. When there is an electric field E present in the semiconductor, it is necessary to augment the previous force equation with terms describing the interaction with the charge. In the long-wavelength limit of the electric field E, the force equations then become

$$-m\omega^2 u_{2r} = m(d^2 u_{2r}/dt^2) = \alpha(u_{2r+1} + u_{2r-1} - 2u_{2r}) + e^* E$$
$$= \alpha(e^{i2qa} + 1)u_{2r-1} - 2\alpha u_{2r} + e^* E \qquad (2.8)$$

and

$$-M\omega^2 u_{2r+1} = M(d^2 u_{2r+1}/dt^2) = \alpha(u_{2r+2} + u_{2r} - 2u_{2r+1}) - e^* E$$
$$= \alpha(1 + e^{-i2qa})u_{2r+2} - 2\alpha u_{2r+1} - e^* E. \qquad (2.9)$$

Regarding the phonon displacements, in the long-wavelength limit there is no need to distinguish between the different sites for a given mass type since all atoms of the same mass are displaced by the same amount. In this limit, $q \to 0$. Denoting the displacements on even-numbered sites by u_1 and those on odd-numbered sites by u_2, in the long-wavelength limit the force equations reduce to

$$-m\omega^2 u_1 = 2\alpha(u_2 - u_1) + e^* E \qquad (2.10)$$

and

$$-M\omega^2 u_2 = 2\alpha(u_1 - u_2) - e^* E. \qquad (2.11)$$

Adding these equations demonstrates that $-m\omega^2 u_1 - M\omega^2 u_2 = 0$ and it is clear that $mu_1 = -Mu_2$; thus

$$-m\omega^2 u_1 = 2\alpha\left(-\frac{m}{M}u_1 - u_1\right) + e^* E \qquad (2.12)$$

and

$$-M\omega^2 u_2 = 2\alpha\left(u_1 + \frac{m}{M}u_1\right) - e^*E; \tag{2.13}$$

accordingly,

$$-(\omega^2 - \omega_0^2)u_1 = e^*E/m \tag{2.14}$$

and

$$-(\omega^2 - \omega_0^2)u_2 = -e^*E/M \tag{2.15}$$

where $\omega_0^2 = 2\alpha(1/m + 1/M)$ is the resonant frequency squared, in the absence of Coulomb effects; that is, for $e^* = 0$. The role of e^* in shifting the phonon frequency will be discussed further in the next section.

Clearly, the electric polarization P produced by such a polar diatomic lattice is given by

$$P = \frac{Ne^*u}{\epsilon(\infty)} = \frac{Ne^*(u_1 - u_2)}{\epsilon(\infty)} = \frac{1}{\epsilon(\infty)}\frac{Ne^{*2}}{(\omega_0^2 - \omega^2)}\left(\frac{1}{m} + \frac{1}{M}\right)E, \tag{2.16}$$

where $u = u_1 - u_2$, N is the number of pairs per unit volume, and e^* is as defined previously. This equation may be rewritten to show that it describes a driven oscillator:

$$(\omega_0^2 - \omega^2)u = e^*\left(\frac{1}{m} + \frac{1}{M}\right)E. \tag{2.17}$$

2.3.2 Displacement patterns for phonons

As discussed in subsection 2.3.1, in the limit $q \to 0$ the displacements, u_1 and u_2, of the optical modes satisfy $-mu_1 = Mu_2$ and the amplitudes of the two types of mass have opposite signs. That is, for the optical modes the atoms vibrate out of phase, and so with their center of mass fixed. For the acoustic modes, the maximum frequency is $(2\alpha/M)^{1/2}$. This maximum frequency occurs at the zone edge so that, near the center of the zone, ω is much less than $(2\alpha/M)^{1/2}$. From subsection 2.3.1, the ratio A_2/A_1 may be expressed as

$$\frac{A_2}{A_1} = \frac{2\alpha\cos qa}{2\alpha - M\omega^2} = \frac{2\alpha - m\omega^2}{2\alpha\cos qa}, \tag{2.18}$$

and it is clear that the ratio of the displacement amplitudes is approximately equal to unity for acoustic phonons near the center of the Brillouin zone. Thus, in contrast to the optical modes, the acoustic modes are characterized by in-phase motion of the different masses m and M. Typical mode patterns for zone-center acoustic and optical modes are depicted in Figures 2.3(a), (b). The transverse modes are illustrated here since the longitudinal modes are more difficult to depict graphically. The higher-frequency optical modes involve out-of-plane oscillations of adjacent ions, while the lower-frequency acoustic modes are characterized by motion of adjacent ions on the same sinusoidal curve.

2.3.3 Polaritons

In the presence of a transverse electric field, transverse optical (TO) phonons of a polar medium couple strongly to the electric field. When the wavevectors and frequencies of the electric field are in resonance with those of the TO phonon, a coupled phonon–photon field is necessary to describe the system. The quantum of this coupled field is known as the polariton. The analysis of subsection 2.3.1 may be generalized to apply to the case of transverse displacements. In particular, for a transverse field E, the oscillator equation takes the form

$$(\omega_{TO}^2 - \omega^2)P = \frac{Ne^{*2}}{\epsilon(\infty)}\left(\frac{1}{m} + \frac{1}{M}\right)E, \tag{2.19}$$

where ω_0^2 of subsection 2.3.1 has been designated $\omega_{TO}^2 = 2\alpha(1/m + 1/M)$ since the resonant frequency in the absence of Coulomb effects, $e^{*2} = 0$, corresponds to

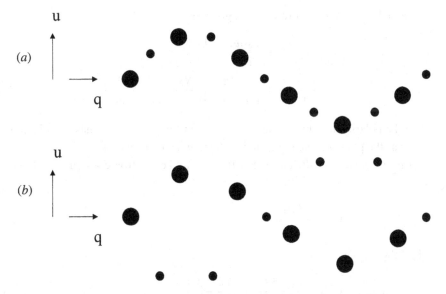

Figure 2.3. Transverse displacements of heavy ions (large disks) and light ions (small disks) for (a) transverse acoustic modes, and (b) transverse optical modes propagating in the q-direction.

the transverse optical frequency. As will become apparent later in this section, the LO phonon frequency squared differs from the TO phonon frequency squared by an amount proportional to e^{*2}.

According to the electromagnetic wave equation, $\partial^2 D/\partial t^2 = c^2 \nabla^2 E$, where $D = E + 4\pi P$, the dispersion relation describing the coupling of the field E of the electromagnetic wave to the electric polarization P of the TO phonon is

$$c^2 q^2 E = \omega^2 (E + 4\pi P) \tag{2.20}$$

or, alternatively,

$$4\pi \omega^2 P = (c^2 q^2 - \omega^2) E, \tag{2.21}$$

where waves of the form $e^{i(qr-\omega t)}$ have been assumed. The driven oscillator equation and the electromagnetic wave equation have a joint solution when the determinant of the coefficients of the fields E and P vanishes,

$$\begin{vmatrix} \omega^2 - c^2 q^2 & 4\pi \omega^2 \\ \dfrac{Ne^{*2}}{\epsilon(\infty)} \left(\dfrac{1}{m} + \dfrac{1}{M} \right) & -(\omega_{TO}^2 - \omega^2) \end{vmatrix} = 0 \tag{2.22}$$

At $q = 0$, there are two roots: $\omega = 0$ and

$$\omega^2 = \omega_{TO}^2 + 4\pi \frac{Ne^{*2}}{\epsilon(\infty)} \left(\frac{1}{m} + \frac{1}{M} \right) = \omega_{LO}^2. \tag{2.23}$$

The dielectric function $\epsilon(\omega)$ is then given by

$$\epsilon(\omega) = \frac{D(\omega)}{E(\omega)} = 1 + \frac{4\pi P_e(\omega)}{E(\omega)} + \frac{4\pi P(\omega)}{E(\omega)}$$

$$= 1 + \frac{4\pi P_e(\omega)}{E(\omega)} + \frac{4\pi}{(\omega_{TO}^2 - \omega^2)} \frac{Ne^{*2}}{\epsilon(\infty)} \left(\frac{1}{m} + \frac{1}{M} \right), \tag{2.24}$$

where the polarization due to the electronic contribution, $P_e(\omega)$, has been included as well as the polarization associated with the ionic contribution, $P(\omega)$.

As is customary, the dielectric constant due to the electronic response is denoted by

$$\epsilon(\infty) = 1 + \frac{4\pi P_e(\omega)}{E(\omega)}, \tag{2.25}$$

and it follows that

$$\epsilon(\omega) = \epsilon(\infty) + \frac{4\pi}{(\omega_{TO}^2 - \omega^2)} \frac{Ne^{*2}}{\epsilon(\infty)} \left(\frac{1}{m} + \frac{1}{M} \right). \tag{2.26}$$

The so-called static dielectric constant $\epsilon(0)$ is then given by

$$\epsilon(0) = \epsilon(\infty) + \frac{4\pi}{\omega_{TO}^2} \frac{Ne^{*2}}{\epsilon(\infty)} \left(\frac{1}{m} + \frac{1}{M} \right). \tag{2.27}$$

From these last two results it follows straightforwardly that

$$\epsilon(\omega) = \epsilon(\infty) + \frac{[\epsilon(0) - \epsilon(\infty)]\omega_{TO}^2}{(\omega_{TO}^2 - \omega^2)}$$

$$= \epsilon(\infty) + \frac{\epsilon(0) - \epsilon(\infty)}{1 - \omega^2/\omega_{TO}^2}. \tag{2.28}$$

From electromagnetic theory it is known that the dielectric function $\epsilon(\omega)$ must vanish for any longitudinal electromagnetic disturbance to propagate. Accordingly, the frequency of the LO phonons, ω_{LO}, must be such that $\epsilon(\omega_{LO}) = 0$; from the last equation, this condition implies that

$$\epsilon(\omega_{LO}) = 0 = \epsilon(\infty) + \frac{\epsilon(0) - \epsilon(\infty)}{1 - \omega_{LO}^2/\omega_{TO}^2} \tag{2.29}$$

or, equivalently,

$$\omega_{LO} = \left[\frac{\epsilon(0)}{\epsilon(\infty)} \right]^{1/2} \omega_{TO}. \tag{2.30}$$

It then follows that

$$\epsilon(\omega) = \epsilon(\infty) + \frac{\epsilon(0) - \epsilon(\infty)}{1 - \omega^2/\omega_{TO}^2} = \epsilon(\infty) + \frac{(\omega_{LO}/\omega_{TO})^2\epsilon(\infty) - \epsilon(\infty)}{1 - \omega^2/\omega_{TO}^2}$$

$$= \epsilon(\infty) \left\{ 1 + \frac{(\omega_{LO}/\omega_{TO})^2 - 1}{1 - \omega^2/\omega_{TO}^2} \right\}$$

$$= \epsilon(\infty) \left(\frac{\omega_{TO}^2 - \omega^2}{\omega_{TO}^2 - \omega^2} + \frac{\omega_{LO}^2 - \omega_{TO}^2}{\omega_{TO}^2 - \omega^2} \right)$$

$$= \epsilon(\infty) \frac{\omega_{LO}^2 - \omega^2}{\omega_{TO}^2 - \omega^2}, \tag{2.31}$$

or alternatively

$$\frac{\epsilon(\omega)}{\epsilon(\infty)} = \frac{\omega_{LO}^2 - \omega^2}{\omega_{TO}^2 - \omega^2}. \tag{2.32}$$

In the special case where $\omega = 0$, this relation reduces to the celebrated Lyddane–Sachs–Teller relationship

$$\frac{\epsilon(0)}{\epsilon(\infty)} = \frac{\omega_{LO}^2}{\omega_{TO}^2}. \tag{2.33}$$

When $\omega = \omega_{LO}$ the dielectric constant vanishes, $\epsilon(\omega_{LO}) = 0$; as stated above, this condition is familiar from electromagnetics as a requirement for the propagation of a longitudinal electromagnetic wave. That is, a longitudinal electromagnetic wave propagates only at frequencies where the dielectric constant vanishes; accordingly, ω_{LO} is identified as the frequency of the LO phonon. From the relation

$$\omega_{TO}^2 + 4\pi \frac{Ne^{*2}}{\epsilon(\infty)} \left(\frac{1}{m} + \frac{1}{M} \right) = \omega_{LO}^2,$$

it follows that $\omega_{TO} = \omega_{LO}$ for zone-center phonons in materials with $e^* = 0$; this is just as observed in non-polar materials such as silicon. In polar materials such as GaAs there is a gap between ω_{TO} and ω_{LO}, associated with the Coulomb energy density arising from e^*. When $\omega = \omega_{TO}$, $\epsilon(\omega_{TO})^{-1} = 0$ and the pole in $\epsilon(\omega)$ reflects the fact that electromagnetic waves with the frequency of the TO phonon are absorbed. Throughout the interval $(\omega_{TO}, \omega_{LO})$, $\epsilon(\omega)$ is negative and electromagnetic waves do not propagate.

2.3.4 Macroscopic theory of polar modes in cubic crystals

As was apparent in subsections 2.3.1 and 2.3.3, polar-optical phonon vibrations produce electric fields and electric polarization fields that may be described in terms of Maxwell's equations and the driven-oscillator equations. Loudon (1964) advocated a model of optical phonons based on these macroscopic fields that has had great utility in describing the properties of optical phonons in so-called uniaxial crystals such as würtzite crystals. The Loudon model for uniaxial crystals will be developed more fully in Chapters 3 and 7. In this section, the concepts underlying the Loudon model will be discussed in the context of cubic crystals.

From the pair of Maxwell's equations,

$$\nabla \times \mathbf{E} + \frac{1}{c}\frac{\partial \mathbf{B}}{\partial t} = 0 \qquad \text{and} \qquad \nabla \times \mathbf{B} - \frac{1}{c}\frac{\partial \mathbf{D}}{\partial t} = \mathbf{J}, \tag{2.34}$$

it follows that

$$\nabla \times (\nabla \times \mathbf{E}) + \frac{1}{c}\frac{\partial(\nabla \times \mathbf{B})}{\partial t} = \nabla(\nabla \cdot \mathbf{E}) - \nabla^2 \mathbf{E} + \frac{1}{c^2}\frac{\partial^2 \mathbf{D}}{\partial^2 t} = 0, \tag{2.35}$$

where the source current, $\mathbf{J}$, has been taken to equal zero. Then since $\nabla \cdot \mathbf{D} = \nabla \cdot \mathbf{E} + 4\pi \nabla \cdot \mathbf{P} = 4\pi\rho = 0$, it follows that

$$-4\pi \nabla(\nabla \cdot \mathbf{P}) - \nabla^2 \mathbf{E} + \frac{1}{c^2}\frac{\partial^2 \mathbf{E}}{\partial^2 t} + 4\pi \frac{1}{c^2}\frac{\partial^2 \mathbf{P}}{\partial^2 t} = 0. \tag{2.36}$$

Assuming that P and E both have spatial and time dependences of the form $e^{i(\mathbf{q}\cdot\mathbf{r}-\omega t)}$, this last result takes the form

$$\mathbf{E} = \frac{-4\pi[\mathbf{q}(\mathbf{q} \cdot \mathbf{P}) - \omega^2 \mathbf{P}/c^2]}{q^2 - \omega^2/c^2}. \tag{2.37}$$

The condition $\mathbf{q} \cdot \mathbf{P} = 0$ corresponds to the transverse wave; in this case,

$$\mathbf{E} = \frac{-4\pi \omega^2 \mathbf{P}/c^2}{q^2 - \omega^2/c^2}.$$

(2.38)

From Appendix A, $\mathbf{E}$ and $\mathbf{P}$ are also related through

$$\mathbf{P} = \frac{1}{4\pi} \left\{ \frac{[\epsilon(0) - \epsilon(\infty)]\omega_{TO}^2}{\omega_{TO}^2 - \omega^2} + [\epsilon(\infty) - 1] \right\} \mathbf{E};$$

(2.39)

thus

$$\frac{q^2 - \omega^2/c^2}{\omega^2/c^2} = \frac{[\epsilon(0) - \epsilon(\infty)]\omega_{TO}^2}{\omega_{TO}^2 - \omega^2} + [\epsilon(\infty) - 1],$$

(2.40)

or, equivalently,

$$\frac{q^2 c^2}{\omega^2} = \frac{\omega_{TO}^2 \epsilon(0) - \omega^2 \epsilon(\infty)}{\omega_{TO}^2 - \omega^2}.$$

(2.41)

For longitudinal waves, $\mathbf{q} \cdot \mathbf{P} = qP$, so that $\mathbf{q} = (q/P)\mathbf{P}$, and it follows that

$$\begin{aligned}
\mathbf{E} &= \frac{4\pi \omega^2 \mathbf{P}/c^2}{q^2 - \omega^2/c^2} - \frac{4\pi q P \mathbf{q}}{q^2 - \omega^2/c^2} \\
&= \frac{4\pi \omega^2 \mathbf{P}/c^2}{q^2 - \omega^2/c^2} - \frac{4\pi q^2 \mathbf{P}}{q^2 - \omega^2/c^2} \\
&= \frac{4\pi}{q^2 - \omega^2/c^2} \left(\frac{\omega^2}{c^2} - q^2 \right) \mathbf{P} \\
&= -4\pi \mathbf{P}.
\end{aligned}$$

(2.42)

Then

$$\begin{aligned}
\mathbf{P} &= \frac{1}{4\pi} \left\{ \frac{[\epsilon(0) - \epsilon(\infty)]\omega_{TO}^2}{\omega_{TO}^2 - \omega^2} + [\epsilon(\infty) - 1] \right\} \mathbf{E} \\
&= -\left\{ \frac{[\epsilon(0) - \epsilon(\infty)]\omega_{TO}^2}{\omega_{TO}^2 - \omega^2} + [\epsilon(\infty) - 1] \right\} \mathbf{P}
\end{aligned}$$

(2.43)

or, equivalently,

$$\omega = \omega_{TO} \left[\frac{\epsilon(0)}{\epsilon(\infty)} \right]^{1/2} = \omega_{LO},$$

and the Lyddane–Sachs–Teller relation is recovered once again! In Chapter 3, we shall return to the Loudon model to describe uniaxial crystals of the würtzite type.

Chapter 3

Phonons in bulk würtzite crystals

Next when I cast mine eyes and see that brave vibration, each
way free; O how that glittering taketh me.
Robert Herrick, 1648

3.1 Basic properties of phonons in würtzite structure

The GaAlN-based semiconductor structures are of great interest in the electronics
and optoelectronics communities because they possess large electronic bandgaps
suitable for fabricating semiconductor lasers with wavelengths in the blue and
ultraviolet as well as electronic devices designed to work at elevated operating
temperatures. These III-V nitrides occur in both zincblende and würtzite structures.
In this chapter, the würtzite structures will be considered rather than the zincblende
structures, since the treatment of the phonons in these würtzite structures is more
complicated than for the zincblendes. Throughout the remainder of this book,
phonon effects in nanostructures will be considered for both the zincblendes and
würtzites. This chapter focuses on the basic properties of phonons in bulk würtzite
structures as a foundation for subsequent discussions on phonons in würtzite
nanostructures.

The crystalline structure of a würtzite material is depicted in Figure 3.1. As in
the zincblendes, the bonding is tetrahedral. The würtzite structure may be generated
from the zincblende structure by rotating adjacent tetrahedra about their common
bonding axis by an angle of 60 degrees with respect to each other. As illustrated in
Figure 3.1, würtzite structures have four atoms per unit cell.

The total number of normal vibrational modes for a unit cell with s atoms in
the basis is $3s$. As for cubic materials, in the long-wavelength limit there are three
acoustic modes, one longitudinal and two transverse. Thus, the total number of
optical modes in the long-wavelength limit is $3s - 3$. These optical modes must,
of course, appear with a ratio of transverse to longitudinal optical modes of two.

The numbers of the various long-wavelength modes are summarized in Table 3.1.

For the zincblende case, $s = 2$ and there are six modes: one LA, two TA, one LO and two TO. For the würtzite case, $s = 4$ and there are 12 modes: one LA, two TA, three LO and six TO. In the long-wavelength limit the acoustic modes are simple translational modes. The optical modes for a würtzite structure are depicted in Figure 3.2.

From Figure 3.2 it is clear that the A_1 and E_1 modes will produce large electric polarization fields when the bonding is ionic. Such large polarization fields result in strong carrier–optical-phonon scattering. These phonon modes are known as infrared active. As we shall see in Chapter 5, the fields associated with these infrared modes may be derived from a potential describing the carrier–phonon interaction of

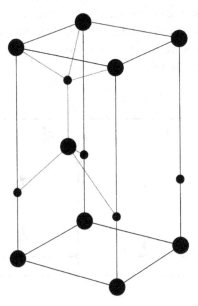

Figure 3.1. Unit cell of the hexagonal würtzite crystal.

Table 3.1. Phonon modes associated with a unit cell having s atoms in the basis.

Type of mode	Number of modes
Longitudinal acoustic (LA)	1
Transverse acoustic (TA)	2
All acoustic modes	3
Longitudinal optical (LO)	$s - 1$
Transverse optical (TO)	$2s - 2$
All optical modes	$3s - 3$
All modes	$3s$

such modes. In Chapter 5, this carrier–phonon interaction potential will be identified as the Fröhlich interaction. The dispersion relations for the 12 phonon modes of the würtzite structure are depicted in Figure 3.3.

The low-frequency behavior of these modes near the Γ point makes it apparent that three of these 12 modes are acoustic modes. This behavior is, of course, consistent with the number of acoustic modes identified in Table 3.1.

3.2   Loudon model of uniaxial crystals

As discussed in subsection 2.3.4, Loudon (1964) advanced a model for uniaxial crystals that provides a useful description of the longitudinal optical phonons in würtzite crystals. In Loudon's model of uniaxial crystals such as GaN or AlN, the angle between the c-axis and q is denoted by θ, and the isotropic dielectric constant of the cubic case is replaced by dielectric constants for the directions parallel and perpendicular to the c-axis, $\epsilon_{\parallel}(\omega)$ and $\epsilon_{\perp}(\omega)$ respectively. That is,

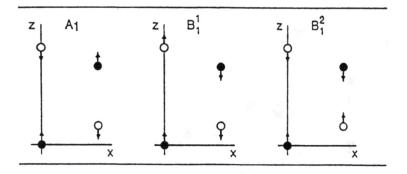

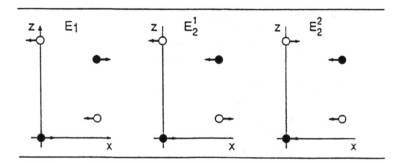

Figure 3.2. Optical phonons in würtzite structure. From Gorczyca *et al.* (1995), American Physical Society, with permission.

$$\epsilon(\omega) = \begin{pmatrix} \epsilon_\perp(\omega) & 0 & 0 \\ 0 & \epsilon_\perp(\omega) & 0 \\ 0 & 0 & \epsilon_\parallel(\omega) \end{pmatrix} \qquad (3.1)$$

with

$$\epsilon_\perp(\omega) = \epsilon_\perp(\infty)\frac{\omega^2 - \omega_{LO,\perp}^2}{\omega^2 - \omega_{TO,\perp}^2} \quad \text{and} \quad \epsilon_\parallel(\omega) = \epsilon_\parallel(\infty)\frac{\omega^2 - \omega_{LO,\parallel}^2}{\omega^2 - \omega_{TO,\parallel}^2},$$

$$(3.2)$$

as required by the Lyddane–Sachs–Teller relation. The c-axis is frequently taken to be in the z-direction and the dielectric constant is then sometimes labeled by the z-coordinate; that is, $\epsilon_\parallel(\omega) = \epsilon_z(\omega)$. Figure 3.4 depicts the two dielectric constants for GaN as well as those for AlN.

In such a uniaxial crystal, there are two types of phonon wave: (a) ordinary waves where for any θ both the electric field **E** and the polarization **P** are perpendicular to the c-axis and **q** simultaneously, and (b) extraordinary waves, for which the orientation of **E** and **P** with respect to **q** and the c-axis is more complicated. As discussed in subsection 2.3.4, the ordinary wave has E_1 symmetry, is transverse, and is polarized in the $\perp$-plane. There are two extraordinary waves, one associated with the $\perp$-polarized vibrations and having A_1 symmetry and the other associated with $\parallel$-polarized vibrations and having E_1 symmetry. For $\theta = 0$, one of these modes is the A_1(LO) mode and the other is the E_1(TO) mode. As θ varies between 0 and $\pi/2$, these modes evolve to the A_1(TO) and E_1(TO) modes respectively. For values of θ intermediate between 0 and $\pi/2$ they are mixed and do not have purely LO or

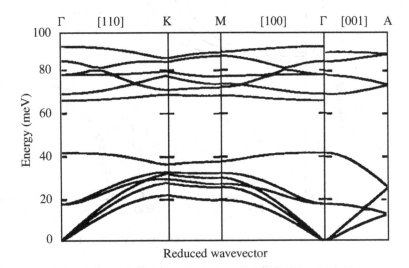

Figure 3.3. Phonon dispersion curves for GaN crystal of würtzite structure. From Nipko *et al.* (1998), American Institute of Physics, with permission.

TO character or A_1 or E_1 symmetry (Loudon, 1964). For würtzite structures at the Γ point, it will be obvious in Chapter 7 that only three of the nine optical phonon modes, the $A_1(Z)$ and $E_1(X, Y)$ modes, produce significant carrier–optical-phonon scattering rates. These are the so-called infrared-active modes. For the case of würtzite structures, Loudon's model of uniaxial crystals is based upon generalizing Huang's equations, equations (A.8) and (A.9) of Appendix A, and the relationship of subsection 2.3.4, equation (2.43). Specifically, for each of these equations there is a set of two more equations, one in terms of quantities along the c-axis and the other in terms of quantities perpendicular to the c-axis:

$$(\omega_{TO,\perp}^2 - \omega^2)\mathbf{u}_\perp = \left(\frac{V}{4\pi\mu N}\right)^{1/2} \sqrt{\epsilon_\perp(0) - \epsilon_\perp(\infty)}\, \omega_{TO,\perp}\mathbf{E}_\perp, \qquad (3.3)$$

$$(\omega_{TO,\parallel}^2 - \omega^2)\mathbf{u}_\parallel = \left(\frac{V}{4\pi\mu N}\right)^{1/2} \sqrt{\epsilon_\parallel(0) - \epsilon_\parallel(\infty)}\, \omega_{TO,\parallel}\mathbf{E}_\parallel, \qquad (3.4)$$

$$\mathbf{P}_\perp = \left(\frac{\mu N}{4\pi V}\right)^{1/2} \sqrt{\epsilon_\perp(0) - \epsilon_\perp(\infty)}\, \omega_{TO,\perp}\mathbf{u}_\perp + \left[\frac{\epsilon_\perp(\infty) - 1}{4\pi}\right]\mathbf{E}_\perp, \quad (3.5)$$

$$\mathbf{P}_\parallel = \left(\frac{\mu N}{4\pi V}\right)^{1/2} \sqrt{\epsilon_\parallel(0) - \epsilon_\parallel(\infty)}\, \omega_{TO,\parallel}\mathbf{u}_\parallel + \left[\frac{\epsilon_\parallel(\infty) - 1}{4\pi}\right]\mathbf{E}_\parallel, \qquad (3.6)$$

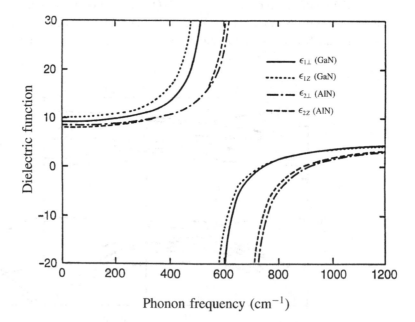

Figure 3.4. Dielectric constants for GaN, $\epsilon_{1\perp}$(GaN) and ϵ_{1z}(GaN), and for AlN, $\epsilon_{2\perp}$(AlN) and ϵ_{2z}(AlN). From Lee *et al.* (1998), American Physical Society, with permission.

$$E_\perp = \frac{-4\pi[\mathbf{q}_\perp(\mathbf{q}\cdot\mathbf{P}) - \omega^2\mathbf{P}_\perp/c^2]}{q^2 - \omega^2/c^2}, \tag{3.7}$$

$$E_\parallel = \frac{-4\pi[\mathbf{q}_\parallel(\mathbf{q}\cdot\mathbf{P}) - \omega^2\mathbf{P}_\parallel/c^2]}{q^2 - \omega^2/c^2}. \tag{3.8}$$

Eliminating $\mathbf{u}_\perp$ and $\mathbf{u}_\parallel$ in the first four of these equations yields

$$\begin{aligned}
\mathbf{P}_\perp &= \frac{1}{4\pi}\left\{\frac{[\epsilon_\perp(0) - \epsilon_\perp(\infty)]\omega^2_{TO,\perp}}{\omega^2_{TO,\perp} - \omega^2} + [\epsilon_\perp(\infty) - 1]\right\}\mathbf{E}_\perp \\
&= \frac{1}{4\pi}A_\perp\mathbf{E}_\perp
\end{aligned} \tag{3.9}$$

and

$$\begin{aligned}
\mathbf{P}_\parallel &= \frac{1}{4\pi}\left\{\frac{[\epsilon_\parallel(0) - \epsilon_\parallel(\infty)]\omega^2_{TO,\parallel}}{\omega^2_{TO,\parallel} - \omega^2} + [\epsilon_\parallel(\infty) - 1]\right\}\mathbf{E}_\parallel \\
&= \frac{1}{4\pi}A_\parallel\mathbf{E}_\parallel,
\end{aligned} \tag{3.10}$$

where $A_\perp$ and $A_\parallel$ may be written as

$$A_\perp = \frac{\omega^2_{LO,\perp} - \omega^2}{\omega^2_{TO,\perp} - \omega^2}\epsilon_\perp(\infty) - 1, \tag{3.11}$$

$$A_\parallel = \frac{\omega^2_{LO,\parallel} - \omega^2}{\omega^2_{TO,\parallel} - \omega^2}\epsilon_\parallel(\infty) - 1, \tag{3.12}$$

upon using the Lyddane–Sachs–Teller relations

$$\omega_{TO,\perp}\left[\frac{\epsilon_\perp(0)}{\epsilon_\perp(\infty)}\right]^{1/2} = \omega_{LO,\perp} \quad \text{and} \quad \omega_{TO,\parallel}\left[\frac{\epsilon_\parallel(0)}{\epsilon_\parallel(\infty)}\right]^{1/2} = \omega_{LO,\parallel}. \tag{3.13}$$

For the ordinary wave, $\mathbf{E}_\parallel = 0$, $\mathbf{P}_\parallel = 0$, and $\mathbf{q}\cdot\mathbf{P} = 0$, so that the derivation of subsection 2.3.4 now gives

$$\frac{q^2c^2}{\omega^2} = \frac{\omega^2_{TO,\perp}\epsilon_\perp(0) - \omega^2\epsilon_\perp(\infty)}{\omega^2_{TO,\perp} - \omega^2}. \tag{3.14}$$

For the ordinary mode it also follows that $\mathbf{u}_\parallel = \mathbf{u}_\perp = 0$.

For the extraordinary wave, $q_\perp = q\sin\theta$ and $q_\parallel = q\cos\theta$, where θ is the angle between q and the c-axis. Then, it follows that

$$\mathbf{q}\cdot\mathbf{P} = (q\sin\theta, q\cos\theta)\cdot(P_\perp, P_\parallel) = qP_\perp\sin\theta + qP_\parallel\cos\theta. \tag{3.15}$$

Thus,

$$q_\perp(\mathbf{q} \cdot \mathbf{P}) = q^2 (P_\perp \sin^2 \theta + P_\parallel \sin \theta \cos \theta),$$

$$q_\parallel(\mathbf{q} \cdot \mathbf{P}) = q^2 (P_\perp \sin \theta \cos \theta + P_\parallel \cos^2 \theta). \tag{3.16}$$

In the limit where retardation effects are neglected, $c \to \infty$ and it follows that

$$E_\perp = \frac{-4\pi [q_\perp(\mathbf{q} \cdot \mathbf{P}) - \omega^2 P_\perp/c^2]}{q^2 - \omega^2/c^2} \to -4\pi (P_\perp \sin^2 \theta + P_\parallel \sin \theta \cos \theta)$$

$$= -\sin^2 \theta \, A_\perp E_\perp - \sin \theta \cos \theta \, A_\parallel E_\parallel, \tag{3.17}$$

and

$$E_\parallel = \frac{-4\pi [q_\parallel(\mathbf{q} \cdot \mathbf{P}) - \omega^2 P_\parallel/c^2]}{q^2 - \omega^2/c^2} \to -4\pi (P_\perp \sin \theta \cos \theta + P_\parallel \cos^2 \theta)$$

$$= -\sin \theta \cos \theta \, A_\perp E_\perp - \cos^2 \theta \, A_\parallel E_\parallel. \tag{3.18}$$

These equations may be written as

$$\begin{pmatrix} 1 + \sin^2 \theta \, A_\perp & \sin \theta \cos \theta \, A_\parallel \\ \sin \theta \cos \theta \, A_\perp & 1 + \cos^2 \theta \, A_\parallel \end{pmatrix} \begin{pmatrix} E_\perp \\ E_\parallel \end{pmatrix} = 0, \tag{3.19}$$

and it follows that the condition for non-trivial solutions to exist is

$$1 + \sin^2 \theta \, A_\perp + \cos^2 \theta \, A_\parallel = \frac{\omega_{\mathrm{LO},\perp}^2 - \omega^2}{\omega_{\mathrm{TO},\perp}^2 - \omega^2} \epsilon_\perp(\infty) \sin^2 \theta$$

$$+ \frac{\omega_{\mathrm{LO},\parallel}^2 - \omega^2}{\omega_{\mathrm{TO},\parallel}^2 - \omega^2} \epsilon_\parallel(\infty) \cos^2 \theta$$

$$= \epsilon_\perp(\omega) \sin^2 \theta + \epsilon_\parallel(\omega) \cos^2 \theta = 0 \tag{3.20}$$

or equivalently

$$\epsilon_\perp(\omega) q_\perp^2 + \epsilon_\parallel(\omega) q_\parallel^2 = 0. \tag{3.21}$$

Since the high-frequency electronic response of a medium should not depend strongly on the crystalline structure, it is usually assumed (Loudon, 1964) that $\epsilon_\perp(\infty) \approx \epsilon_\parallel(\infty)$. Thus

$$\frac{\omega_{\mathrm{LO},\perp}^2 - \omega^2}{\omega_{\mathrm{TO},\perp}^2 - \omega^2} \sin^2 \theta + \frac{\omega_{\mathrm{LO},\parallel}^2 - \omega^2}{\omega_{\mathrm{TO},\parallel}^2 - \omega^2} \cos^2 \theta = 0 \tag{3.22}$$

or equivalently

$$\omega^4 - (\omega_1^2 + \omega_2^2)\omega^2 + \omega_{\mathrm{TO},\perp}^2 \omega_{\mathrm{LO},\parallel}^2 \cos^2 \theta + \omega_{\mathrm{LO},\perp}^2 \omega_{\mathrm{TO},\parallel}^2 \sin^2 \theta = 0, \tag{3.23}$$

where

$$
\begin{aligned}
\omega_1^2 &= \omega_{TO,\parallel}^2 \sin^2\theta + \omega_{TO,\perp}^2 \cos^2\theta, \\
\omega_2^2 &= \omega_{LO,\parallel}^2 \cos^2\theta + \omega_{LO,\perp}^2 \sin^2\theta.
\end{aligned}
\tag{3.24}
$$

When $\left|\omega_{TO,\parallel} - \omega_{TO,\perp}\right|$ is very much less than $\omega_{LO,\parallel} - \omega_{TO,\parallel}$ and $\omega_{LO,\perp} - \omega_{TO,\perp}$ this equation has roots

$$
\omega^2 = \frac{1}{2}\left\{(\omega_1^2 + \omega_2^2) \pm [(\omega_1^2 - \omega_2^2) + 2\Delta\omega^2(\theta)]\right\},
\tag{3.25}
$$

where

$$
\Delta\omega^2(\theta) = 2\frac{(\omega_{LO,\parallel}^2 - \omega_{LO,\perp}^2)(\omega_{TO,\parallel}^2 - \omega_{TO,\perp}^2)}{\omega_2^2 - \omega_1^2}\sin^2\theta\cos^2\theta;
\tag{3.26}
$$

thus

$$
\begin{aligned}
\omega^2 &= \omega_{TO,\parallel}^2 \sin^2\theta + \omega_{TO,\perp}^2 \cos^2\theta \\
&\quad - \frac{(\omega_{LO,\parallel}^2 - \omega_{LO,\perp}^2)(\omega_{TO,\parallel}^2 - \omega_{TO,\perp}^2)}{\omega_2^2 - \omega_1^2}\sin^2\theta\cos^2\theta \\
&\approx \omega_{TO,\parallel}^2 \sin^2\theta + \omega_{TO,\perp}^2 \cos^2\theta
\end{aligned}
\tag{3.27}
$$

and

$$
\begin{aligned}
\omega^2 &= \omega_{LO,\parallel}^2 \cos^2\theta + \omega_{LO,\perp}^2 \sin^2\theta \\
&\quad + \frac{(\omega_{LO,\parallel}^2 - \omega_{LO,\perp}^2)(\omega_{TO,\parallel}^2 - \omega_{TO,\perp}^2)}{\omega_2^2 - \omega_1^2}\sin^2\theta\cos^2\theta \\
&\approx \omega_{LO,\parallel}^2 \cos^2\theta + \omega_{LO,\perp}^2 \sin^2\theta.
\end{aligned}
\tag{3.28}
$$

3.3 Application of Loudon model to III–V nitrides

The conditions $\left|\omega_{TO,\parallel} - \omega_{TO,\perp}\right| \ll \omega_{LO,\parallel} - \omega_{TO,\parallel}$ and $\left|\omega_{TO,\parallel} - \omega_{TO,\perp}\right| \ll \omega_{LO,\perp} - \omega_{TO,\perp}$ are satisfied reasonably well for a number of würtzite materials including the III–V nitrides. Indeed, for GaN, $\epsilon(\infty) = 5.26$, $\omega_{LO,\perp} = 743$ cm^{-1}, $\omega_{LO,\parallel} = 735$ cm^{-1}, $\omega_{TO,\perp} = 561$ cm^{-1}, and $\omega_{TO,\parallel} = 533$ cm^{-1} (Azuhata et al., 1995). For AlN, $\epsilon(\infty) = 5.26$, $\omega_{LO,\perp} = 916$ cm^{-1}, $\omega_{LO,\parallel} = 893$ cm^{-1}, $\omega_{TO,\perp} = 673$ cm^{-1}, and $\omega_{TO,\parallel} = 660$ cm^{-1} (Perlin et al., 1993). For these and other würtzite crystals (Hayes and Loudon, 1978), Table 3.2 summarizes the various frequency differences appearing in the previously stated frequency conditions.

As is clear from Table 3.2, the inequalities assumed in Section 3.2 are reasonably well satisfied for both GaN and AlN as well as for the other materials listed. The infrared-active modes in these III-V nitrides are the $A_1(\text{LO})$, $A_1(\text{TO})$, $E_1(\text{LO})$, and $E_1(\text{TO})$ modes and the frequencies associated with these modes, $\omega_{A_1(\text{LO})}$, $\omega_{A_1(\text{TO})}$, $\omega_{E_1(\text{LO})}$, and $\omega_{E_1(\text{TO})}$ are given by $\omega_{\text{LO},\parallel}$, $\omega_{\text{TO},\parallel}$, $\omega_{\text{LO},\perp}$, and $\omega_{\text{TO},\perp}$, respectively. Let us consider the case of GaN in more detail. From the results of Section 3.2, it follows immediately that

$$\frac{E_\perp}{E_\parallel} = -\frac{\sin\theta\cos\theta\, A_\parallel}{1 + \sin^2\theta\, A_\perp} = \frac{\sin\theta\cos\theta\, A_\parallel}{\cos^2\theta\, A_\parallel} = \frac{\sin\theta}{\cos\theta}, \tag{3.29}$$

and

$$\begin{aligned}
\frac{u_\perp}{u_\parallel} &= \frac{\omega_{\text{TO},\parallel}^2 - \omega^2}{\omega_{\text{TO},\perp}^2 - \omega^2}\left[\frac{\epsilon_\perp(0) - \epsilon_\perp(\infty)}{\epsilon_\parallel(0) - \epsilon_\parallel(\infty)}\right]^{1/2}\left(\frac{\omega_{\text{TO},\perp}}{\omega_{\text{TO},\parallel}}\right)\frac{E_\perp}{E_\parallel} \\
&= \frac{\omega_{\text{TO},\parallel}^2 - \omega^2}{\omega_{\text{TO},\perp}^2 - \omega^2}\left[\frac{\epsilon_\perp(0) - \epsilon_\perp(\infty)}{\epsilon_\parallel(0) - \epsilon_\parallel(\infty)}\right]^{1/2}\left(\frac{\omega_{\text{TO},\perp}}{\omega_{\text{TO},\parallel}}\right)\frac{\sin\theta}{\cos\theta}.
\end{aligned} \tag{3.30}$$

Since $\mathbf{q} = (q_\perp, q_\parallel) = (q\sin\theta, q\cos\theta)$, the first of these relations illustrates the fact that $\mathbf{E}\parallel\mathbf{q}$, as expected from $q^2\mathbf{E} = -4\pi\mathbf{q}(\mathbf{q}\cdot\mathbf{P})$; this last equality follows from $\nabla\cdot(\mathbf{E}+4\pi\mathbf{P}) = 0$. The ratio $u_\perp/u_\parallel$ may be estimated for GaN for the transverse-like modes, with $\omega^2 = \omega_{\text{TO},\parallel}^2\sin^2\theta + \omega_{\text{TO},\perp}^2\cos^2\theta$, as

$$\frac{u_\perp}{u_\parallel} = -\left[\frac{\epsilon_\perp(0) - \epsilon_\perp(\infty)}{\epsilon_\parallel(0) - \epsilon_\parallel(\infty)}\right]^{1/2}\left(\frac{\omega_{\text{TO},\perp}}{\omega_{\text{TO},\parallel}}\right)\frac{\cos\theta}{\sin\theta} \approx -0.95\frac{\cos\theta}{\sin\theta}, \tag{3.31}$$

and for the longitudinal-like modes, with $\omega^2 = \omega_{\text{LO},\parallel}^2\cos^2\theta + \omega_{\text{LO},\perp}^2\sin^2\theta$, as

$$\begin{aligned}
\frac{u_\perp}{u_\parallel} &= \frac{\omega_{\text{TO},\parallel}^2 - \omega^2}{\omega_{\text{TO},\perp}^2 - \omega^2}\left[\frac{\epsilon_\perp(0) - \epsilon_\perp(\infty)}{\epsilon_\parallel(0) - \epsilon_\parallel(\infty)}\right]^{1/2}\left(\frac{\omega_{\text{TO},\perp}}{\omega_{\text{TO},\parallel}}\right)\frac{\sin\theta}{\cos\theta} \\
&= \frac{\omega_{\text{TO},\parallel}^2 - \omega_{\text{LO}}^2}{\omega_{\text{TO},\perp}^2 - \omega_{\text{LO}}^2}\left(\frac{\epsilon_\perp(0) - \epsilon_\perp(\infty)}{\epsilon_\parallel(0) - \epsilon_\parallel(\infty)}\right)^{1/2}\left[\frac{\omega_{\text{TO},\perp}}{\omega_{\text{TO},\parallel}}\right]\frac{\sin\theta}{\cos\theta} \\
&= 1.07\frac{\sin\theta}{\cos\theta}.
\end{aligned} \tag{3.32}$$

Table 3.2. Difference frequencies in cm^{-1} for GaN and AlN as well as for other würtzite crystals.

Würtzite	$\|\omega_{\text{TO},\parallel} - \omega_{\text{TO},\perp}\|$	$\omega_{\text{LO},\parallel} - \omega_{\text{TO},\parallel}$	$\omega_{\text{LO},\perp} - \omega_{\text{TO},\perp}$
GaN	27	211	186
AlN	59	279	243
AgI	0	18	18
BeO	44	403	375
CdS	9	71	64
ZnO	33	199	178
ZnS	0	76	76

where ω_{LO}^2 is taken to be equal to both $\omega_{\text{LO},\parallel}^2$ and $\omega_{\text{LO},\perp}^2$ since $\omega_{\text{LO},\parallel}^2 \approx \omega_{\text{LO},\perp}^2$.

The properties of uniaxial crystals derived in this section and in Section 3.2 will be used extensively in Chapter 7 to determine the Fröhlich potentials in würtzite nanostructures.

Chapter 4

Raman properties of bulk phonons

When you measure what you are speaking about and express it
in numbers, you know something about it; but when you
cannot measure it, when you cannot express it in numbers, your
knowledge is of a meagre and unsatisfactory kind; it may be the
beginning of knowledge but you have scarcely in your thoughts
advanced to the stage of science, whatever the matter may be.
Lord Kelvin, 1889

4.1 Measurements of dispersion relations for bulk samples

This chapter deals with the application of Raman scattering techniques to measure
basic properties of phonons in dimensionally confined systems. It is, however,
appropriate at this point to emphasize that non-Raman techniques such as neutron
scattering (Waugh and Dolling, 1963) have been used for many years to determine
the phonon dispersion relations for bulk semiconductors. Indeed, for thermal
neutrons the de Broglie wavelengths are comparable to the phonon wavelengths.
For bulk samples, neutron scattering cross sections are large enough to facilitate
the measurement of phonon dispersion relations. This is generally not the case for
quantum wells, quantum wires, and quantum dots, where Raman and micro-Raman
techniques are needed to make accurate measurements of dispersion relations
in structures of such small volume. Further comparisons of neutron and Raman
scattering measurements of phonon dispersion relations are found in Section 7.5.

4.2 Raman scattering for bulk zincblende and würtzite structures

Raman scattering has been a very effective experimental technique for observing
phonons; it involves measuring the frequency shift between the incident and

scattered photons. It is a three-step process: the incident photon of frequency ω_i is absorbed; the intermediate electronic state which is thus formed interacts with phonons or other elementary excitations of energy via several mechanisms, creating or annihilating them; finally, the scattered photon, of different energy ω_s, is emitted. Energy and momentum are conserved and are given by the following equations:

$$\hbar\omega_i = \hbar\omega_s \pm \hbar\Omega, \tag{4.1}$$

$$\mathbf{k}_i = \mathbf{k}_s \pm \mathbf{q}. \tag{4.2}$$

Since the momenta of the incident and scattered photons are small compared with the reciprocal lattice vectors, only excitations with $\mathbf{q} \simeq 0$ take part in the Raman process illustrated in Figure 4.1. In the case of Raman scattering in semiconductors, the absorption of photons gives rise to electron–hole pairs; hence the intensity of the Raman scattering and the resonances reflect the underlying electronic structure of the material. The Raman intensity, $I(\omega_i)$, is given by

$$I(\omega_i) \propto \omega_s^4 \, |\boldsymbol{\varepsilon}_s \mathsf{T} \boldsymbol{\varepsilon}_i|^2 \sum_{\alpha,\beta} \frac{1}{(E_\alpha - \hbar\omega_i)(E_\beta - \hbar\omega_s)} \tag{4.3}$$

where the ω_i and ω_s are the frequencies of the incoming photon and of the scattered photon respectively, E_α and E_β are the energies of the intermediate states, T the Raman tensor, and $\boldsymbol{\varepsilon}_i$ and $\boldsymbol{\varepsilon}_s$ are the incident and scattered polarization vectors. The summation is over all possible intermediate states. In general, for semiconductors there may be the following real intermediate states: Bloch states, which form the conduction or valence bands, exciton states and in-gap impurity states. In equation (4.3), the second factor gives the Raman selection rules, which come about from symmetry considerations of the interactions involved in a Raman process. The selection rules are conveniently summarized in the form of Raman tensors. These selection rules are essential tools for determining crystal orientation and quality.

Details of the theoretical description of Raman scattering and these effects in the vicinity of the critical points of the semiconductor are given in excellent books and reviews elsewhere (Loudon, 1964; Hayes and Loudon, 1978; Cardona, 1975; Cardona and Güntherodt, 1982a, b, 1984, 1989, 1991) and will not be repeated here.

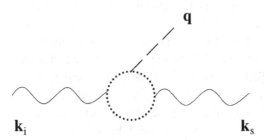

Figure 4.1. Diagrammatic representation of the Raman process. The broken line represents the phonon, the wavy lines represent the photons, and the dotted line represents the electronic state.

Instead we will summarize key results in zincblende and würtzite crystals both for the bulk case and, in Chapter 7, for quantum wells and superlattices. While first-rate articles and book chapters exist for the results of the zincblende structures, the work on the nitrides, with their würtzite structure, is more recent and hence in this book we will cover the latter results in more detail.

4.2.1   Zincblende structures

The features that can be observed in a Raman experiment for particular values of incident and scattered polarization can be determined from the symmetry properties of the second-order susceptibility for the excitation concerned as well as from the spatial symmetry of the scattering medium. The cubic zincblende structure has a space-group symmetry T_d^2, and there is one three-fold Raman active mode of the T_2 representation. The optic mode is polar so that the macroscopic field lifts the degeneracy, producing a non-degenerate longitudinal mode that is at a higher frequency than the two transverse modes. The allowed light-scattering symmetries, as indicated by the second-order susceptibilities for the zincblende structure are given below by appropriate matrices for the tensor T in the T_2 representation:

$$\begin{pmatrix} 0 & 0 & 0 \\ 0 & 0 & d \\ 0 & d & 0 \end{pmatrix} \qquad R(x) \text{ mode,}$$

$$\begin{pmatrix} 0 & 0 & d \\ 0 & 0 & 0 \\ d & 0 & 0 \end{pmatrix} \qquad R(y) \text{ mode,} \qquad\qquad (4.4)$$

$$\begin{pmatrix} 0 & d & 0 \\ d & 0 & 0 \\ 0 & 0 & 0 \end{pmatrix} \qquad R(z) \text{ mode.}$$

Raman scattering has been used now for several decades as a characterization tool in understanding, for example, crystal structure and quality, impurity content, strain, interface disorder, and the effects of alloying and sample preparation. Much work has been done in this class of cubic zincblende crystals since the first laser measurements of Hobden and Russell (1964) in zincblende GaP. The prototypical system that has been studied extensively is GaAs, and comprehensive reviews are available (Loudon and Hayes, 1978; Cardona, 1975; Cardona and Güntherodt, 1982a, b, 1984, 1989, 1991). Frequencies of the LO and TO modes, ω_{LO} and ω_{TO} respectively, for some of these systems are listed in Table 4.1.

4.2.2 Würtzite structures

In the last several years Raman scattering has also contributed a great deal to the advances in understanding of the III-V nitride materials. The wealth of experiments and information collected over the past 25 years on the GaAs-based material systems is now starting to be duplicated in the nitride system, albeit somewhat slowly, as the growth techniques and material systems continue to improve.

GaN-, AlN- and InN-based materials are highly stable in the hexagonal würtzite structure although they can be grown in the zincblende phase and unintentional phase separation and coexistence may occur. The würtzite crystal structure belongs to the space group C_{6v}^4 and group theory predicts zone-center optical modes are $A_1, 2B_1, E_1$ and $2E_2$. The A_1 and E_1 modes and the two E_2 modes are Raman active while the B modes are silent. The A and E modes are polar, resulting in a splitting of the LO and the TO modes (Hayes and Loudon, 1978). The Raman tensors for the würtzite structure are as follows:

$$
\begin{pmatrix} a & 0 & 0 \\ 0 & a & 0 \\ 0 & 0 & b \end{pmatrix} \qquad A_1(z) \text{ mode}
$$

$$
\begin{pmatrix} 0 & 0 & c \\ 0 & 0 & 0 \\ c & 0 & 0 \end{pmatrix} \qquad E_1(x) \text{ mode}
$$

(4.5)

$$
\begin{pmatrix} 0 & 0 & 0 \\ 0 & 0 & c \\ 0 & c & 0 \end{pmatrix} \qquad E_1(y) \text{ mode}
$$

$$
\begin{pmatrix} f & 0 & 0 \\ 0 & -f & 0 \\ 0 & 0 & 0 \end{pmatrix} \begin{pmatrix} 0 & -f & 0 \\ -f & 0 & 0 \\ 0 & 0 & 0 \end{pmatrix} \qquad E_2 \text{ mode.}
$$

The vibrational modes in würtzite structures are given in Figure 3.3. Details of the frequencies are given in Table 4.2.

Table 4.1. Frequencies in cm^{-1} of the LO and TO modes for zincblende crystals.

	ω_{LO} (cm^{-1})	ω_{TO} (cm^{-1})
AlN	902	655
GaAs	292	269
GaN	740	554
GaP	403	367
InP	345	304
ZnS	352	271

Following some early work (Manchon *et al.*, 1970; Lemos *et al.*, 1972; Burns *et al.*, 1973) there has been a number of more recent experiments (Murugkar *et al.*, 1995; Cingolani *et al.*, 1986; Azuhata *et al.*, 1995) identifying the Raman modes in these nitride materials. The early work was mainly on crystals in the form of needles and platelets and the more recent work has been on epitaxial layers grown on sapphire, on 6H-SiC, and on ZnO as well as some more unusual substrates. Table 4.2 gives the Raman modes as well as the scattering geometry in which they were observed in the experiments of Azuhata *et al.* (1995). Experiments on AlN and InN crystallites and films, particularly for the latter material, are more scarce, reflecting the difficulties in achieving good growth qualities for these materials. In uniaxial materials, when the long-range electrostatic field interactions of the polar phonons dominate the short-range field of the vibrational force constants, phonons of mixed symmetry can be observed (Loudon, 1964) under specific conditions of propagation direction and polarization. They have been seen in the case of AlN (Bergman *et al.*, 1999).

4.3   Lifetimes in zincblende and würtzite crystals

Phonon–carrier interactions have an impact on semiconductor device performance and, hence, a knowledge of the phonon lifetimes is important. Phonon lifetimes demonstrate the effects of anharmonic interactions as well as scattering via point defects and impurities. Anharmonic interactions (Klemens, 1958; Klemens, 1966; Borer *et al.*, 1971; Debernardi, 1998; Menéndez and Cardona, 1984; Ridley, 1996) include the decay of phonons into other normal modes with the conservation of energy and momentum. For a three-phonon decay process, a phonon of frequency ω_1 and wavevector $\mathbf{q}_1$ decays into two phonons of frequencies ω_2 and ω_3, with wavevectors $\mathbf{q}_2$ and $\mathbf{q}_3$ respectively, such that $\omega_1 = \omega_2 + \omega_3$ and $\mathbf{q}_1 = \mathbf{q}_2 + \mathbf{q}_3$.

The investigation of the dynamical behavior of the vibrational modes provides a direct measure of the electron–phonon interaction. The measurement of the decay

Table 4.2. Frequencies in cm^{-1} of the vibrational modes in some würtzite structures.

	ω_{AlN}	ω_{CdS}	ω_{GaN}	ω_{InN}	ω_{ZnO}
E_2^1	252	44	144		101
E_2^2	660	252	569	495	437
$A_1(TO)$	614	228	533		380
$A_1(LO)$	893	305	735	596	574
$E_1(TO)$	673	235	561		407
$E_1(LO)$	916	305	743		583

of the optical modes, which involves the anharmonic effects mentioned previously, will be discussed here. Other processes that give experimental information on the electron–phonon interaction include the generation of optical phonons by high-energy carriers, intervalley scattering between different minima in the conduction band, and carrier–carrier scattering; these are reported by Kash and Tsang (1991) for the prototypical system of GaAs.

Measurements of phonon linewidths for Raman and infrared measurements in GaAs, ZnSe, and GaP give phonon lifetimes of 2–10 ps (von der Linde, 1980; Menéndez and Cardona, 1984). For systems that are not far from equilibrium, the lifetimes of the phonons can be described by anharmonic processes. The decay of an optical phonon is frequently via pairs of acoustic phonons or via one acoustic phonon and one optical phonon of appropriate energies and momenta (Cowley, 1963; Klemens, 1966). The first measurements with continuous-wave pumping (Shah *et al.*, 1970) of highly non-equilibrium LO phonons in GaAs yielded estimates of LO-phonon lifetimes of approximately 5 ps at room temperature. This was consistent with values obtained from linewidth studies. von der Linde (1980) used time-resolved Raman scattering to obtain directly the time decay of non-equilibrium LO phonons. They obtained a value of 7 ps for GaAs LO phonons at 77 K. Subsequent experiments by Kash *et al.* (1985) led to the conclusion that the LO phonon lifetime in GaAs was limited by its anharmonic decay into two acoustic phonons.

Kash *et al.* (1987, 1988) and Tsen and Morkoç (1988a, b) used time-resolved Raman scattering for the alloy system AlGaAs. The results for the lifetimes are similar to those for pure GaAs; here, though, the phonon linewidths are broadened owing to the disorder of the alloys and these inhomogeneous broadening effects need to be considered. Secondly, although the dispersion relations of AlAs are different from those of GaAs there is a similarity in decay times that is interesting and unexpected. Tsen (1992) and Tsen *et al.* (1989) reported on the use of time-resolved Raman studies of non-equilibrium LO phonons in GaAs-based structures.

Tsen *et al.* (1996, 1997, 1998) have studied the electron–phonon interactions in GaN of würtzite structure via picosecond and sub-picosecond Raman spectroscopy. Results on undoped GaN with an electron density of $n = 5 \times 10^{16}$ cm^{-3} showed that the relaxation mechanism of the hot electrons is via the emission of LO phonons and that the Fröhlich interaction is much stronger than the deformation-potential interaction in that material. The measured lifetime was found to be 3 ps at 300 K and 5 ps at 5–25 K (Tsen *et al.*, 1996, 1997, 1998). The electron–LO-phonon scattering rate was seen to be an order of magnitude larger than that for GaAs and was attributed to the much larger ionicity in GaN. These experiments also indicated that the longitudinal phonons decay into a TO and an LO phonon or two TO phonons.

Raman investigations of phonon lifetimes have been reported by Bergman *et al.* (1999) in GaN, AlN, and ZnO würtzite crystals. These lifetimes were obtained from measured Raman linewidths using the uncertainty relation, after correcting for instrument broadening (Di Bartolo, 1969). These results demonstrate that the E_2^1 mode has a lifetime of 10 ps, an order of magnitude greater than that of the E_2^2, $E_1(\text{TO})$, $A_1(\text{TO})$ and $A_1(\text{LO})$ modes. This result was found to be true for samples of high-quality GaN, AlN, ZnO as well as for AlN with a high level of impurities. An explanation of the relative long lifetime of the E_2^2 phonons was given in terms of factors including energy conservation constraints, density of final states, and anharmonic interaction coefficients. The E_2^2 mode lies at the lowest energy of the optical phonon modes in the würtzite dispersion curves (Nipko et al., 1998; Nipko and Loong, 1998; Hewat, 1970) and only the acoustic phonons provide channels of decay. At the zone edges, the acoustic phonons are equal to or larger than those of the E_2^2 mode. Thus, for energy conservation to hold, the E_2^2 phonons have to decay to acoustic phonons at the zone center, where their density is low.

4.4   Ternary alloys

The phonons of the ternary alloys AB_xC_{1-x} formed from the binaries AB and AC crystals in the III-V as well as the II-VI semiconductors have been studied for some time (Chang and Mitra, 1968). The III-nitrides have been studied more recently and the alloys of the würtzite materials show some interesting features (Hayashi et al., 1991; Behr et al., 1997; Cros et al., 1997; Demangeot et al., 1998; Wisniewski et al., 1998). The AB_xC_{1-x} mixed crystals of the zincblende materials fall into two main groups when classified according to the characteristics of the phonons. These two classes are generally referred to as one-mode or two-mode behavior, where 'one-mode' refers to the situation where the frequency of the AB phonons gradually approaches the frequency of the AC phonons as the x-value of the alloy increases. In the two-mode situation, the phonon frequencies are distinct and in the limit of $x = 0$ (1) the AC (AB) phonon frequency is a local mode in the AB (AC) crystal. Intermediate behavior has also been observed for certain crystals (Lucovski and Chen, 1970). While there is no general agreement, several criteria for phonon-mode behavior based on the mass differences of the atoms have been proposed (Chang and Mitra, 1968). Typically, when the frequencies of the phonons in the AB and the AC binary crystals are very different a two-mode behavior is expected; otherwise, a one-mode behavior is seen. There is more uncertainty as well as a smaller number of reports in the case of the würtzite nitrides. Hayashi et al. (1991) reported studies on AlGaN würtzite films in the range $0 < x < 0.15$. The E_2, $E_1(\text{TO})$, $E_1(\text{LO})$ and $A_1(\text{TO})$ modes were investigated and, in the composition range studied, one-mode behavior was observed. Similar results were obtained by Behr et al. (1997) in a narrow composition range. The E_2 mode was seen to be unaffected by a change in composition. Cros et al. (1997) studied the AlGaN alloys over the whole concentration range. They concluded that the E_2 mode exhibits two-mode character, while the $A_1(\text{LO})$ mode is one-mode; the results for the

A_2(TO) mode were inconclusive. Studies by Demangeot *et al.* (1998) concluded that the A_1(LO), A_1(TO), and E_1(TO) modes all exhibit one-mode behavior. However, infrared reflectance experiments indicate that the E_1(TO) mode displays two-mode behavior (Wisniewski *et al.*, 1998).

Yu *et al.* (1998) extended the modified random-element isodisplacement model developed for zincblende structures by including the additional phonon modes and the anisotropy of the würtzite structure. According to this model, the E_1 and the A_1 phonon modes should show one-mode behavior. In zincblende AlGaN crystals, the results of Harima *et al.* (1999) indicate that the LO-phonon shows one-mode behavior while the TO mode shows two-mode behavior.

Raman experiments using two ultraviolet wavelengths were performed by Alexson *et al.* (2000) on InGaN in the range $0 < x < 0.5$. They investigated the A_1 and the E_2 phonons. These studies show a one-mode behavior of the A_1(LO) phonon while the E_2 phonon demonstrates a two-mode characteristic.

The fact that the E_2 mode behaves differently from the E_1 and A_1 modes is not surprising, when one considers the specific atoms giving rise to the vibrations. This in fact emerges from experiments on GaN würtzite films from natural GaN as well as from GaN containing the isotope ^{15}N reported by Zhang *et al.* (1997). All the A_1 and the E_1 modes observed in the ^{15}N isotope were seen to shift to lower frequencies. Niether the E_2^1 nor E_2^2 mode showed a similar shift; the E_2^1 mode was essentially unaffected by the different isotopic mass. The authors thus concluded that the E_2^1 vibration is due to the motion of the Ga atoms, which are heavy, alone and, thus, that there is no frequency response to an isotropic change in the nitrogen mass, which is considerably lighter.

4.5   Coupled plasmon–phonon modes

A coupling of the LO phonons to the plasma oscillations of the free carriers – these oscillations are known as plasmons – occurs when an appreciable free-carrier concentration is present in a polar semiconductor. These coupled phonon–plasmon modes may be observed by Raman scattering, so providing information about the free-carrier density of a given sample.

The free electrons scatter light weakly, although in solids the effect is enhanced by band structure effects. Detailed accounts are provided by Platzman and Wolff (1973), Yafet (1966), and Klein (1975). Mooradian and Wright (1966) made the first observation of plasmons in n-GaAs using a Nd:YAG laser. These results demonstrated the coupling between the LO phonon and the plasmon via the interaction of the longitudinal electric fields produced by each of these excitations. Other systems for which plasmon–phonon scattering has been observed include zincblende structures such as InSb, GaP, and InAs (Hon and Faust, 1973; Patel and Slusher, 1968). Various würtzite crystals have been investigated, for example CdS

(Scott *et al.*, 1969), SiC (Klein *et al.*, 1972) and, more recently, the nitrides discussed in the next paragraph.

Here, we summarize the experimental Raman scattering results of phonon–plasmon coupled modes in GaN, which as grown tends to be an n-type material (Edgar, 1994); a higher carrier concentration is achieved via intentionally doping the material. Due to this fairly high carrier concentration, the plasmons in GaN are considered to be overdamped, similarly to those in SiC (Klein *et al.*, 1972). Experiments by Kozawa *et al.* (1998) in GaN films with a relatively low concentration of carriers show a broadening and a weakening of the intensity of the Raman features as well as a shift to higher frequencies. The carrier concentrations obtained by fitting the Raman lineshape as in Klein *et al.* (1972), Hon and Faust (1973), and Irmer *et al.* (1983) give values similar to those obtained from Hall measurements (Kozawa *et al.*, 1998). Other experiments on GaN to study these effects at lower free-carrier concentrations have been carried out by Wetzel *et al.* (1996) and Ponce *et al.* (1996) and show similar results. Kirillov *et al.* (1996) studied the effect on the Raman modes in the high-carrier-concentration limit. This study found that with these higher carrier concentrations the Raman features corresponding to the upper branch of the phonon–plasmon coupled modes are too broad to extract meaningful information. Demangeot *et al.* (1997) carried out a study of the lower branch of the coupled phonon–plasmon mode for GaN for higher carrier concentrations. The broad Raman peak observed could be fitted by the models referenced above (Klein *et al.*, 1972; Hon and Faust, 1973).

Chapter 5

Occupation number representation

Oh mighty-mouthed inventor of harmonies.
Alfred, Lord Tennyson, 1863

5.1 Phonon mode amplitudes and occupation numbers

In the study of carrier–phonon interactions in nanostructures it is convenient to use the so-called phonon-number-occupation basis. In this basis the phonon system is modeled by the Hamiltonian for a simple harmonic oscillator. Specifically, the familiar conjugate variables of position and momentum are replaced by creation and annihilation operators. These creation and annihilation operators act on states each having a given number of phonons. In particular, the creation operator acting on a state of n_q phonons of wavevector q increases the number of phonons to $n_q + 1$ and the phonon annihilation operator acting on a state of n_q phonons of wavevector q decreases the number of phonons to $n_q - 1$.

In some applications – such as those involving the ground state of a Bose–Einstein condensation – the creation and annihilation operators are essential to describing the physical properties of the system. However, in the applications considered in this book, these operators are used merely as a convenient way of keeping track of the number of phonons before and after a carrier–phonon scattering event; they do not introduce new physics. Nevertheless, it is important to understand these operators since they are used widely by the semiconductor community in modeling the electronic and optical properties of bulk semiconductors as well as nanostructures.

The Hamiltonian describing the harmonic oscillator associated with a phonon mode of wavevector q is

$$H_q = \frac{p_q^2}{2m} + \frac{1}{2}m\omega_q^2 u_q^2, \tag{5.1}$$

where m is the mass of the oscillator, ω_q is the frequency of the phonon, u_q is the displacement associated with it, and p_q is its momentum. Introducing the operators, a_q and $a_q^\dagger$,

$$a_q = \sqrt{\frac{m\omega_q}{2\hbar}}u_q + i\sqrt{\frac{1}{2\hbar m\omega_q}}p_q \tag{5.2}$$

and

$$a_q^\dagger = \sqrt{\frac{m\omega_q}{2\hbar}}u_q - i\sqrt{\frac{1}{2\hbar m\omega_q}}p_q, \tag{5.3}$$

it is straightforward to show that

$$a_q^\dagger a_q = \left(\sqrt{\frac{m\omega_q}{2\hbar}}u_q - i\sqrt{\frac{1}{2\hbar m\omega_q}}p_q\right)\left(\sqrt{\frac{m\omega_q}{2\hbar}}u_q + i\sqrt{\frac{1}{2\hbar m\omega_q}}p_q\right)$$

$$= \frac{1}{2}\frac{m\omega_q^2}{\hbar\omega_q}u_q^2 + \frac{1}{2m}\frac{1}{\hbar\omega_q}p_q^2 + \frac{i}{2\hbar}[u_q, p_q]. \tag{5.4}$$

Here the commutator $[u_q, p_q] \equiv u_q p_q - p_q u_q = i\hbar$, from the properties of the quantum mechanical operators u_q and p_q. Thus

$$\frac{p_q^2}{2m} + \frac{1}{2}m\omega_q^2 u_q^2 = \hbar\omega_q\left(a_q^\dagger a_q + \frac{1}{2}\right). \tag{5.5}$$

Since the energy of a quantum-mechanical harmonic oscillator is $\hbar\omega_q(n_q + \frac{1}{2})$, where n_q is the number of phonons having wavevector q, it is clear that $N_q = a_q^\dagger a_q$ operating on an eigenstate of N_q phonons $|N_q\rangle$ has eigenvalue N_q. That is, $N_q|N_q\rangle = n_q|N_q\rangle$. Moreover, by calculating $a_q a_q^\dagger$ in the same manner used to derive an expression for $a_q^\dagger a_q$, it follows that $a_q a_q^\dagger - a_q^\dagger a_q = [a_q, a_q^\dagger] = 1$, $[a_q, N_q] = a_q a_q^\dagger a_q - a_q^\dagger a_q a_q = [a_q, a_q^\dagger]a_q = a_q$ and $[a_q^\dagger, N_q] = a_q^\dagger a_q^\dagger a_q - a_q^\dagger a_q a_q^\dagger = a_q^\dagger[a_q^\dagger, a_q] = -a_q^\dagger$.

Accordingly,

$$N_q(a_q^\dagger|N_q\rangle) = (a_q^\dagger N_q + a_q^\dagger)|N_q\rangle = (n_q + 1)a_q^\dagger|N_q\rangle \tag{5.6}$$

and

$$N_q(a_q|N_q\rangle) = (a_q N_q - a_q)|N_q\rangle = (n_q - 1)a_q|N_q\rangle. \tag{5.7}$$

Thus, $a_q^\dagger$ acting on $|N_q\rangle$ gives a new eigenstate with eigenvalue increased by 1 and a_q acting on $|N_q\rangle$ gives a new eigenstate with eigenvalue decreased by 1; that is,

$$a_q^\dagger |N_q\rangle = \sqrt{n_q + 1}|N_q + 1\rangle \tag{5.8}$$

and

$$a_q |N_q\rangle = \sqrt{n_q}|N_q - 1\rangle, \tag{5.9}$$

where the eigenvalues are consistent with the relation $N_q|N_q\rangle = n_q|N_q\rangle$ with $N_q = a_q^\dagger a_q$. The eigenstates $|N_q\rangle$ are orthonormal, so that

$$\langle N_q'||N_q\rangle \equiv \langle N_q'|N_q\rangle = \delta_{N_q',N_q} \tag{5.10}$$

where δ_{N_q',N_q} is the Kronecker delta function.

Thus, the only non-zero matrix elements of $a_q^\dagger$ are those that couple states $\langle N_q'|$ and $|N_q\rangle$ for which $N_q' = N_q + 1$:

$$\langle N_q'|a_q^\dagger|N_q\rangle = \sqrt{n_q + 1}\,\delta_{N_q',N_q+1}. \tag{5.11}$$

Likewise,

$$\langle N_q'|a_q|N_q\rangle = \sqrt{n_q}\,\delta_{N_q',N_q-1}. \tag{5.12}$$

The phonon occupation number n_q may be determined at a temperature T from the relation $\langle \epsilon \rangle = \hbar\omega_q(n_q + \frac{1}{2})$, where the average energy $\langle \epsilon \rangle$ is calculated for the case where the eigenenergies are those of a harmonic oscillator, $E_n = \hbar\omega_q(n + \frac{1}{2})$. Then taking the weighting factor to be the Boltzmann factor $f(E_n) = e^{-E_n/k_BT}$, we find that

$$\langle \epsilon \rangle = \sum E_n f(E_n) \Big/ \sum f(E_n), \tag{5.13}$$

where the sums are over all n from 0 to ∞. Then

$$\begin{aligned}
\langle \epsilon \rangle &= \left(\sum \hbar\omega_q n e^{-\hbar\omega_q n/k_BT}\right) \Big/ \left(\sum e^{-\hbar\omega_q n/k_BT}\right) + \frac{\hbar\omega_q}{2} \\
&= \left(\sum \hbar\omega_q n x^n\right) \Big/ \left(\sum x^n\right) + \frac{\hbar\omega_q}{2},
\end{aligned} \tag{5.14}$$

where $x = e^{-\hbar\omega_q/k_BT}$. Then

$$\begin{aligned}
\langle \epsilon \rangle &= \left(\hbar\omega_q x \frac{d}{dx}\sum x^n\right) \Big/ \left(\sum x^n\right) + \frac{\hbar\omega_q}{2} \\
&= \hbar\omega_q \frac{x/(1-x)^2}{1/(1-x)} + \frac{\hbar\omega_q}{2} \\
&= \frac{\hbar\omega_q x}{1-x} + \frac{\hbar\omega_q}{2} \\
&= \frac{\hbar\omega_q}{e^{-\hbar\omega_q/k_BT} - 1} + \frac{\hbar\omega_q}{2},
\end{aligned} \tag{5.15}$$

and it follows that

$$n_q = \frac{1}{e^{-\hbar\omega_q/k_BT} - 1},\tag{5.16}$$

which is known as the Bose–Einstein distribution. Equations (5.10)–(5.12) and (5.16) are used frequently in calculating carrier–phonon scattering rates in nanostructures. Such calculations will be the subject of following chapters. In these calculations the carrier wavefunctions and the phonon eigenstates are written as products to express the total wavefunctions for the system of carriers and phonons. When matrix elements are evaluated between the final and initial states of the system, the phonon eigenstates and the phonon operators are grouped together since they commute with the carrier wavefunctions and operations. Thus, carrier and phonon matrix elements always appear as products and are evaluated separately. This procedure will be illustrated in Chapters 8, 9, and 10. In Sections 5.2–5.4 we will relate the phonon displacement amplitudes to the interaction Hamiltonians describing the dominant carrier–phonon interaction processes. In these calculations and those of Chapters 8, 9, and 10 it will be convenient to express the normal-mode phonon displacement u_q in terms of the phonon creation and annihilation operators, $a_q^\dagger$ and a_q respectively. By adding equations (5.11) and (5.12) it will be seen readily that

$$u_q = \sqrt{\frac{\hbar}{2m\omega_q}}(a_q + a_q^\dagger).\tag{5.17}$$

In calculations of carrier–phonon scattering probabilities, the normal-mode phonon displacement $\mathbf{u}(\mathbf{r})$ will appear in various linear forms in the Hamiltonians entering the matrix elements being evaluated. In evaluating these matrix elements, it will be important to keep track of not only the phonon occupation numbers n_q emerging from the phonon matrix elements but also the necessary conservation of momentum and energy for each scattering process. In these matrix elements either phonon creation or phonon absorption will take place but not both; therefore, as is manifest from equations (5.11) and (5.12), only one of the two terms appearing in equation (5.17) will contribute to the process under consideration.

$\mathbf{u}(\mathbf{r})$ is, of course, a Fourier series over the modes u_q. In phonon absorption processes the phonon appears as an incoming wave and the factor $e^{i(\mathbf{q}\cdot\mathbf{r}-\omega t)}$ multiplies the amplitudes associated with the phonon fields. Likewise, in phonon emission processes the phonon appears as an outgoing wave and the factor $e^{i(-\mathbf{q}\cdot\mathbf{r}-\omega t)}$ multiplies the amplitudes associated with the phonon fields. These factors are essential in ensuring proper conservation of momentum and energy and it is convenient to include them along with the associated creation or annihilation operator. Indeed, they appear naturally in the Fourier decomposition of $\mathbf{u}(\mathbf{r})$.

Moreover, each incoming or outgoing phonon will be associated with a unit polarization vector; these unit polarization vectors will be denoted by $\hat{\mathbf{e}}_{\mathbf{q},j}$ for incoming waves and by $\hat{\mathbf{e}}^*_{\mathbf{q},j}$ for outgoing waves. The factor $e^{-i\omega t}$ is common to both incoming and outgoing phonons and it is generally included – along with factors of $E/\hbar$ associated with the carrier phases – in the integral over time that appears in the Fermi golden rule. As a means of including the phase factors to ensure proper accounting of energy and momentum as well as the appropriate unit polarization vectors, (5.17) will now be written as a sum over all wavevectors q; the appropriate non-temporal phase factors appear as multipliers of the corresponding phonon operators:

$$
\mathbf{u}(\mathbf{r}) = \frac{1}{\sqrt{N}} \sum_q \sum_{j=1,2,3} \sqrt{\frac{\hbar}{2m\omega_q}} \left(a_q e^{i\mathbf{q}\cdot\mathbf{r}} \hat{\mathbf{e}}_{\mathbf{q},j} + a_q^\dagger e^{-i\mathbf{q}\cdot\mathbf{r}} \hat{\mathbf{e}}^*_{\mathbf{q},j} \right)
$$

$$
= \frac{1}{\sqrt{N}} \sum_q \sum_{j=1,2,3} \sqrt{\frac{\hbar}{2m\omega_q}} \hat{\mathbf{e}}_{\mathbf{q},j} (a_q + a_{-q}^\dagger) e^{i\mathbf{q}\cdot\mathbf{r}} \equiv \sum_q \mathbf{u}(\mathbf{q}) e^{i\mathbf{q}\cdot\mathbf{r}}.
\tag{5.18}
$$

where q is summed over all wavevectors in the Brillouin zone and N is the number of unit cells in the sample. Chapter 7 will treat the role of dimensional confinement in modifying the optical and acoustic phonon modes in nanostructures. Two major modifications result: the phase space is restricted and the plane-wave nature of the phonon is modified. Both of these effects will be treated by appropriate modifications of equation (5.18): the changes in the phase space due to dimensional confinement will be addressed by modifying the sum over q and the dimensional confinement of the plane wave will be described by introducing suitable envelope functions. Equations (5.8)–(5.18) will find many applications here and in Chapters 8, 9 and 10.

The factor $\sqrt{\hbar/2m\omega_q}$ in the final expression for $\mathbf{u}(\mathbf{r})$ ensures that the desired Hamiltonian is consistent with the phonon mode amplitude. It is convenient in practical calculations of $\mathbf{u}(\mathbf{r})$ to cast the amplitude constraints implied by $[u_q, p_q] = i\hbar$ in the form of an integral:

$$
\int \mathbf{u}^*(\mathbf{r})\mathbf{u}(\mathbf{r})d^3\mathbf{r} = \sum_q \sum_{q'} \frac{\hbar}{2Nm} \frac{1}{\sqrt{\omega_q \omega_{q'}}} \hat{\mathbf{e}}_q \cdot \hat{\mathbf{e}}_{q'} \int e^{-i(\mathbf{q}'-\mathbf{q})\cdot\mathbf{r}} d^3\mathbf{r}
$$

$$
= \frac{1}{nm} \sum_q \frac{\hbar}{2\omega_q},
\tag{5.19}
$$

where $n = N/V$ and the integral is performed through use of the identity $\int e^{-i(\mathbf{q}'-\mathbf{q})\cdot\mathbf{r}}d^3\mathbf{r} = V\delta_{q,q'}$. For a single mode q, the so-called phonon normalization condition then becomes

$$
\int [\sqrt{nm}\mathbf{u}^*(\mathbf{r})] \cdot [\sqrt{nm}\mathbf{u}(\mathbf{r})]d^3\mathbf{r} = \frac{\hbar}{2\omega_q}.
\tag{5.20}
$$

In terms of $\mathbf{u}(\mathbf{q})$, the normalization condition is then

$$
[\sqrt{nm}\mathbf{u}^*(\mathbf{q})] \cdot [\sqrt{nm}\mathbf{u}(\mathbf{q})] = \frac{\hbar}{2\omega_q} \frac{1}{V}.
\tag{5.21}
$$

In the literature, quantized fields – such as the displacement field $\mathbf{u}(\mathbf{r})$ – are expressed in terms of a_q and $a_q^\dagger$ in a number of different ways. Indeed, it is possible to perform canonical transformations that essentially exchange the roles of u_q and p_q. Specifically, from the definitions of a_q and $a_q^\dagger$ in terms of u_q and p_q, it follows that

$$u_q' = \sqrt{\frac{\hbar}{2m\omega_q}}(a_q' + a_q'^\dagger) \quad \text{and} \quad \frac{p_q'}{m\omega_q} = -i\sqrt{\frac{\hbar}{2m\omega_q}}(a_q' - a_q'^\dagger),$$

$$(5.22)$$

where the primes have been added in preparation for the following substitutions: $a_q' \to -ia_q$ and $a_q'^\dagger \to +ia_q^\dagger$. With these substitutions it follows that

$$u_q' \to \frac{p_q}{m\omega_q} \quad \text{and} \quad \frac{p_q'}{m\omega_q} \to -u_q.$$

$$(5.23)$$

Thus, the canonical transformation $a_q' \to -ia_q$ has the result that the roles of u_q and $p_q/m\omega_q$ are interchanged. Clearly, this canonical transformation leaves the harmonic-oscillator Hamiltonian unchanged. This is, of course, true for this Hamiltonian whether expressed in terms of u_q and p_q or in terms of a_q and $a_q^\dagger$. As a result of this canonical transformation, the quantized fields – such as the displacement field $\mathbf{u}(\mathbf{r})$ – may be expressed in terms of a_q and $a_q^\dagger$ in a number of different ways. Indeed, the literature testifies to the fact that all the different forms are used widely. Evidently, factors of $\pm i$ as multipliers of a_q and $a_q^\dagger$ do not change the matrix element of any such field times its complex conjugate, since $(\pm i)(\pm i)^* = 1$; thus, quantum-mechanical transition rates are not affected by this transformation, as should clearly be the case. There is one other invariance that is used frequently to rewrite fields expressed in terms of a_q and $a_q^\dagger$ in the most convenient form for a particular application. Namely, in expressions where a sum or integral over positive and negative values of q is present, substitutions of the type $a_q e^{i\mathbf{q}\cdot\mathbf{r}} \to a_{-q}e^{-i\mathbf{q}\cdot\mathbf{r}}$ leave the expression unchanged if all other multiplicative factors contain even powers of q. Again, the quantum-mechanical transition rates are not affected by such a change of variable.

5.2   Polar-optical phonons: Fröhlich interaction

One of the most important carrier–phonon scattering mechanisms in semiconductors occurs when charge carriers interact with the electric polarization, $\mathbf{P}(\mathbf{r})$, produced by the relative displacement of positive and negative ions. In low-defect polar semiconductors such as GaAs, InP, and GaN, carrier scattering in polar semiconductors at room temperature is dominated by this polar–optical-phonon (POP) scattering mechanism. The POP–carrier interaction is referred to as the Fröhlich interaction, after H. Fröhlich, who formulated the first qualitatively correct formal description. In this book, the potential energy associated with the Fröhlich interaction will be denoted by $\phi_{Fr}(\mathbf{r})$. Clearly the polarization $\mathbf{P}$ associated with polar-optical phonons and the potential energy associated with the Fröhlich interaction, $\phi_{Fr}(\mathbf{r})$, are related by

$$\nabla^2 \phi_{Fr}(\mathbf{r}) = 4\pi e \nabla \cdot \mathbf{P}(\mathbf{r}). \tag{5.24}$$

In terms of the phonon creation and annihilation operators of Section 5.1, $\mathbf{P}(\mathbf{r})$ may be written as

$$\mathbf{P}(\mathbf{r}) = \zeta \sum_{j=1,2,3} \int \frac{d^3\mathbf{q}}{(2\pi)^3} \left(a_{\mathbf{q}} e^{i\mathbf{q}\cdot\mathbf{r}} \mathbf{e}_{\mathbf{q},j} + a_{\mathbf{q}}^\dagger e^{-i\mathbf{q}\cdot\mathbf{r}} \mathbf{e}_{\mathbf{q},j}^* \right) \tag{5.25}$$

where $\mathbf{e}_{\mathbf{q},j}$ represents the polarization vector associated with $\mathbf{P}(\mathbf{r})$ and $\mathbf{q}$ is the phonon wavevector; then, it follows that

$$4\pi \nabla \cdot \mathbf{P}(\mathbf{r}) = 4\pi i \zeta \sum_{j=1,2,3} \int \frac{d^3\mathbf{q}}{(2\pi)^3} \left(a_{\mathbf{q}} e^{i\mathbf{q}\cdot\mathbf{r}} \mathbf{q} \cdot \mathbf{e}_{\mathbf{q},j} - a_{\mathbf{q}}^\dagger e^{-i\mathbf{q}\cdot\mathbf{r}} \mathbf{q} \cdot \mathbf{e}_{\mathbf{q},j}^* \right).$$

$$\tag{5.26}$$

Consider the case of a polar crystal with two atoms per unit cell, such as GaAs. Clearly, the dominant contribution to $\mathbf{P}(\mathbf{r})$ results from the phonon modes in which the normal distance between the planes of positive and negative charge varies. Such modes are obviously the LO modes since in the case of LO modes $\mathbf{e}_{\mathbf{q},j}$ is parallel to $\mathbf{q}$. However, TO phonons produce displacements of the planes of charge such that they remain at fixed distances from each other; that is, the charge planes 'slide' by each other but the normal distance between planes of opposite charge does not change. So, TO modes make negligible contributions to $\mathbf{P}(\mathbf{r})$. For TO phonons, $\mathbf{e}_{\mathbf{q},j} \cdot \mathbf{q} = 0$. Accordingly,

$$4\pi \nabla \cdot \mathbf{P}(\mathbf{r}) = 4\pi i \zeta \int \frac{d^3\mathbf{q}}{(2\pi)^3} \left(a_{\mathbf{q}} e^{i\mathbf{q}\cdot\mathbf{r}} q - a_{\mathbf{q}}^\dagger e^{-i\mathbf{q}\cdot\mathbf{r}} q \right), \tag{5.27}$$

and the potential energy associated with the Fröhlich interaction, $\phi_{Fr}(\mathbf{r})$, is given by

$$H_{Fr} = \phi_{Fr}(\mathbf{r}) = -4\pi i e \zeta \int \frac{d^3\mathbf{q}}{(2\pi)^3} \frac{1}{q} \left(a_{\mathbf{q}} e^{i\mathbf{q}\cdot\mathbf{r}} - a_{\mathbf{q}}^\dagger e^{-i\mathbf{q}\cdot\mathbf{r}} \right), \tag{5.28}$$

where $\phi_{Fr}(\mathbf{r})$, has been denoted by H_{Fr}, the Fröhlich interaction Hamiltonian since $\phi_{Fr}(\mathbf{r})$ is the only term contributing to it.

The dependence of $\phi_{Fr}(\mathbf{r})$ on q^{-1} is familiar from the Coulomb interaction; the coupling constant ζ remains to be determined.

From subsection 2.3.3, the electric polarization $\mathbf{P}(\mathbf{r})$ may be written as

$$
\begin{aligned}
\mathbf{P}(\mathbf{r}) &= \frac{Ne^*}{\epsilon(\infty)}\mathbf{u}_\mathbf{q}(\mathbf{r}) \\
&= \frac{Ne^*}{\epsilon(\infty)}\frac{1}{\sqrt{N}}\sum_q \sum_{j=1,2,3}\sqrt{\frac{\hbar}{2\left(\dfrac{mM}{m+M}\right)\omega_{\mathrm{LO}}}} \\
&\quad \times \left(a_\mathbf{q}e^{i\mathbf{q}\cdot\mathbf{r}}\mathbf{e}_{\mathbf{q},j} + a_\mathbf{q}^\dagger e^{-i\mathbf{q}\cdot\mathbf{r}}\mathbf{e}_{\mathbf{q},j}^*\right),
\end{aligned} \tag{5.29}
$$

where the division by $\epsilon(\infty)$ accounts for screening, and the normal-mode expression (5.18) has been used for $\mathbf{u}(\mathbf{r})$. By noticing that

$$
\frac{1}{\sqrt{V}}\sum_q \leftrightarrow \int \frac{d^3q}{(2\pi)^3} \tag{5.30}
$$

and by comparing expressions (5.25) and (5.29), it follows that

$$
\zeta = \frac{Ne^*}{\epsilon(\infty)}\frac{1}{\sqrt{N}}\sqrt{\frac{\hbar}{2\left(\dfrac{mM}{m+M}\right)\omega_{\mathrm{LO}}}}. \tag{5.31}
$$

However, from (2.26) evaluated for $\omega = \omega_0$, it follows that

$$
\begin{aligned}
\frac{Ne^{*2}}{\epsilon(\infty)^2}\left(\frac{1}{m}+\frac{1}{M}\right) &= \frac{\omega_{\mathrm{LO}}^2 - \omega_{\mathrm{TO}}^2}{4\pi\epsilon(\infty)} \\
&= \frac{\omega_{\mathrm{LO}}^2}{4\pi}\left[\frac{1}{\epsilon(\infty)}-\frac{1}{\epsilon(\infty)}\frac{\omega_{\mathrm{TO}}^2}{\omega_{\mathrm{LO}}^2}\right] \\
&= \frac{\omega_{\mathrm{LO}}^2}{4\pi}\left[\frac{1}{\epsilon(\infty)}-\frac{1}{\epsilon(0)}\right],
\end{aligned} \tag{5.32}
$$

so that

$$
\zeta = \frac{Ne^*}{\epsilon(\infty)}\frac{1}{\sqrt{N}}\sqrt{\frac{\hbar}{2\left(\dfrac{mM}{m+M}\right)\omega_{\mathrm{LO}}}} = \sqrt{\frac{\hbar}{2\omega_{\mathrm{LO}}}\frac{\omega_{\mathrm{LO}}^2}{4\pi}\left[\frac{1}{\epsilon(\infty)}-\frac{1}{\epsilon(0)}\right]}, \tag{5.33}
$$

and

$$
\begin{aligned}
H_{\mathrm{Fr}} = \phi_{\mathrm{Fr}}(\mathbf{r}) &= -4\pi i e\zeta \int \frac{d^3q}{(2\pi)^3}\frac{1}{q}\left(a_\mathbf{q}e^{i\mathbf{q}\cdot\mathbf{r}} - a_\mathbf{q}^\dagger e^{-i\mathbf{q}\cdot\mathbf{r}}\right) \\
&= -i\sqrt{\frac{2\pi e^2\hbar\omega_{\mathrm{LO}}}{V}\left[\frac{1}{\epsilon(\infty)}-\frac{1}{\epsilon(0)}\right]}\sum_q \frac{1}{q}\left(a_\mathbf{q}e^{i\mathbf{q}\cdot\mathbf{r}} - a_\mathbf{q}^\dagger e^{-i\mathbf{q}\cdot\mathbf{r}}\right). \quad (5.34)
\end{aligned}
$$

5.3 Acoustic phonons and deformation-potential interaction

The deformation-potential interaction arises from local changes in the crystal's energy bands arising from the lattice distortion created by a phonon. The deformation-potential interaction, introduced by Bardeen and Shockley, is one of the most important interactions in modern semiconductor devices and it has its origin in the displacements caused by phonons. Indeed, the displacements associated with a phonon set up a strain field in the crystal. In the simple case of a one-dimensional lattice, the energy of the conduction band, E_c, or the energy of the valence band, E_v, will change by an amount

$$E_{c,v} = E_{c,v}(a) - E_{c,v}(a + u), \tag{5.35}$$

where a is the lattice constant and u is the displacement produced by the phonon mode. Since $a \gg u$, it follows that

$$\Delta E_{c,v}(a) = (dE_{c,v}(a)/da)u. \tag{5.36}$$

Thus the phonon displacement field u produces a local change in the band energy; the energy associated with the change is known as the deformation potential and it represents one of the major scattering mechanisms in non-polar semiconductors. Indeed, the deformation-potential interaction is a dominant source of electron energy loss in silicon-based electronic devices. The three-dimensional generalization of $\Delta E_{c,v}$ is

$$\Delta E_{c,v}(a) = (dE_{c,v}(a)/dV)\Delta V, \tag{5.37}$$

where V is a volume element and ΔV is the change in the volume element due to the phonon field. For an isotropic medium $\Delta V/V = \nabla \cdot \mathbf{u}$ and the last expression becomes,

$$\Delta E_{c,v}(a) = V(dE_{c,v}(a)/dV)\nabla \cdot \mathbf{u}, \tag{5.38}$$

which is usually written as

$$H_{\text{def}}^{c,v} = \Delta E_{c,v}(a) = E_1^{c,v} \nabla \cdot \mathbf{u}. \tag{5.39}$$

The superscripts on $H_{\text{def}}^{c,v}$ and $E_1^{c,v}$ are necessary since the deformation potential for electrons is different from that for holes. Chapter 9 provides a discussion of the case where the medium is not assumed to be isotropic.

5.4 Piezoelectric interaction

The piezoelectric interaction occurs in all polar crystals lacking an inversion symmetry. In the general case, the application of an external strain to a piezoelectric crystal will produce a macroscopic polarization as a result of the displacements of ions. Thus an acoustic phonon mode will drive a macroscopic polarization in a piezoelectric crystal. In rectangular coordinates, the polarization created by the piezoelectric interaction in cubic crystals, including zincblende crystals, may be written as

$$\mathbf{P} = \left\{ \tfrac{1}{2}e_{x4}(\partial w/\partial y + \partial v/\partial z), \ \tfrac{1}{2}e_{x4}(\partial u/\partial z + \partial w/\partial x), \ \tfrac{1}{2}e_{x4}(\partial u/\partial y + \partial v/\partial x) \right\},$$
$$(5.40)$$

where e_{x4} is the piezoelectric coupling constant and, as will be described in Section 7.2, the factors multiplying e_{x4} are the components of the strain tensor that contribute to the piezoelectric polarization in a zincblende crystal. As with the Fröhlich and deformation-potential interactions, phonons play an essential role in producing piezoelectric interactions. Piezoelectric interactions will be discussed further in Chapter 9.

Chapter 6

Anharmonic coupling of phonons

With a name like yours, you might be any shape almost.
Lewis Carroll, Through the Looking Glass, 1872

6.1   Non-parabolic terms in the crystal potential for ionically bonded atoms

The crystal potential may be expanded in powers of the displacements of the ions from their equilibrium positions to yield a sum over quadratic and higher-order terms. The quadratic terms, of course, represent the harmonic modes considered at length in Chapter 5. The cubic and higher-order terms, containing products of three or more displacements, are generally known as the anharmonic terms. These anharmonic terms lead to modifications of the harmonic modes in the quadratic approximation of the harmonic oscillator. Indeed, while the harmonic approximation may be used to describe the phonon dispersion relations it cannot describe the decay of phonon modes: the anharmonic interaction is necessary to describe the decay of a phonon into other phonons. The leading term in the anharmonic interaction is the cubic term and it may be written as

$$H'_{q,j;q',j';q'',j''} = \frac{1}{\sqrt{N}} P(q, j; q', j'; q'', j'') \mathbf{u}_{q,j} \mathbf{u}_{q',j'} \mathbf{u}_{q'',j''}, \qquad (6.1)$$

where q, q', and q'' are the phonon wavevectors and j, j', and j'' designate the polarizations of the phonon modes participating in the anharmonic interaction. N is the number of unit cells in the crystal, and $P(q, j; q', j'; q'', j'')$ describes the anharmonic coupling. The normal process where $q - q' - q'' = 0$ is taken into account in this analysis but the umklapp process where $q - q' - q'' = b \neq 0$ is ignored since the normal process is expected to dominate for small phonon wavevectors. From Section 5.1, the displacement is given by

$$\mathbf{u}(\mathbf{r}) = \frac{1}{\sqrt{N}} \sum_{q} \sum_{j=1,2,3} \sqrt{\frac{\hbar}{2m\omega_{q,j}}} \left(a_q e^{i\mathbf{q}\cdot\mathbf{r}} \hat{\mathbf{e}}_{\mathbf{q},j} + a_q^\dagger e^{-i\mathbf{q}\cdot\mathbf{r}} \hat{\mathbf{e}}_{\mathbf{q},j}^* \right)$$

$$= \frac{1}{\sqrt{N}} \sum_{q} \sum_{j=1,2,3} \mathbf{u}_{q,j} \tag{6.2}$$

and, accordingly, the normal-mode coordinates $\mathbf{u}_{q,j}$ are

$$\mathbf{u}_{q,j} = \sqrt{\frac{\hbar}{2m\omega_{q,j}}} \left(a_q e^{i\mathbf{q}\cdot\mathbf{r}} \hat{\mathbf{e}}_{\mathbf{q},j} + a_q^\dagger e^{-i\mathbf{q}\cdot\mathbf{r}} \hat{\mathbf{e}}_{\mathbf{q},j}^* \right), \tag{6.3}$$

where, as defined in Chapter 5, the unit polarization vectors are denoted by $\hat{\mathbf{e}}_{\mathbf{q},j}$ for incoming waves and by $\hat{\mathbf{e}}_{\mathbf{q},j}^*$ for outgoing waves, and the phonon creation and annihilation operators are given by $a_q^\dagger$ and a_q, respectively.

6.2   Klemens' channel for the decay process LO → LA(1) + LA(2)

Klemens (1966) considered the anharmonic decay of an optical phonon into two longitudinal acoustic phonons, LA(1) and LA(2). The Klemens channel, LO → LA(1) + LA(2), is now recognized as the dominant LO phonon decay channel in a number of cubic crystals including GaAs (Tsen, 1992). For the Klemens channel,

$$q_{LO} = q'_{LA(1)} + q''_{LA(2)}. \tag{6.4}$$

Writing $q = q_{LO}$, $q' = q'_{LA(1)}$, and $q'' = q''_{LA(2)}$, the anharmonic interaction is

$$H'_{q,j;q',j';q'',j''} = \frac{1}{\sqrt{N}} \sqrt[3]{\frac{\hbar}{2m}} \left(\frac{1}{\omega_q \omega_{q'} \omega_{q''}} \right)^{1/2} P(q,q',q'') a_q a_{q'}^\dagger a_{q''}^\dagger e^{i(q-q'-q'')}. \tag{6.5}$$

The Klemens process is depicted in Figure 6.1 for bulk GaAs.

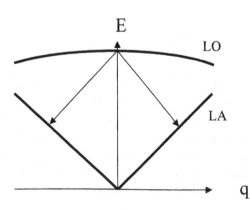

Figure 6.1. Decay of a near-zone-center LO phonon into two acoustic phonons for idealized dispersion curves (thick lines). The arrows begin at the energy and wavevector of the initial LO phonon and terminate at the final state LA phonon energies and wavevectors. Momentum and energy are both conserved in this process.

To determine $P(q, q', q'')$ it is necessary to characterize the anharmonic coupling. Direct measurement of the anharmonic potential is difficult and few *ab initio* calculations have been undertaken (Tua and Mahan, 1982; Tua, 1981). Keating (1966) used the theory of elasticity and exploited the relationship between the third-order elastic coefficients and the anharmonic terms of the crystal potential to determine $P(q, q', q'')$.

The theory of elasticity will be discussed further in Chapter 7 as the basis for a continuum model for acoustic phonons. Bhatt *et al.* (1994) used Keating's approach to estimate $P(q, q', q'')$ as the strain energy density U_a associated with the anharmonic third-order potential. They gave expressions for U_a in terms of the third-order elastic constants and the linear components of the strain variables. U_a is related to a Grüneisen constant and is treated frequently as a parameter. Herein, $P(q, q', q'')$ will be denoted as U_a.

6.3 LO phonon lifetime in bulk cubic materials

For the Klemens channel, LO $\rightarrow$ LA(1) + LA(2), the Fermi golden rule predicts a transition rate

$$\Gamma = \frac{2\pi}{\hbar} \sum_{q',q''} |M|^2 \delta(\hbar\omega_q - \hbar\omega_{q'} - \hbar\omega_{q''}), \tag{6.6}$$

where the matrix element satisfies

$$|M|^2 = \frac{\hbar^3 U_a^3}{8Nm^3} \left(\frac{1}{\omega_q \omega_{q'} \omega_{q''}} \right) n_q (n_{q'} + 1)(n_{q''} + 1)\delta_{q,q'+q''} \tag{6.7}$$

and, as derived in Section 5.1, the Bose–Einstein occupation number is given by

$$n_q = \frac{1}{e^{-\hbar\omega_q/k_B T} - 1}. \tag{6.8}$$

To calculate the net rate of phonon decay through the Klemens channel it is necessary to consider not only the decay rate Γ but also the generation rate for the reverse process, G. The total loss rate for the LO phonons is then given by $-\Gamma + G$. Writing Γ as $\Gamma' n_q (n_{q'} + 1)(n_{q''} + 1)$ it follows that G is given by $\Gamma'(n_q + 1)n_{q'}n_{q''}$ and the inverse lifetime takes the form (Bhatt *et al.*, 1994; Klemens, 1966; Ferry, 1991)

$$\frac{1}{\tau} = \sum_{q',q''} \frac{\pi\hbar^2 U_a^3}{4Nm^3} \left(\frac{1}{\omega_q \omega_{q'} \omega_{q''}} \right)(1 + n_{q'} + n_{q''})$$
$$\times \delta_{q,q'+q''} \delta(\hbar\omega_q - \hbar\omega_{q'} - \hbar\omega_{q''}). \tag{6.9}$$

In the general case the sum over final states must include a sum over j' and j'' as well. Bhatt *et al.* (1994) applied this result to calculate the lifetime of the bulk LO phonon in GaAs as a function of temperature and as a function of phonon wavevector along the $\langle 100 \rangle$ and $\langle 111 \rangle$ directions at selected temperatures; these results are shown in Figures 6.2 and 6.3.

6.4   Phonon lifetime effects in carrier relaxation

The energy loss rate of a carrier in a polar semiconductor is determined by both the rate at which the carrier's energy is lost by phonon emission and the rate at which the carrier gains energy from phonon absorption. This latter rate can be significant in dimensionally confined structures – such as quantum-well lasers – since the phonons emitted by energetic carriers can accumulate in these structures. Indeed, the phonon densities in dimensionally confined semiconductor devices may well be above those of the equilibrium phonon population and there is a large probability that these non-equilibrium phonons will be reabsorbed. The net loss of energy by a carrier – also known as the carrier relaxation rate – depends on

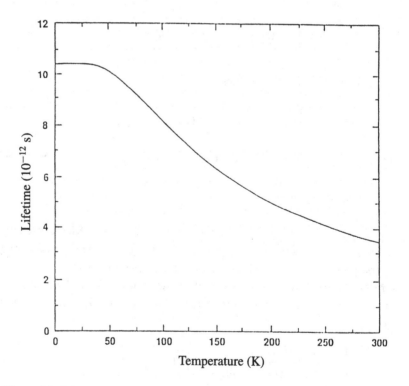

Figure 6.2. LO phonon lifetime in GaAs as a function of temperature. From Bhatt *et al.* (1994), American Institute of Physics, with permission.

the rates for both phonon absorption and emission. Clearly, the lifetimes of the optical phonons are important in determining the total energy loss rate for such carriers. As discussed in the last section, the LO phonons in GaAs and many other polar materials decay into acoustic phonons through the Klemens channel. Over a wide range of temperatures and phonon wavevectors, the lifetimes of LO phonons in GaAs vary from a few picoseconds to about 10 ps (Bhatt *et al.*, 1994), as illustrated by Figures 6.2 and 6.3. Longitudinal-optical phonon lifetimes for other polar semiconductors are also of this magnitude. As a result of the Klemens channel, the 'hot' LO phonons decay into acoustic phonons in times of the order of 10 ps. These LO phonons undergoing decay into acoustic phonons are not available for absorption by the carriers and, as a result of the Klemens channel, the carrier energy relaxation rate is influenced strongly by the LO phonon lifetime. Indeed, it follows that the time required to modulate a semiconductor quantum-well laser through the switching of the electronic current being used to pump the laser is limited by the time it takes the carrier to thermalize in the active quantum-well region of the laser.

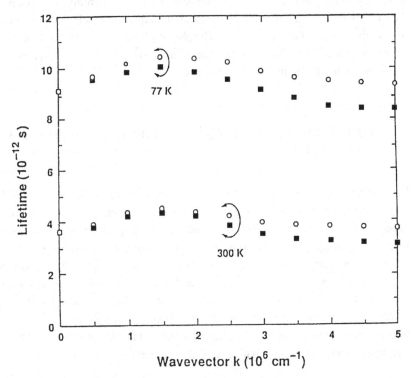

Figure 6.3. Wavevector dependence of the lifetime of bulk LO phonons in GaAs for wavevectors along the $\langle 100 \rangle$ (squares) and $\langle 111 \rangle$ (circles) directions at 77 K (upper set of points) and 300 K (lower set of points). From Bhatt *et al.* (1994), American Institute of Physics, with permission.

Moreover, this thermalization time of carriers in a polar semiconductor quantum well such as in GaAs is determined by the LO phonon emission rate minus the LO phonon absorption rate. Since there is generally a large probability that such non-equilibium LO phonons in the quantum well will be reabsorbed, the net rate of carrier energy loss is reduced considerably from the thermalization rate that could be achieved if the carrier lost energy through the sequential emission of LO phonons without reabsorption. This effect is referred to frequently as the phonon bottleneck. In fact, the thermalization times of semiconductor quantum-well lasers are generally tens of picoseconds – in accord with the LO phonon lifetime – and not the fractions of picoseconds characteristic of the carrier–LO-phonon emission times for many polar semiconductors over a wide range of carrier energies. To reduce the phonon-bottleneck effect in quantum-well lasers, Zhang *et al.* (1996) conceived the tunneling injection laser, in which the carrier tunnels into the active region of the quantum-well laser with energies only one or two times the LO-phonon energy. As a result of this ingenious design, the carriers can thermalize by emitting only one or two LO phonons and the probability of significant LO phonon absorption is reduced greatly. These tunneling injection lasers have intrinsic switching speeds close to 100 GHz, which is several times faster than for the semiconductor quantum-well lasers, where the carriers have to be captured and thermalize in quantum wells having well depths several times the LO phonon energy. The phonon-bottleneck effect will be discussed further in subsection 8.3.4 in connection with carrier energy loss in quantum wires and in Section 10.1 on intersubband lasers.

6.5   Anharmonic effects in würtzite structures: the Ridley channel

Longitudinal-optical phonon decay in GaN cannot proceed by the Klemens' channel since the LO phonon energy is more than twice the energy of any of the available acoustic phonons. Ridley (1996) observed that a four-phonon decay process would be much slower than the rates typical of polar semiconductors and proposed that a three-phonon decay channel where the LO phonon decays into a TO phonon and an LA phonon is possible. The Ridley channel, LO → TO + LA, competes with the channel LO → TO + TA but it is expected that the Ridley channel will dominate, owing to symmetry considerations (Ridley, 1996). Through an analysis similar to that presented in Sections 6.1–6.3, Ridley showed that the net rate of annihilation of LO phonons via the Ridley channel in GaN is given by

$$
\frac{1}{\tau_{\mathrm{GaN}}} = \frac{\Gamma^2}{2\pi\rho_{\mathrm{LO}}\omega_{\mathrm{LO}}v_{\mathrm{LA}}^3}\hbar\omega_{\mathrm{TO}}(\omega_{\mathrm{LO}} - \omega_{\mathrm{TO}})^3
$$
$$
\times [n(\omega_{\mathrm{TO}}) + n(\omega_{\mathrm{LO}} - \omega_{\mathrm{TO}}) + 1], \tag{6.10}
$$

where ρ_{LO} is the reduced-mass density, v_{LA} is the group velocity, and q_{LA} is the LA wavevector associated with the LA frequency $\omega_{LO} - \omega_{TO}$; see also Ridley and Gupta (1991). By taking $\Gamma \approx 10^8$ cm^{-1} as for GaAs, Ridley obtained a zero-temperature rate of 5.0×10^{11} s^{-1}. A full understanding of phonon decay in the wide-bandgap III-V nitride materials is yet to be obtained. However, it is likely that the optical-phonon lifetimes in the III-V nitrides will play an important role in the design of electronic and optoelectronic devices fabricated from them. The Raman analysis of anharmonic phonon decay rates given in Chapter 4 provides essential information for the systematic design of such devices.

Chapter 7

Continuum models for phonons

There is not a single thing, however small, in the world that
does not depend on something that is higher ... for everything
is interdependent.

Zohar, Kabbalah, Book 1, 156b

7.1   Dielectric continuum model of phonons

The dielectric continuum model of optical phonons in polar materials is based on
the concept that the associated lattice vibrations produce an electric polarization
$\mathbf{P}(\mathbf{r})$ that is describable in terms of the equations of electrostatics for a medium
of dielectric constant $\epsilon(\omega)$ (Fuchs and Kliewer, 1965, 1966a, b, c; Engelman and
Ruppin, 1968a, b, c; Ruppin and Engelman, 1970; Licari and Evrard, 1977; Wendler,
1985; Mori and Ando, 1989). The volume of the structure is assumed to be L^3
($-L/2 \leq x, y, z \leq +L/2$) with periodic boundary conditions. The potential $\Phi(\mathbf{r})$
associated with $\mathbf{P}(\mathbf{r})$ is given by (Kim and Stroscio, 1990)

$$\nabla^2 \Phi(\mathbf{r}) = 4\pi \nabla \cdot \mathbf{P}(\mathbf{r}) \tag{7.1}$$

and the electric field $\mathbf{E}(\mathbf{r})$ is given by

$$\mathbf{E}(\mathbf{r}) = -\nabla \Phi(\mathbf{r}). \tag{7.2}$$

Moreover, $\mathbf{E}(\mathbf{r})$ and $\mathbf{P}(\mathbf{r})$, in medium n are related through the dielectric susceptibility, $\varkappa_n(\omega)$:

$$\mathbf{P}(\mathbf{r}) = \varkappa_n(\omega)\mathbf{E}(\mathbf{r}) \tag{7.3}$$

where

$$\varkappa_n(\omega) = [\epsilon_n(\omega) - 1]/4\pi. \tag{7.4}$$

Using the results of Chapter 5, the Lyddane–Sachs–Teller relation for medium n
may be written as

$$\epsilon_n(\omega) = \epsilon_n(\infty)\frac{\omega^2 - \omega_{LO,n}^2}{\omega^2 - \omega_{TO,n}^2}, \tag{7.5}$$

for a binary polar semiconductor AB and as

$$\epsilon_n(\omega) = \epsilon_n(\infty)\frac{\omega^2 - \omega_{LO,n,a}^2}{\omega^2 - \omega_{TO,n,a}^2}\frac{\omega^2 - \omega_{LO,n,b}^2}{\omega^2 - \omega_{TO,n,b}^2}, \tag{7.6}$$

for a ternary polar material $A_y B_{1-y} C$, where the subscript a denotes frequencies associated with the dipole pairs AC and the subscript b denotes frequencies associated with the dipole pairs BC. As in subsection 2.3.1, the displacement field is related to the fields $\mathbf{E}(\mathbf{r})$ and $\mathbf{P}(\mathbf{r})$ through the driven oscillator equation and through the effective charge, e_n^* : for a binary medium n,

$$-\mu_n\omega^2\mathbf{u}_n(\mathbf{r}) = -\mu_n\omega_{0n}^2\mathbf{u}_n(\mathbf{r}) + e_n^*\mathbf{E}_{local}(\mathbf{r}),$$
$$\mathbf{P}(\mathbf{r}) = n_n e_n^*\mathbf{u}_n(\mathbf{r}) + n_n\alpha_n\mathbf{E}_{local}(\mathbf{r}), \tag{7.7}$$

where n_n is the number of unit cells in region n, $\mu_n = m_n M_n/(m_n + M_n)$ is the reduced mass and α_n is the electronic polarizability per unit cell and where, by the Lorentz relation,

$$\mathbf{E}_{local}(\mathbf{r}) = \mathbf{E}(\mathbf{r}) + \frac{4\pi}{3}\mathbf{P}(\mathbf{r}). \tag{7.8}$$

Within the virtual-crystal model, for the dipole pairs AC (BC) in a ternary medium m, we have

$$-\mu_{m,a(b)}\omega^2\mathbf{u}_{m,a(b)}(\mathbf{r}) = -\mu_m\omega_{0m,a(b)}^2\mathbf{u}_{m,a(b)}(\mathbf{r}) + e_{m,a(b)}^*\mathbf{E}_{local}(\mathbf{r}),$$
$$\mathbf{P}(\mathbf{r}) = n_m[ye_{m,a}^*\mathbf{u}_{m,a}(\mathbf{r}) + (1-y)e_{m,b}^*\mathbf{u}_{m,b}(\mathbf{r})] + n_m\alpha_m\mathbf{E}_{local}(\mathbf{r}). \tag{7.9}$$

An alternative and useful form of these equations for the case of a binary material results straightforwardly from the relations of Appendix A. Indeed, it is shown in Appendix A, equations (A.8) and (A.9), for a diatomic polar material that

$$\ddot{\mathbf{u}} = -\omega_{TO}^2\mathbf{u} + \left(\frac{V}{4\pi\mu N}\right)^{1/2}\sqrt{\epsilon(0) - \epsilon(\infty)}\,\omega_{TO}\mathbf{E},$$
$$\mathbf{P} = \left(\frac{\mu N}{4\pi V}\right)^{1/2}\sqrt{\epsilon(0) - \epsilon(\infty)}\,\omega_{TO}\mathbf{u} + \left[\frac{\epsilon(\infty) - 1}{4\pi}\right]\mathbf{E}. \tag{7.10}$$

In the first equation it has been assumed that $\ddot{\mathbf{u}}$ has a general form for the time dependence and may not be simply sinusoidal in ω. As will become evident, this pair of equations is well suited as the basis for an alternative method for performing the calculations of subsection 7.3.1. Moreover, it provides a convenient starting point for the derivation of the macroscopic equations describing optical phonons in polar uniaxial materials (Loudon, 1964). Uniaxial materials such as the hexagonal würtzite structures GaN, AlN, and $Ga_x Al_{1-x}N$ have relatively wide bandgaps and are suited for high-temperature electronics and short-wavelength optoelectronic devices. Loudon (1964) introduced a useful model for describing the macroscopic equations of a uniaxial polar crystal by introducing one dielectric constant associated with the direction parallel to the c-axis, $\epsilon_{\parallel}$, and another dielectric constant associated with the direction perpendicular to the c-axis, $\epsilon_{\perp}$. In Loudon's model a separate set of Huang–Born equations is necessary for the phonon mode displacements parallel to the c-axis, $\mathbf{u}_{\parallel}$, and perpendicular to it, $\mathbf{u}_{\perp}$. For a medium denoted by n it then follows straightforwardly that

$$\ddot{\mathbf{u}}_{\perp,n} = -\omega_{TO,\perp,n}^2 \mathbf{u}_{\perp,n} + \sqrt{\frac{V}{4\pi \mu_n N}} \sqrt{\epsilon(0)_{\perp,n} - \epsilon(\infty)_{\perp,n}}\, \omega_{TO,\perp,n} \mathbf{E}_{\perp,n},$$

$$\mathbf{P}_{\perp,n} = \sqrt{\frac{\mu_n N}{4\pi V}} \sqrt{\epsilon(0)_{\perp,n} - \epsilon(\infty)_{\perp,n}}\, \omega_{TO,\perp,n} \mathbf{u}_{\perp,n} + \left[\frac{\epsilon(\infty)_{\perp,n} - 1}{4\pi}\right] \mathbf{E}_{\perp,n},$$

$$\epsilon_{\perp,n}(\omega) = \epsilon_{\perp,n}(\infty)\left(\frac{\omega^2 - \omega_{LO,\perp,n}^2}{\omega^2 - \omega_{TO,\perp,n}^2}\right), \tag{7.11}$$

and

$$\ddot{\mathbf{u}}_{\parallel,n} = -\omega_{TO,\parallel,n}^2 \mathbf{u}_{\parallel,n} + \sqrt{\frac{V}{4\pi \mu_n N}} \sqrt{\epsilon(0)_{\parallel,n} - \epsilon(\infty)_{\parallel,n}}\, \omega_{TO,\parallel,n} \mathbf{E}_{\parallel,n},$$

$$\mathbf{P}_{\parallel,n} = \sqrt{\frac{\mu_n N}{4\pi V}} \sqrt{\epsilon(0)_{\parallel,n} - \epsilon(\infty)_{\parallel,n}}\, \omega_{TO,\parallel,n} \mathbf{u}_{\parallel,n} + \left[\frac{\epsilon(\infty)_{\parallel,n} - 1}{4\pi}\right] \mathbf{E}_{\parallel,n},$$

$$\epsilon_{\parallel,n}(\omega) = \epsilon_{\parallel,n}(\infty)\left(\frac{\omega^2 - \omega_{LO,\parallel,n}^2}{\omega^2 - \omega_{TO,\parallel,n}^2}\right). \tag{7.12}$$

Of course, in Loudon's model these six equations must be supplemented by the following three equations of electrostatics for the case where there is no free charge:

$$\mathbf{E}(\mathbf{r}) = -\nabla\phi(\mathbf{r}),$$

$$\mathbf{D}(\mathbf{r}) = \mathbf{E}(\mathbf{r}) + 4\pi\mathbf{P}(\mathbf{r})$$
$$= \epsilon_{\perp}(\omega)E_{\perp}(\mathbf{r})\hat{\boldsymbol{\rho}} + \epsilon_{\parallel}(\omega)E_{\parallel}(\mathbf{r})\hat{\mathbf{z}}, \tag{7.13}$$

$$\nabla \cdot \mathbf{D}(\mathbf{r}) = 0,$$

where $\hat{\mathbf{z}}$ and $\hat{\boldsymbol{\rho}}$ are the unit vectors in the $\parallel$ and $\perp$ directions respectively.

In the first and third of this set of nine equations, it has been assumed that $\ddot{\mathbf{u}}_{\perp,n}$ and $\ddot{\mathbf{u}}_{\parallel,n}$ have a general form of the dependence and may not be simply sinusoidal in ω; the assumption of sinusoidal time dependence made in Appendix A was not necessary and simply by replacing $-\omega^2\mathbf{u}$ by $\ddot{\mathbf{u}}$ it is straightforward to rederive the results of Appendix A without assuming a sinusoidal time dependence.

The above set of nine equations provides a convenient basis for describing carrier–optical-phonon scattering in würtzite crystals. Indeed, using the relations for the displacement perpendicular (parallel) to the c-axis,

$$\mathbf{u}(\mathbf{r})_{\perp(\parallel)} = \frac{1}{\sqrt{N}} \sum_q \sum_{j=1,2,3} \sqrt{\frac{\hbar}{2m\omega_q}} \hat{\mathbf{e}}_{\mathbf{q},j,\perp(\parallel)} (a_q + a_{-q}^{\dagger}) e^{i\mathbf{q}\cdot\mathbf{r}}, \qquad (7.14)$$

which follow from the results of Section 5.1, and defining

$$\phi(\mathbf{r})_{\perp(\parallel)} = \sum_q \phi(q)_{\perp(\parallel)} e^{i\mathbf{q}\cdot\mathbf{r}} \qquad (7.15)$$

$$\mathbf{E}(\mathbf{r})_{\perp(\parallel)} = -\nabla\phi(\mathbf{r})_{\perp(\parallel)} = -i\mathbf{q} \sum_q \phi(q)_{\perp(\parallel)} e^{i\mathbf{q}\cdot\mathbf{r}}, \qquad (7.16)$$

it follows, by taking $m = \mu_n$, $\omega^2 = \omega_q^2$ and assuming a sinusoidal dependence for $\mathbf{u}(\mathbf{r},\mathbf{t})$, that (Lee *et al.*, 1997)

$$(\omega_{\mathrm{TO},\perp(\parallel),n}^2 - \omega_q^2) \sqrt{\frac{\hbar}{2\mu_n N \omega_q}} \hat{\mathbf{e}}_{\mathbf{q},j,\perp(\parallel)} (a_q + a_{-q}^{\dagger})$$

$$= \sqrt{\frac{V}{4\pi\mu_n N}} \sqrt{\epsilon(0)_{\perp(\parallel),n} - \epsilon(\infty)_{\perp(\parallel),n}} \, \omega_{\mathrm{TO},\perp(\parallel),n}(-i)q_{\perp(\parallel)}\phi(q),$$

$$\qquad (7.17)$$

with $q_{\perp} = q\sin\theta$, where θ is the angle between q and the c-axis, which is taken as the z-axis. Moreover, $q_{\parallel} = q\cos\theta$ and $\hat{\mathbf{e}}_{\mathbf{q},j,\perp}^2 + \hat{\mathbf{e}}_{\mathbf{q},j,\parallel}^2 = 1$. Hence

$$\sqrt{\frac{2\pi\hbar}{V\omega_q}} \hat{\mathbf{e}}_{\mathbf{q},j,\perp(\parallel)} (a_q + a_{-q}^{\dagger})$$

$$= \frac{\sqrt{\epsilon(0)_{\perp(\parallel),n} - \epsilon(\infty)_{\perp(\parallel),n}} \, \omega_{\mathrm{TO},\perp(\parallel),n}}{\omega_{\mathrm{TO},\perp(\parallel),n}^2 - \omega_q^2} (-i)q_{\perp(\parallel)}\phi(q), \qquad (7.18)$$

so that

$$\left(\frac{2\pi\hbar}{V\omega_q}\right)(a_q + a_{-q}^{\dagger})^2$$

$$= -q^2 \left\{ \frac{[\epsilon(0)_{\perp} - \epsilon(\infty)_{\perp}]\omega_{\mathrm{TO},\perp}^2}{(\omega_{\perp,\mathrm{TO}}^2 - \omega_q^2)^2} \sin^2\theta \right\} \phi^2(q),$$

$$-q^2 \left\{ \frac{[\epsilon(0)_{\parallel} - \epsilon(\infty)_{\parallel}]\omega_{\mathrm{TO},\parallel}^2}{(\omega_{\mathrm{TO},\parallel}^2 - \omega_q^2)^2} \cos^2\theta \right\} \phi^2(q), \qquad (7.19)$$

and

$$\phi(q) = -i\sqrt{\frac{2\pi\hbar}{Vq^2\omega_q}} (a_q + a_{-q}^{\dagger})(\omega_{\mathrm{TO},\perp}^2 - \omega_q^2)(\omega_{\mathrm{TO},\parallel}^2 - \omega_q^2)$$

$$\times \left\{ [\epsilon(0)_{\perp} - \epsilon(\infty)_{\perp}]\omega_{\mathrm{TO},\perp}^2(\omega_{\mathrm{TO},\parallel}^2 - \omega_q^2)^2 \sin^2\theta \right.$$

$$\left. + [\epsilon(0)_{\parallel} - \epsilon(\infty)_{\parallel}]\omega_{\mathrm{TO},\parallel}^2(\omega_{\mathrm{TO},\perp}^2 - \omega_q^2)^2 \cos^2\theta \right\}^{-1/2}. \qquad (7.20)$$

Thus

$$H = \sum_q (-e)\phi(q) e^{i\mathbf{q}\cdot\mathbf{r}}(a_q + a_{-q}^\dagger)$$

$$= i \sum_q \sqrt{\frac{2\pi e^2 \hbar}{V \omega_q}} \frac{1}{q} e^{i\mathbf{q}\cdot\mathbf{r}}(a_q + a_{-q}^\dagger)(\omega_{\perp,\text{TO}}^2 - \omega_q^2)(\omega_{\text{TO},\|}^2 - \omega_q^2)$$

$$\times \{[\epsilon(0)_\perp - \epsilon(\infty)_\perp]\omega_{\text{TO},\perp}^2(\omega_{\text{TO},\|}^2 - \omega_q^2)^2 \sin^2\theta$$

$$+ [\epsilon(0)_\| - \epsilon(\infty)_\|]\omega_{\text{TO},\|}^2(\omega_{\perp,\text{TO}}^2 - \omega_q^2)^2 \cos^2\theta\}^{-1/2}$$

$$= i \sum_q \left\{ \frac{4\pi e^2 \hbar V^{-1}}{(\partial/\partial\omega)[\epsilon(\omega)_\perp \sin^2\theta + \epsilon(\omega)_\| \cos^2\theta]} \right\}^{1/2} \frac{1}{q} e^{i\mathbf{q}\cdot\mathbf{r}}(a_q + a_{-q}^\dagger).$$

$$(7.21)$$

In the isotropic case, $\epsilon(\omega)_\perp = \epsilon(\omega)_\|$ and this result must reduce to the expression, (5.34), obtained in Section 5.2 for the interaction Hamiltonian describing the carrier–LO-phonon interactions. Indeed, since the general form of the Lyddane–Sachs–Teller relation implies that

$$\frac{\omega_{\text{TO}}^2 - \omega_{\text{LO}}^2}{\omega_{\text{TO}}\sqrt{\epsilon(0) - \epsilon(\infty)}} = -\omega_{\text{LO}}\left[\frac{1}{\epsilon(\infty)} - \frac{1}{\epsilon(0)}\right]^{1/2}, \qquad (7.22)$$

the Hamiltonian for the uniaxial case reduces to (5.34) upon taking $\epsilon(\omega)_\perp = \epsilon(\omega)_\|$ and $\omega_q = \omega_{\text{LO}}$.

7.2 Elastic continuum model of phonons

As will become clear, the elastic continuum model of acoustic phonons provides an adequate description of acoustic phonons for nanostructures having confined dimensions of about two atomic monolayers. A simple and illustrative application of the elastic continuum model is found in the case of a longitudinal acoustic mode propagating in a quasi-one-dimensional structure. Consider an element dx located along this structure between x and $x+dx$. Let $u(x, t)$ be the elastic displacement at x along the axis of the one-dimensional structure; that is, $u(x, t)$ describes the uniform longitudinal displacement of the element dx. In the elastic continuum model the dynamics of the mass-containing element, dx, are described in terms of Newton's laws. Indeed, defining the strain as $e = du/dx$ and the stress, $T(x)$, as the force per unit area in the quasi-one-dimensional structure of area A, it follows from Hooke's law that

$$T = Ye, \qquad (7.23)$$

where Y is a proportionality constant known as Young's modulus. The force equation describing the dynamics of the element dx of density $\rho(x)$ is given by

Newton's second law:

$$\rho(x)A\,dx\frac{\partial^2 u(x,t)}{\partial t^2} = [T(x+dx) - T(x)]A, \qquad (7.24)$$

where $\rho A\,dx$ is the mass associated with the element dx and $\partial^2 u/\partial t^2$. By Hooke's law

$$T(x+dx) - T(x) = \left(\frac{\partial T}{\partial x}\right)dx = \left(Y\frac{\partial e}{\partial x}\right)dx = \left(Y\frac{\partial^2 u}{\partial x^2}\right)dx, \qquad (7.25)$$

and it follows that

$$\frac{\partial^2 u}{\partial x^2} = \left(\frac{\rho(x)}{Y}\right)\frac{\partial^2 u}{\partial t^2}. \qquad (7.26)$$

Seeking solutions of this one-dimensional wave equation of the form $u(x) = \xi e^{i(qx-\omega t)}$, where $q = 2\pi/\lambda$ and ω is the angular frequency of the wave, it follows that the dispersion relation for the longitudinal acoustic (LA) mode is $\rho\omega^2 = Yq^2$ or $\omega = v_l q$, where $v_l = \sqrt{Y/\rho}$. The longitudinal sound speed, v_l, has typical values $(3\text{–}5) \times 10^5$ cm s^{-1} and for $\rho = 4$ g cm^{-3} it follows that Y must have an order of magnitude of 10^{12} g cm s^{-2}.

The three-dimensional generalization of these results may be accomplished through the replacements (Auld, 1973) $u(x) \rightarrow \mathbf{u}(x, y, z) = (u, v, w)$ and $T = Ye \rightarrow T = c : S$ with $T_i = c_{ij}S_j$. In this generalization, Young's modulus is replaced by a 6×6 matrix of elastic constants c_{ij}; T is replaced by a six-component object T_i; e is replaced by a six-component object S_j. For the cubic, zincblende, and würtzite crystals the most general form of the stress–strain relation, $T_{ij} = c_{ijkl}S_{kl}$, where i, j, l, k run over x, y, z, may be represented by $T_i = c_{ij}S_j$. In this last result, i and j run over the integers 1–6; $1 \equiv xx, 2 \equiv yy, 3 \equiv zz, 4 \equiv yz$ or $zy, 5 \equiv xz$ or zx, and $6 \equiv xy$ or yx. The resulting forms for S_j are

$$S_1 = S_{xx} = \frac{\partial u}{\partial x}, \qquad S_2 = S_{yy} = \frac{\partial v}{\partial y}, \qquad S_3 = S_{zz} = \frac{\partial w}{\partial z},$$

$$S_4 = S_{yz} = S_{zy} = \frac{1}{2}\left(\frac{\partial w}{\partial y} + \frac{\partial v}{\partial z}\right),$$

$$S_5 = S_{xz} = S_{zx} = \frac{1}{2}\left(\frac{\partial u}{\partial z} + \frac{\partial w}{\partial x}\right), \qquad (7.27)$$

$$S_6 = S_{xy} = S_{yx} = \frac{1}{2}\left(\frac{\partial u}{\partial y} + \frac{\partial v}{\partial x}\right).$$

For T_i the forms are

$$T_1 = T_{xx}, \qquad T_2 = T_{yy}, \qquad T_3 = T_{zz},$$
$$T_4 = T_{yz} = T_{zy}, \qquad T_5 = T_{xz} = T_{zx}, \qquad T_6 = T_{xy} = T_{yx}. \qquad (7.28)$$

For the elastic energy to be single valued $c_{ij} = c_{ji}$, and it follows that only 21 distinct elements are necessary to define the 6×6 matrix c_{ij}. Nanostructures of widespread interest in modern electronics and optoelectronics are generally fabricated from zincblende and würtzite crystals. For cubic crystals, including zincblende crystals, the matrix c_{ij} is of the form

$$
\begin{pmatrix}
c_{11} & c_{12} & c_{12} & 0 & 0 & 0 \\
c_{12} & c_{11} & c_{12} & 0 & 0 & 0 \\
c_{12} & c_{12} & c_{11} & 0 & 0 & 0 \\
0 & 0 & 0 & c_{44} & 0 & 0 \\
0 & 0 & 0 & 0 & c_{44} & 0 \\
0 & 0 & 0 & 0 & 0 & c_{44}
\end{pmatrix}
\tag{7.29}
$$

and for würtzite crystals c_{ij} is of the form

$$
\begin{pmatrix}
c_{11} & c_{12} & c_{13} & 0 & 0 & 0 \\
c_{12} & c_{11} & c_{13} & 0 & 0 & 0 \\
c_{13} & c_{13} & c_{33} & 0 & 0 & 0 \\
0 & 0 & 0 & c_{44} & 0 & 0 \\
0 & 0 & 0 & 0 & c_{44} & 0 \\
0 & 0 & 0 & 0 & 0 & (c_{11} - c_{12})/2
\end{pmatrix}.
\tag{7.30}
$$

For a cubic medium such as the zincblende crystal, only three independent elastic constants, c_{11}, c_{12}, and c_{44}, are needed to specify all the c_{ij}. For an isotropic cubic medium $c_{12} = c_{11} - 2c_{44}$, and only two constants λ and μ are necessary to define the c_{ij}:

$$
\lambda = c_{12} = c_{13} = c_{21} = c_{23} = c_{31} = c_{32},
$$

$$
\mu = c_{44} = c_{55} = c_{66} = \tfrac{1}{2}(c_{11} - c_{12}),
\tag{7.31}
$$

$$
\lambda + 2\mu = c_{11} = c_{22} = c_{33}.
$$

The constants λ and μ are known as Lamé's constants. Thus, in the cubic case three independent constants replace Y: c_{11}, which relates the compressive stress to the strain along the same direction, [100]; c_{44}, which relates the shear stress and the strain in the same direction; and c_{12}, which relates the compressive stress in one direction and the strain in another direction.

For the isotropic case, it follows that

$$
\begin{aligned}
T_{xx} &= \lambda(S_{xx} + S_{yy} + S_{zz}) + 2\mu S_{xx} = \lambda\Delta + 2\mu S_{xx}, \\
T_{yy} &= \lambda(S_{xx} + S_{yy} + S_{zz}) + 2\mu S_{yy} = \lambda\Delta + 2\mu S_{yy}, \\
T_{zz} &= \lambda(S_{xx} + S_{yy} + S_{zz}) + 2\mu S_{zz} = \lambda\Delta + 2\mu S_{zz}, \\
T_{yz} &= \mu S_{yz}, \qquad T_{zx} = \mu S_{zx}, \qquad T_{xy} = \mu S_{xy},
\end{aligned}
\tag{7.32}
$$

where $\Delta = \partial u/\partial x + \partial v/\partial y + \partial w/\partial z$ represents the dilatation of the medium. Then, it is straightforward to show that the three-dimensional generalization of $\rho(x)\partial^2 u/\partial t^2 = Y\partial^2 u/\partial x^2 = \partial T/\partial x$ is given by the equations

$$\rho\frac{\partial^2 u}{\partial t^2} = \frac{\partial T_{xx}}{\partial x} + \frac{\partial T_{yx}}{\partial y} + \frac{\partial T_{zx}}{\partial z} = (\lambda + \mu)\frac{\partial \Delta}{\partial x} + \mu \nabla^2 u,$$

$$\rho\frac{\partial^2 v}{\partial t^2} = \frac{\partial T_{xy}}{\partial x} + \frac{\partial T_{yy}}{\partial y} + \frac{\partial T_{zy}}{\partial z} = (\lambda + \mu)\frac{\partial \Delta}{\partial y} + \mu \nabla^2 v, \qquad (7.33)$$

$$\rho\frac{\partial^2 w}{\partial t^2} = \frac{\partial T_{xz}}{\partial x} + \frac{\partial T_{yz}}{\partial y} + \frac{\partial T_{zz}}{\partial z} = (\lambda + \mu)\frac{\partial \Delta}{\partial z} + \mu \nabla^2 w,$$

where, as usual,

$$\nabla^2 \equiv \frac{\partial^2}{\partial x^2} + \frac{\partial^2}{\partial y^2} + \frac{\partial^2}{\partial z^2}. \qquad (7.34)$$

Two alternative forms of the three-dimensional force equations are encountered frequently in the literature. The first of these is derived by writing the components of $\mathbf{u}(x, y, z) = (u_1, u_2, u_3)$ as u_α, $\alpha = 1, 2, 3$; it then follows that the three force equations may be rewritten as

$$\rho\frac{\partial^2 u_\alpha}{\partial t^2} = \frac{\partial T_{\alpha\beta}}{\partial r_\beta}, \qquad (7.35)$$

where

$$T_{\alpha\beta} = \lambda S_{\alpha\alpha}\delta_{\alpha\beta} + 2\mu S_{\alpha\beta}. \qquad (7.36)$$

In these equations, the subscripts α and β run over 1, 2, 3 (corresponding to x, y, z). A repeated index in a term implies summation. $\delta_{\alpha\beta}$ is the Kronecker delta function. In a second alternative form the three force equations are written straightforwardly as the single vector equation

$$\frac{\partial^2 \mathbf{u}}{\partial t^2} = c_t^2 \nabla^2 \mathbf{u} + (c_l^2 - c_t^2)\,\mathrm{grad}\,\Delta, \qquad (7.37)$$

where it is now clear that

$$\Delta = \nabla \cdot \mathbf{u} = \mathrm{div}\,\mathbf{u}. \qquad (7.38)$$

Here, c_t and c_l are the transverse and longitudinal sound speeds and we have

$$c_t^2 = \frac{\lambda}{\rho} \qquad \text{and} \qquad c_l^2 = \frac{\lambda + 2\mu}{\rho}. \qquad (7.39)$$

In physical acoustics the solutions for the displacement fields are frequently specified in terms of two potential functions, a scalar potential ϕ and a vector potential $\boldsymbol{\Psi} = (\Psi_x, \Psi_y, \Psi_z)$, through

$$u = \frac{\partial \phi}{\partial x} + \frac{\partial \Psi_x}{\partial y} - \frac{\partial \Psi_y}{\partial z},$$

$$v = \frac{\partial \phi}{\partial y} + \frac{\partial \Psi_x}{\partial z} - \frac{\partial \Psi_z}{\partial x}, \qquad (7.40)$$

$$w = \frac{\partial \phi}{\partial z} + \frac{\partial \Psi_y}{\partial x} - \frac{\partial \Psi_x}{\partial y},$$

where ϕ and Ψ_i, $i = x, y, z$, satisfy

$$\nabla^2 \phi = \frac{1}{c_l^2} \frac{\partial^2 \phi}{\partial t^2}, \qquad c_l^2 = \frac{\lambda + 2\mu}{\rho},$$

$$\nabla^2 \Psi_i = \frac{1}{c_t^2} \frac{\partial^2 \Psi_i}{\partial t^2}, \qquad i = x, y, z, \qquad c_t^2 = \frac{\lambda}{\rho}.$$

(7.41)

The scalar potential ϕ corresponds to the 'irrotational' part of the solution and the vector potential corresponds to any remaining 'rotational' fields. In the literature the irrotational solutions are also referred to as the longitudinal, compressional, or dilatational solutions. Moreover, seismologists frequently refer to these solutions as P waves. Likewise, the rotational vector-potential solutions based on Ψ_i are identified as transverse, shear, distortional, or equivoluminal solutions. In seismology these solutions are commonly identified as S waves.

Herein, the irrotational solutions will generally be referred to as longitudinal modes and the corresponding sound speed will be denoted by c_l. Likewise, the rotational fields will be denoted as transverse modes and the associated sound speed will be denoted by c_t. The principal interest in this book is on using the longitudinal and transverse solutions of the elastic continuum model to describe the longitudinal acoustic (LA) and transverse acoustic (TA) phonons in nanostructures. Hence, the notation adopted herein is that corresponding most naturally to the solid-state community's descriptions of phonons as longitudinal and transverse.

Every year, experimental observations of acoustic modes in nanostructures are being reported in the literature in increasing numbers. Examples of such experimental studies include the works of Nabity and Wybourne (1990a, b), Seyler and Wybourne (1992), and Sun *et al.* (1999). Wybourne and his coworkers report measurements of confined acoustic phonons in very thin metallic foils and wires on dielectric substrates. Sun *et al.* (1999) presented data on folded acoustic modes in semiconductor superlattices. In view of such observations, Section 7.6 presents extensions of the elastic continuum theory of this section to the case of acoustic modes in dimensionally confined structures and Chapter 9 focuses on carrier–acoustic-phonon scattering rates in both bulk materials and dimensionally confined nanostructures.

7.3   Optical modes in dimensionally confined structures

The dielectric continuum model has been applied to describe the properties of dimensionally confined optical phonons in many electronic and optoelectronics devices fabricated from semiconductor nanostructures (Dutta and Stroscio, 1998, 2000; Mitin *et al.*, 1999). These include quantum wells, superlattices, quantum wires, and quantum dots. To illustrate the basic features of the dielectric continuum model of optical phonons, the case of confinement in just one dimension – as in a quantum well or superlattice – is considered first. In addition, the dielectric continuum model will be compared with other continuum models, including the hydrodynamic model and the reformulated dielectric continuum model. These models predict different sets of confined optical phonon modes but each model predicts the same carrier–phonon scattering rate as long as it includes a complete set of orthogonal phonon modes (Nash, 1992). Following a comparison of these models, a microscopic treatment of confined phonon modes will be discussed.

7.3.1 Dielectric continuum model for slab modes: normalization of interface modes

The dielectric continuum model predicts a set of confined optical phonon modes commonly referred to as the slab modes. These slab modes may be determined by applying the dielectric continuum model and by imposing electrostatic boundary conditions at each heterointerface. The normal-mode frequencies and orthogonal confined phonon modes are obtained through the simultaneous solution of the equations arising from the dielectric continuum model, subject to the boundary conditions that the potential, $\Phi(\mathbf{r})$ and the normal component of $\mathbf{D}(\mathbf{r})$ are continuous at each heterointerface. Taking the heterointerfaces to be normal to the z-direction, the electrostatic potential $\Phi_i(\mathbf{r})$ in the region $\mathbf{R}_i = (z_i, z_{i+1})$ and its two-dimensional Fourier transform $\Phi_i(\mathbf{q}, z)$ are related by

$$\Phi_i(\mathbf{r}) = \sum_q \Phi_i(\mathbf{q}, z) \, e^{-i\mathbf{q}\cdot\rho}, \tag{7.42}$$

where $\rho \equiv (x, y)$ and $\mathbf{q}$ is the two-dimensional wavevector in the xy-plane, that is, $\mathbf{q} = q_x\hat{\mathbf{x}} + q_y\hat{\mathbf{y}}$, where $\hat{\mathbf{x}}$ and $\hat{\mathbf{y}}$ are unit vectors. Then

$$\mathbf{E}_i(\mathbf{r}) = -\nabla\Phi_i(\mathbf{r}) = \sum_q \mathbf{E}_i(\mathbf{q}, z) \, e^{-i\mathbf{q}\cdot\rho},$$

$$\mathbf{P}_i(\mathbf{r}) = \varkappa_i(\omega)\mathbf{E}_i(\mathbf{r}) = \sum_q \mathbf{P}_i(\mathbf{q}, z) \, e^{-i\mathbf{q}\cdot\rho}. \tag{7.43}$$

Following the concepts of Section 5.1, the mode normalization condition requires that the energy of a phonon of mode $\mathbf{q}$ is $\hbar\omega_q$; for the case of a single interface at $z = 0$ separating two layers n and m, this condition is

$$\int_0^\infty L^2\big[\sqrt{(\mu_n n_n)}\,\mathbf{u}_n(\mathbf{q},z)\big]^* \cdot \big[\sqrt{(\mu_n n_n)}\,\mathbf{u}_n(\mathbf{q},z)\big]dz$$

$$+\,y\int_{-\infty}^0 L^2\big[\sqrt{(\mu_m n_m)}\,\mathbf{u}_{m,a}(\mathbf{q},z)\big]^* \cdot \big[\sqrt{(\mu_m n_m)}\,\mathbf{u}_{m,a}(\mathbf{q},z)\big]dz$$

$$+\,(1-y)\int_{-\infty}^0 L^2\big[\sqrt{(\mu_m n_m)}\,\mathbf{u}_{m,b}(\mathbf{q},z)\big]^* \cdot \big[\sqrt{(\mu_m n_m)}\,\mathbf{u}_{m,b}(\mathbf{q},z)\big]dz$$

$$=\frac{\hbar}{2\omega_q}. \tag{7.44}$$

To illustrate the normalization procedure, let us consider one of the classes of optical phonon modes existing for this one-heterointerface structure. The wave equations for the fields of relevance here admit both oscillating and exponential solutions. In particular, let us consider the solutions having an exponential character. For these modes, known as the interface (IF) modes, we take

$$\Phi(\mathbf{r}) = \sum_q \Phi(\mathbf{q},z)\,e^{-i\mathbf{q}\cdot\boldsymbol{\rho}} = \sum_q c\,e^{-q|z|}e^{-i\mathbf{q}\cdot\boldsymbol{\rho}} \tag{7.45}$$

so that

$$\mathbf{E}_i(\mathbf{q},z) = \begin{cases} c(iqe^{-qz}\hat{\mathbf{q}} - qe^{-qz}\hat{\mathbf{z}}) & z \geq 0, \\ c(iqe^{qz}\hat{\mathbf{q}} + qe^{qz}\hat{\mathbf{z}}) & z \leq 0, \end{cases}$$

$$\mathbf{P}_i(\mathbf{q},z) = \begin{cases} \varkappa_i c(iqe^{-qz}\hat{\mathbf{q}} - qe^{-qz}\hat{\mathbf{z}}) & z \geq 0, \\ \varkappa_i c(iqe^{qz}\hat{\mathbf{q}} + qe^{qz}\hat{\mathbf{z}}) & z \leq 0. \end{cases} \tag{7.46}$$

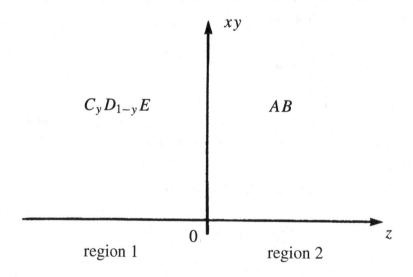

Figure 7.1. A ternary–binary structure. From Kim and Stroscio (1990), American Institute of Physics, with permission.

where $\hat{\mathbf{q}}$ is the unit vector specifying the direction of $\mathbf{q} \equiv (q_x, q_y)$. Let material n be a binary layer filling the space $z \geq 0$, region 1, and material m be a ternary layer filling the space $z \leq 0$, region 2, as illustrated in Figure 7.1. Then, for the right-hand medium, material 1, using the notation of Appendix B, it follows that

$$-\mu_1 \omega^2 \mathbf{u}_1(\mathbf{q}, z) = -\mu_1 \omega_{01}^2 \mathbf{u}_1(\mathbf{q}, z) + e_1^* \mathbf{E}_{\text{local}}(\mathbf{r}),$$

$$\mathbf{P}_1(\mathbf{q}, z) = n_1 e_n^* \mathbf{u}_1(\mathbf{q}, z) + n_1 \alpha_1 \frac{\mu_1(\omega_{01}^2 - \omega^2)}{e_1^*} \mathbf{u}_1(\mathbf{q}, z)$$

$$= n_1 e_1^* \left[1 + \frac{\alpha_1 \mu_1(\omega_{01}^2 - \omega^2)}{e_1^{*2}} \right] \mathbf{u}_1(\mathbf{q}, z) \qquad (7.47)$$

$$= \varkappa_1 c(iqe^{-qz}\hat{\mathbf{q}} - qe^{-qz}\mathbf{z});$$

thus

$$\mathbf{u}_1(\mathbf{q}, z) = \frac{\varkappa_1 c(iqe^{-qz}\hat{\mathbf{q}} - qe^{qz}\hat{\mathbf{z}})}{n_1 e_1^* [1 + \alpha_1 \mu_1(\omega_{01}^2 - \omega^2)/e_1^{*2}]}. \qquad (7.48)$$

For material 2, there are two driven-oscillator equations, one for the AC pair, denoted by a, and one for the BC pair, denoted by b:

$$-\mu_{2,a(b)}^2 \omega^2 \mathbf{u}_{2,a(b)}(\mathbf{q}, z) = -\mu_{2,a(b)} \omega_{02,a(b)}^2 \mathbf{u}_{2,a(b)}(\mathbf{q}, z)$$
$$+ e_{2,a(b)}^* \mathbf{E}_{\text{local}}(\mathbf{q}, z), \qquad (7.49)$$

and the electric polarization in the virtual-crystal approximation is

$$\mathbf{P}_2(\mathbf{q}, z) = n_2[(1-y)e_{2b}^* \mathbf{u}_{2b}(\mathbf{q}, z) + ye_{2a}^* \mathbf{u}_{2a}(\mathbf{q}, z)] + n_2 \alpha_2 \mathbf{E}_{\text{local}}(\mathbf{q}, z)$$

$$= n_2 \left[\frac{ye_{2a}^{*2}}{\mu_{2,a}(\omega_{02,a}^2 - \omega^2)} + \frac{(1-y)e_{2b}^{*2}}{\mu_{2,b}(\omega_{02,b}^2 - \omega^2)} + \alpha_2 \right] \mathbf{E}_{\text{local}}(\mathbf{q}, z)$$

$$= \varkappa_2 c(iqe^{qz}\hat{\mathbf{q}} + qe^{qz}\hat{\mathbf{z}}). \qquad (7.50)$$

Then

$$\mathbf{E}_{\text{local}}(\mathbf{q}, z) = \frac{\varkappa_2 c(iqe^{qz}\hat{\mathbf{q}} + qe^{qz}\hat{\mathbf{z}})}{n_2 \left[\dfrac{ye_{2a}^{*2}}{\mu_{2,a}(\omega_{02,a}^2 - \omega^2)} + \dfrac{(1-y)e_{2b}^{*2}}{\mu_{2,b}(\omega_{02,b}^2 - \omega^2)} + \alpha_2 \right]},$$

$$\mathbf{u}_{2,a(b)}(\mathbf{q}, z) = \frac{e_{2,a(b)}^*}{\mu_{2,a(b)}(\omega_{02,a(b)}^2 - \omega^2)}$$

$$\times \frac{\varkappa_2 c(iqe^{qz}\hat{\mathbf{q}} + qe^{qz}\hat{\mathbf{z}})}{n_2 \left[\dfrac{ye_{2a}^{*2}}{\mu_{2,a}(\omega_{02,a}^2 - \omega^2)} + \dfrac{(1-y)e_{2b}^{*2}}{\mu_{2,b}(\omega_{02,b}^2 - \omega^2)} + \alpha_2 \right]}. \qquad (7.51)$$

With these expressions for $\mathbf{u}_1(\mathbf{q}, z)$ and $\mathbf{u}_{2,a(b)}(\mathbf{q}, z)$, the normalization condition, with $n = 1$ and $m = 2$, yields

$$\left(\frac{\hbar}{2\omega}\right)\frac{1}{L^2 c^2} = \frac{\mu_1 n_1 \chi_1^2 q}{\{n_1 e_1^*[1 + \alpha_1 \mu_1 e_1^{*-2}(\omega_{01}^2 - \omega^2)]\}^2}$$

$$+ \mu_{2a} n_2 y \left[\frac{e_{2a}^*}{\mu_{2,a}}(\omega_{02,a}^2 - \omega^2)\right]^2 q$$

$$\times \left\{\frac{\chi_2^2}{n_2^2}\left[\frac{y e_{2a}^{*2}}{\mu_{2,a}(\omega_{02,a}^2 - \omega^2)} + \frac{(1-y)e_{2b}^{*2}}{\mu_{2,b}(\omega_{02,b}^2 - \omega^2)} + \alpha_2\right]^2\right\}$$

$$+ \mu_{2b} n_2 (1-y)\left[\frac{e_{2b}^*}{\mu_{2,b}}(\omega_{02,b}^2 - \omega^2)\right]^2 q$$

$$\times \left\{\frac{\chi_2^2}{n_2^2}\left[\frac{y e_{2a}^{*2}}{\mu_{2,a}(\omega_{02,a}^2 - \omega^2)} + \frac{(1-y)e_{2b}^{*2}}{\mu_{2,b}(\omega_{02,b}^2 - \omega^2)} + \alpha_2\right]^2\right\},$$

$$(7.52)$$

where the integral in the first term has been performed using

$$\int_0^\infty dz(iqe^{-qz}\hat{\mathbf{q}} - qe^{-qz}\mathbf{z})^* \cdot (iqe^{-qz}\hat{\mathbf{q}} - qe^{-qz}\hat{\mathbf{z}}) = \int_0^\infty dz\, 2q^2 e^{-2qz} = q,$$

$$(7.53)$$

and the second and third integrals have been performed using

$$\int_{-\infty}^0 dz(iqe^{qz}\hat{\mathbf{q}} + qe^{qz}\hat{\mathbf{z}})^* \cdot (iqe^{qz}\hat{\mathbf{q}} + qe^{qz}\hat{\mathbf{z}}) = \int_0^\infty dz\, 2q^2 e^{-2qz} = q. \qquad (7.54)$$

Thus the normalization constant c is determined. It is convenient to rewrite this expression using conditions derived by Wendler (1985). These conditions are discussed in Appendix B. The conditions that are useful at this point in our derivation are

$$\epsilon_n(\infty) = 1 + 4\pi \frac{n_n \alpha_n}{1 - \frac{4}{3}\pi n_n \alpha_n},$$

$$\omega_{\text{LO},n}^2 = \omega_{0,n}^2 + \frac{2}{3}\omega_{\text{plasma},n}^2 \frac{1}{1 + \frac{8}{3}\pi n_n \alpha_n},$$

$$\omega_{\text{TO},n}^2 = \omega_{0,n}^2 + \frac{1}{3}\omega_{\text{plasma},n}^2 \frac{1}{1 - \frac{4}{3}\pi n_n \alpha_n},$$

$$\omega_{\text{LO},n}^2 - \omega_{\text{TO},n}^2 = \frac{2}{3}\omega_{\text{plasma},n}^2 \frac{1}{1 + \frac{8}{3}\pi n_n \alpha_n} + \frac{1}{3}\omega_{\text{plasma},n}^2 \frac{1}{1 - \frac{4}{3}\pi n_n \alpha_n} \qquad (7.55)$$

$$= \frac{\omega_{\text{plasma},n}^2}{(1 + \frac{8}{3}\pi n_n \alpha_n)(1 - \frac{4}{3}\pi n_n \alpha_n)}.$$

Here the subscript n represents either material 1 or material 2. In these relations, the plasma frequency squared, $\omega_{\text{plasma},n}^2$, is given by

$$\omega_{plasma,n,a(b)}^2 = 4\pi n_n e_{n,a(b)}^{*2}/\mu_{n,a(b)}. \tag{7.56}$$

As may be verified algebraically, the Lyddane–Sachs–Teller relations of subsection 2.3.3 are satisfied by these frequencies (Wendler, 1985).

With these results of Wendler, a straightforward but lengthy derivation yields

$$
\begin{aligned}
c &= -\left[\frac{\hbar}{2\omega}\left(\frac{8\pi\omega}{L^2 q}\right)\right]^{1/2}\left[\frac{\partial E_1(\omega)}{\partial\omega} + \frac{\partial E_2(\omega)}{\partial\omega}\right]^{-1/2}\\
&= -\left(\frac{4\pi\hbar}{L^2 q}\right)^{1/2}\left[\frac{\partial E_1(\omega)}{\partial\omega} + \frac{\partial E_2(\omega)}{\partial\omega}\right]^{-1/2}.
\end{aligned}
\tag{7.57}
$$

Thus,

$$
\begin{aligned}
\Phi(\mathbf{r}) &= \sum_q \Phi(\mathbf{q}, z)\, e^{-i\mathbf{q}\cdot\rho} = \sum_q c e^{-q|z|} e^{-i\mathbf{q}\cdot\rho}\\
&= -\sum_q \left(\frac{4\pi\hbar}{L^2 q}\right)^{1/2}\left[\frac{\partial E_1(\omega)}{\partial\omega} + \frac{\partial E_2(\omega)}{\partial\omega}\right]^{-1/2} e^{q|z|} e^{-i\mathbf{q}\cdot\rho},
\end{aligned}
\tag{7.58}
$$

where, as discussed previously,

$$
\begin{aligned}
\epsilon_1(\omega) &= \epsilon_1(\infty)\left(\frac{\omega^2 - \omega_{LO,1}^2}{\omega^2 - \omega_{TO,1}^2}\right),\\
\epsilon_2(\omega) &= \epsilon_2(\infty)\left(\frac{\omega^2 - \omega_{LO,2,a}^2}{\omega^2 - \omega_{TO,2,a}^2}\right)\left(\frac{\omega^2 - \omega_{LO,2,b}^2}{\omega^2 - \omega_{TO,2,b}^2}\right).
\end{aligned}
\tag{7.59}
$$

Finally, multiplying $\Phi(\mathbf{r})$ by $-e$ and introducing a_q and $a_q^{\dagger}$, according to the procedure described in Section 5.1, the interaction Hamiltonian for the interface (IF) optical phonon mode may be written as (Kim and Stroscio, 1990)

$$
H_{IF} = \sum_q \left(\frac{4\pi e^2\hbar}{L^2 q}\right)^{1/2}\left[\frac{\partial E_1(\omega)}{\partial\omega} + \frac{\partial E_2(\omega)}{\partial\omega}\right]^{-1/2} e^{-q|z|} e^{i\mathbf{q}\cdot\rho}(a_q + a_{-q}^{\dagger}),
\tag{7.60}
$$

where $\hat{\mathbf{e}}_{\mathbf{q},j}$ and $\hat{\mathbf{e}}_{\mathbf{q},j}^*$ of equation (6.2) have been taken as unit vectors in the longitudinal direction, since the IF phonon modes considered here are longitudinal optical (LO) phonons. The dispersion relation for this optical phonon mode is given from the requirement that the normal components of the electric displacement field be continuous at $z = 0$, that is, $\epsilon_2(\omega)E_{2,z}\big|_{z=0} = \epsilon_1(\omega)E_{1,z}\big|_{z=0}$. From this condition, it follows immediately that the frequencies of the IF optical phonons must satisfy $\epsilon_1(\omega) + \epsilon_2(\omega) = 0$.

This result is similar to that for a bulk semiconductor, where the optical phonon frequencies must satisfy $\epsilon(\omega) = 0$. Moreover, since this is the condition necessary for the propagation of any longitudinal electromagnetic disturbance, it was expected that the frequencies of longitudinal optical phonons should satisfy this dispersion relation.

In the case of the two-region, single-heterointerface structure, the IF longitudinal optical phonon frequencies depend on both $\epsilon_1(\omega)$ and $\epsilon_2(\omega)$. Therefore, the IF optical phonon mode is a joint mode of both materials. Indeed, the electrostatic phonon potential for this mode falls off exponentially with distance from the heterointerface in both materials and, of course, the IF phonon electrostatic potential has only one common value at the interface. Hence, such an IF optical phonon mode must have one common electrostatic phonon potential throughout the two-region heterostructure. It is therefore not surprising that the frequency of such an interface LO phonon mode depends on the dielectric constants of both materials. Clearly, the IF optical phonon mode described by H_{IF} is a joint mode of both materials. In general, IF optical phonon modes are joint modes of all the materials in a given heterostructure. This property is manifest throughout this book.

Clearly, the IF optical phonon modes do not form a complete set of optical phonon modes for the case of two semi-infinite regions joined at a single hetero-interface. Indeed, these IF modes vanish exponentially as $|z| \rightarrow \infty$ and it is clear that bulk-like optical phonons must exist in regions significantly removed from the heterointerface. These additional modes are known as half-space modes. For a structure with a single heterointerface these half-space modes have been given by Mori and Ando (1989) for the case where the two semi-infinite regions are composed of binary semiconductors. Mori and Ando also gave the full set of optical phonon modes for double-heterointerface structures, where one type of binary semiconductor layer with interface planes situated at $-a/2$ and $+a/2$ is bounded by two semi-infinite regions of a different binary semiconductor. Appendix C provides a summary of the phonon modes of the double-heterointerface structure for three models, the slab modes of the dielectric continuum model with electrostatic boundary conditions, as in Mori and Ando (1989), and two other models discussed widely in the literature. Appendix C also discusses Raman measurements useful in understanding the behavior of these modes (Sood et al., 1985). The modes arising from the second and third models are known as the guided modes and the reformulated (or Huang–Zhu) modes. As discussed in Appendix C, all these sets of phonon modes predict the same intrasubband and intersubband scattering rates provided that each set is composed of a complete, orthogonal set of phonon modes.

7.3.2   Electron–phonon interaction for slab modes

Here, it is instructive to consider an earlier – and intuitively very appealing – theory of electron–phonon interactions in a dielectric slab given by Licari and Evrard (1977). In this theory, a single dielectric slab of infinite extent in the x- and y-directions is situated with its faces at $-a$ and $+a$ and with its surface bounded by a vacuum in the regions with $|z| \geq a$. Within this dielectric slab $\nabla \cdot \mathbf{D} = \mathbf{0}$, where, as usual, $\mathbf{D}(\mathbf{r}) = \epsilon(\omega)\mathbf{E}(\mathbf{r}) = \mathbf{E}(\mathbf{r}) + 4\pi\mathbf{P}(\mathbf{r})$; $\epsilon(\omega)$ is the dielectric constant of the slab and $\mathbf{P}(\mathbf{r})$ is the electric polarization associated with the optical phonons in the slab. Defining a scalar potential through $\mathbf{E}(\mathbf{r}) = -\nabla\phi(\mathbf{r})$ and, since the system is translationally invariant in the xy-plane, taking $\phi(\mathbf{r})$ to be of the form $\phi(\mathbf{r}) = \phi(z)e^{i\mathbf{q}_\| \cdot \rho}$, where $\rho = (x, y)$ and $\mathbf{q}_\| = (q_x, q_y)$, it follows that

$$\epsilon(\omega)\left(\frac{\partial^2}{\partial z^2} - q_\|^2\right)\phi(z) = 0 \tag{7.61}$$

where $q_\|^2 = q_x^2 + q_y^2$. This equation is satisfied when $\epsilon(\omega) = 0$ or when $(\partial^2/\partial^2 z - q_\|^2)\phi(z) = 0$. As shown previously, from the general form of $\epsilon(\omega)$ – as given for example in Appendix A – and the Lydanne–Sachs–Teller relation, the condition $\epsilon(\omega) = 0$ is satisfied for a single-material system when $\omega = \omega_{LO}$. In this case an arbitrary function, $\phi(z)$, is a solution of the wave equation; Licari and Evrard took this solution (as did Fuchs and Kliewer, 1965, and Kliewer and Fuchs, 1966a, b) to be of the form

$$\phi(z) = \sum_{q_z}(\phi_1 \sin q_z z + \phi_2 \cos q_z z), \tag{7.62}$$

inside the slab i.e., in the range $(-a, +a)$. Outside the slab, where $\epsilon = 1$, the solutions have the form $\phi(z) = \phi_\pm \exp(\pm\sqrt{q_x^2 + q_y^2}\, z)$, where the positive sign applies for $z \leq -a$ and the negative sign applies for $z \geq +a$. The constants ϕ_1, ϕ_2, ϕ_+, and ϕ_- are determined by the usual boundary conditions that the tangential component of $\mathbf{E}$ and the normal component of $\mathbf{D}$ are continuous at $z = \pm a$. From these conditions it is seen that $\phi_\pm = 0$ and it is thus clear that for this mode $\phi(z)$, $\mathbf{E}(\mathbf{r})$, and $\mathbf{D}(\mathbf{r})$ are zero in the regions surrounding the slab; in particular $\phi(z)$ vanishes at the surfaces of the layer, where $z = \pm a$. For z in the range $(-a, +a)$, the boundary conditions may be satisfied by taking either $\phi_1 = 0$ or $\phi_2 = 0$, so that there are two solutions corresponding to the two polarization vectors:

$$\mathbf{P}_+^m(\mathbf{r}) = \frac{\phi_2}{4\pi a}e^{i\mathbf{q}_\| \cdot \rho}\, i\left(i\mathbf{q}_\| a \cos\frac{m\pi}{2a}z - \hat{\mathbf{z}}\frac{m\pi}{2}\sin\frac{m\pi}{2a}z\right)$$

$$m = 1, 3, 5, \ldots ,$$

$$\mathbf{P}_-^m(\mathbf{r}) = \frac{\phi_1}{4\pi a}e^{i\mathbf{q}_\| \cdot \rho}\, i\left(i\mathbf{q}_\| a \sin\frac{m\pi}{2a}z + \hat{\mathbf{z}}\frac{m\pi}{2}\cos\frac{m\pi}{2a}z\right)$$

$$m = 2, 4, 6, \ldots , \tag{7.63}$$

where $\hat{\mathbf{z}}$ is the unit vector in the z-direction. Of course, $\nabla \cdot \mathbf{D}(\mathbf{r}) = 0$ implies that $\nabla^2\phi(\mathbf{r}) = -\nabla \cdot \mathbf{E}(\mathbf{r}) = +4\pi\nabla \cdot \mathbf{P}(\mathbf{r})$. These standing modes are now widely known as the confined optical phonon modes in a slab. They exist for m running from 1 to some maximum number N_{2a}; the values of m must terminate at N_{2a}, the number of unit cells in thickness $2a$, since the continuum model adopted here must fail when the number of half-wavelengths in $2a$ becomes equal to or greater than the number of unit cells in the same thickness.

The remaining solution corresponds to the case where $\epsilon(\omega) \neq 0$ inside the slab

$$\phi(x) = \phi_1 \exp\left(+\sqrt{q_x^2 + q_y^2}\, z\right) + \phi_2 \exp\left(-\sqrt{q_x^2 + q_y^2}\, z\right).$$

As before, when $|z| \geq a$, the solution is $\phi(z) = \phi_\pm \exp(\pm\sqrt{q_x^2 + q_y^2}z)$. The boundary conditions then restrict the modes to the forms

$$
\mathbf{P}_+^0(\mathbf{r}) = -\phi_1\sqrt{q_x^2 + q_y^2}\,\frac{1 - \epsilon}{4\pi} e^{i\mathbf{q}_\parallel \cdot \boldsymbol{\rho}}\, i
$$
$$
\times \left(i\hat{\mathbf{q}}_\parallel \cosh\sqrt{q_x^2 + q_y^2}\, z + \hat{\mathbf{z}}\sinh\sqrt{q_x^2 + q_y^2}\, z\right),
$$

$$
\mathbf{P}_-^0(\mathbf{r}) = -\phi_1\sqrt{q_x^2 + q_y^2}\,\frac{1 - \epsilon}{4\pi} e^{i\mathbf{q}_\parallel \cdot \boldsymbol{\rho}}\, i
\tag{7.64}
$$
$$
\times \left(i\hat{\mathbf{q}}_\parallel \sinh\sqrt{q_x^2 + q_y^2}\, z + \hat{\mathbf{z}}\cosh\sqrt{q_x^2 + q_y^2}\, z\right),
$$

where $\hat{\mathbf{q}}_\parallel$ is the unit two-dimensional wavevector. These last two modes describe the so-called IF optical phonon modes in the polar semiconductor slab of thickness $2a$. The boundary conditions imply that the frequencies for these modes are solutions of

$$
\frac{1 + \epsilon(\omega)}{1 - \epsilon(\omega)} = \pm \exp\left(-2\sqrt{q_x^2 + q_y^2}a\right),
\tag{7.65}
$$

which, using $\epsilon(\omega) = \epsilon(\infty) + [\epsilon(0) - \epsilon(\infty)]/(1 - \omega^2/\omega_{TO}^2)$, may be written as

$$
\omega_\pm^2 = \omega_{TO}^2 \frac{[\epsilon(0) + 1] \mp [\epsilon(0) - 1]\exp\left(-\sqrt{q_x^2 + q_y^2}\, a\right)}{[\epsilon(\infty) + 1] \mp [\epsilon(\infty) - 1]\exp\left(-\sqrt{q_x^2 + q_y^2}\, a\right)},
\tag{7.66}
$$

where the plus sign corresponds to the even mode, the minus sign to the odd mode.

As pointed out by Licari and Evrard (1977), this continuum model is capable of predicting both the confined LO phonons and the interface IF optical phonons because for both of these modes there exists a polarization charge density. In particular, both $\rho' = -\nabla \cdot \mathbf{P}$, the volume charge density, and $\sigma' = -\mathbf{P} \cdot \hat{\mathbf{n}}$, the surface charge density, contribute to the confined LO modes; here, $\hat{\mathbf{n}}$ is the unit vector normal to the surface and pointing into the vacuum. For the IF modes, only σ' makes a contribution. Clearly, in this model the polarization charge acts as the source of the fields associated with these phonon modes. Transverse modes are not predicted by this continuum approach since for such modes $\nabla \cdot \mathbf{P} = 0$ and $\mathbf{P} \cdot \hat{\mathbf{n}} = \mathbf{0}$.

Licari and Evrard (1977) used this model to study the effects of electronic polarizability on the phonon modes and they derive conditions for the slab which are equivalent to Wendler's conditions for the two-layer system described in Appendix B. Licari and Evrard also used their model to construct the normalized polarization eigenvectors and frequencies for the phonon modes of the dielectric slab. Moreover, they constructed the Hamiltonian for the electron–polar-optical-phonon interaction and showed that the correct harmonic oscillator energy is recovered when the eigenvectors of the slab are used to evaluate the Hamiltonian; in particular, it can be shown that the normal modes are consistent with the harmonic oscillator energy of Section 5.1. Finally, Licari and Evrard presented a very enlightening physical derivation of the electron–phonon interaction Hamiltonian for a slab by applying boundary conditions to the electron–phonon interaction Hamiltonian for a bulk semiconductor. Specifically, starting with expression for the bulk Fröhlich interaction, which we take as the expression (5.34) derived in Section 5.2,

$$
\begin{aligned}
H_{Fr} &= -i \left\{ \frac{2\pi e^2 \hbar \omega_{LO}}{V} \left[\frac{1}{\epsilon(\infty)} - \frac{1}{\epsilon(0)} \right] \right\}^{1/2} \sum_{\mathbf{q}} \frac{1}{q} (a_{\mathbf{q}} + a_{-\mathbf{q}}^{\dagger}) e^{-i\mathbf{q}\cdot\mathbf{r}} \\
&= -i \sum_{\mathbf{q}} \left\{ \frac{2\pi e^2 \hbar \omega_{LO}}{V q^2} \left[\frac{1}{\epsilon(\infty)} - \frac{1}{\epsilon(0)} \right] \right\}^{1/2} (a_{\mathbf{q}} + a_{-\mathbf{q}}^{\dagger}) e^{-i\mathbf{q}\cdot\mathbf{r}} \\
&= -i \sum_{\mathbf{q}} V_{\mathbf{q}} (a_{\mathbf{q}} + a_{-\mathbf{q}}^{\dagger}) e^{-i\mathbf{q}\cdot\mathbf{r}},
\end{aligned}
\tag{7.67}
$$

Licari and Evrard took $\mathbf{q} = (\mathbf{q}_{\parallel}, q_z)$ and split the sum over $\mathbf{q}$ into a sum over $\mathbf{q}_{\parallel}$ and a sum over $q_z > 0$:

$$
\begin{aligned}
H_{Fr} = \sum_{\mathbf{q}_{\parallel}, q_z > 0} V_{\mathbf{q}} e^{-i\mathbf{q}_{\parallel}\cdot\mathbf{r}} \big[& e^{-iq_z z} (a_{\mathbf{q}_{\parallel}, q_z} + a_{-\mathbf{q}_{\parallel}, -q_z}^{\dagger}) \\
& + e^{iq_z z} (a_{\mathbf{q}_{\parallel}, -q_z} + a_{-\mathbf{q}_{\parallel}, q_z}^{\dagger}) \big].
\end{aligned}
\tag{7.68}
$$

Then, using $e^{i\theta} = \cos\theta + i\sin\theta$ to write $e^{\pm iq_z z}$ in terms of sines and cosines,

$$
\begin{aligned}
H_{Fr} = \sqrt{2} \sum_{\mathbf{q}_{\parallel}, q_z > 0} V_{\mathbf{q}} e^{-i\mathbf{q}_{\parallel}\cdot\mathbf{r}} \\
\times \{ \cos q_z z [a_+(\mathbf{q}_{\parallel}) + a_+^{\dagger}(-\mathbf{q}_{\parallel})] + \sin q_z z \, [a_-(\mathbf{q}_{\parallel}) + a_-^{\dagger}(-\mathbf{q}_{\parallel})] \},
\end{aligned}
\tag{7.69}
$$

where

$$
a_+(\mathbf{q}_{\parallel}) = \frac{1}{\sqrt{2}} (a_{\mathbf{q}_{\parallel}, q_z} + a_{\mathbf{q}_{\parallel}, -q_z}), \qquad a_-(\mathbf{q}_{\parallel}) = \frac{-i}{\sqrt{2}} (a_{\mathbf{q}_{\parallel}, q_z} - a_{\mathbf{q}_{\parallel}, -q_z}).
\tag{7.70}
$$

The operators $a_+^{\dagger}(-\mathbf{q}_{\parallel})$ and $a_-^{\dagger}(-\mathbf{q}_{\parallel})$ are given by taking the adjoints. These operators describe phonons which propagate as plane waves in the x- and y-directions but as standing modes in the z-direction. Indeed, since $q_z = m\pi/2a$ the Fröhlich Hamiltonian for the two-dimensional slab takes the form

$$H_{Fr} = \left\{\frac{4\pi e^2 \hbar \omega_{LO}}{V}\left[\frac{1}{\epsilon(\infty)} - \frac{1}{\epsilon(0)}\right]\right\}^{1/2} \sum_{\mathbf{q}_\parallel} e^{-i\mathbf{q}_\parallel \cdot \mathbf{r}}$$

$$\times \left\{\sum_{m=1,3,5...} \frac{\cos(m\pi/2a)z}{[q_x^2 + q_y^2 + (m\pi/2a)^2]^{1/2}}[a_{m+}(\mathbf{q}_\parallel) + a_{m+}^\dagger(-\mathbf{q}_\parallel)]\right.$$

$$\left. + \sum_{m=2,4,6...} \frac{\sin(m\pi/2a)z}{[q_x^2 + q_y^2 + (m\pi/2a)^2]^{1/2}}[a_{m-}(\mathbf{q}_\parallel) + a_{m-}^\dagger(-\mathbf{q}_\parallel)]\right\}.$$

$$(7.71)$$

This Hamiltonian vanishes for $z = \pm a$, as it must since the Fröhlich interaction Hamiltonian is given by $-e\phi$, as explained in Section 5.2, and since $\phi(\pm a) = 0$ for the potential describing the fields associated with phonon modes in the dielectric slab. This heuristic derivation makes manifest the fact that the confined phonon modes in the slab located between $-a$ and $+a$ are standing modes with an integer number of half-wavelengths confined within the slab. This Hamiltonian does not contain the contributions of the IF optical phonons in the slab since it satisfies only the boundary conditions for the confined optical phonon modes at $z = \pm a$, namely $H_{Fr}(a) = -e\phi(\pm a) = 0$. As shown by Licari and Evrard (1977), the Fröhlich interaction Hamiltonian for the IF optical phonon modes in the dielectric slab is

$$H_{Fr} = -\left\{\frac{2\pi e^2 \hbar \omega_{TO}}{L^2}[\epsilon(0) - \epsilon(\infty)]\right\}^{1/2}$$

$$\times \sum_{\mathbf{q}_\parallel} e^{-i\mathbf{q}_\parallel \cdot \mathbf{r}}\left(\frac{\sinh 2\sqrt{q_x^2 + q_y^2}\, a}{\sqrt{q_x^2 + q_y^2}}\right)^{1/2} e^{-\sqrt{q_x^2+q_y^2}a}$$

$$\times \left\{G_+\left(\sqrt{q_x^2 + q_y^2}, z'\right)[a_{0+}(\mathbf{q}_\parallel) + a_{0+}^\dagger(-\mathbf{q}_\parallel)]\right.$$

$$\left. + G_-\left(\sqrt{q_x^2 + q_y^2}, z'\right)[a_{0-}(\mathbf{q}_\parallel) + a_{0-}^\dagger(-\mathbf{q}_\parallel)]\right\},$$

$$(7.72)$$

where

$$G_+\left(\sqrt{q_x^2 + q_y^2}, z'\right) = \frac{\cosh\sqrt{q_x^2 + q_y^2}\, z'\big/\cosh\sqrt{q_x^2 + q_y^2}\, a}{[\epsilon(\infty) + 1] - [\epsilon(\infty) - 1]e^{-2\sqrt{q_x^2+q_y^2}\, a}}$$

$$\times \left\{\frac{[\epsilon(\infty) + 1] - [\epsilon(\infty) - 1]e^{-2\sqrt{q_x^2+q_y^2}\, a}}{[\epsilon(0) + 1] - [\epsilon(0) - 1]e^{-2\sqrt{q_x^2+q_y^2}\, a}}\right\}^{1/4},$$

$$(7.73)$$

and

$$G_-\left(\sqrt{q_x^2 + q_y^2}, z'\right) = \frac{\sinh\sqrt{q_x^2 + q_y^2}\, z'\big/\sinh\sqrt{q_x^2 + q_y^2}\, a}{[\epsilon(\infty) + 1] + [\epsilon(\infty) - 1]e^{-2\sqrt{q_x^2+q_y^2}\, a}}$$

$$\times \left\{\frac{[\epsilon(\infty) + 1] + [\epsilon(\infty) - 1]e^{-2\sqrt{q_x^2+q_y^2}\, a}}{[\epsilon(0) + 1] + [\epsilon(0) - 1]e^{-2\sqrt{q_x^2+q_y^2}\, a}}\right\}^{1/4}.$$

$$(7.74)$$

In the result (7.72) the phonon creation and annihilation operators are not summed over m; since there are just two IF optical phonon modes, the subscript m on the creation and annihilation operators for the confined optical phonon modes is replaced by 0 and the plus sign in the subscript corresponds to the even mode while the minus sign corresponds to the odd mode. The modes discussed by Licari and Evrard are recognized as the optical modes of a double-interface heterostructure in the special case where $\epsilon = 1$ outside the region bounded by the two heterointerfaces. As mentioned previously, Appendix C compares the three frequently used complete sets of optical phonon modes for a double-interface heterostructure for the case where all the material layers are polar semiconductors. As may be seen straightforwardly, the modes considered by Licari and Evrard correspond to the slab modes derived with electrostatic boundary condition for the special cases where the quantum well is bounded by a vacuum, so that $\epsilon = 1$.

7.3.3 Slab modes in confined würtzite structures

The so-called slab modes for dimensionally confined würtzite semiconductor structures have been derived using an extension of Loudon's model of uniaxial semiconductors, which was introduced in Chapter 3. The normalization condition of Chapter 5 must be modified to take into account the fact that for uniaxial semiconductor crystals there are separate equations governing $\mathbf{u}_{\perp,n}$ and $\mathbf{u}_{\parallel,n}$, where, as in subsection 7.3.2, the subscripts denote the components normal and parallel to the c-axis of a uniaxial semiconductor with materials properties associated with those of a medium n.

For such a medium, the normalization condition of Section 5.1 for a single mode q (with $\omega_q = \omega$) becomes

$$\int \left\{ \left| \sqrt{n_n \mu_n} \mathbf{u}_{\perp,n}(\mathbf{r}) \right|^2 + \left| \sqrt{n_n \mu_n} \mathbf{u}_{\parallel,n}(\mathbf{r}) \right|^2 \right\} d\mathbf{r} = \frac{\hbar}{2\omega}, \tag{7.75}$$

where μ_n denotes the reduced mass and, as in Section 5.1, $n_n = N_n/V$. Then, defining

$$\mathbf{u}_{\perp,\parallel}(\mathbf{r}) = \sum_q \mathbf{u}_{\perp,\parallel}(\mathbf{q}) e^{i\mathbf{q}\cdot\mathbf{r}}, \tag{7.76}$$

it follows that

$$\left| \sqrt{n_n \mu_n} \mathbf{u}_{\perp,n}(\mathbf{q}) \right|^2 + \left| \sqrt{n_n \mu_n} \mathbf{u}_{\parallel,n}(\mathbf{q}) \right|^2 = \frac{\hbar}{2\omega} \frac{1}{V}, \tag{7.77}$$

provides the necessary generalization to the case of a uniaxial crystal. From Section 7.1, the equations governing $\mathbf{u}_{\perp,n}$ and $\mathbf{u}_{\parallel,n}$ may be written as

$$\mathbf{u}_{\perp(\parallel),n} = \frac{1}{\sqrt{4\pi \mu_n n_n}} \frac{\sqrt{\epsilon(0)_{\perp(\parallel),n} - \epsilon(\infty)_{\perp(\parallel),n}}}{\omega^2_{TO,\perp(\parallel),n} - \omega^2} \omega_{TO,\perp(\parallel),n} \mathbf{E}_{\perp,n}, \tag{7.78}$$

where the time dependence of each displacement has been assumed to be of the form $e^{i\omega t}$. Then, it follows that

$$
\left|\sqrt{n_n \mu_n} \mathbf{u}_{\perp,n}(\mathbf{r})\right|^2 + \left|\sqrt{n_n \mu_n} \mathbf{u}_{\|,n}(\mathbf{r})\right|^2
$$
$$
= \frac{1}{4\pi} \frac{[\epsilon(0)_{\perp,n} - \epsilon(\infty)_{\perp,n}]\omega_{\mathrm{TO},\perp,n}^2}{(\omega_{\mathrm{TO},\perp,n}^2 - \omega^2)^2} \left|\mathbf{E}_{\perp,n}\right|^2
$$
$$
+ \frac{1}{4\pi} \frac{(\epsilon(0)_{\|,n} - \epsilon(\infty)_{\|,n})\omega_{\mathrm{TO},\|,n}^2}{(\omega_{\mathrm{TO},\|,n}^2 - \omega^2)^2} \left|\mathbf{E}_{\|,n}\right|^2 . \tag{7.79}
$$

However, from the generalized Lyddane–Sachs–Teller relation, subsection 2.3.3, it is known that

$$
\epsilon(\omega)_{\perp(\|),n} = \epsilon(\infty)_{\perp(\|),n} \left(\frac{\omega_{\mathrm{LO},\perp(\|),n}^2 - \omega^2}{\omega_{\mathrm{TO},\perp(\|),n}^2 - \omega^2} \right), \tag{7.80}
$$

and it follows that

$$
\frac{1}{2\omega} \frac{\partial \epsilon(\omega)_{\perp(\|),n}}{\partial \omega} = \frac{[\epsilon(0)_{\perp(\|),n} - \epsilon(\infty)_{\perp(\|),n}]}{(\omega_{\mathrm{TO},\perp(\|),n}^2 - \omega^2)^2} \omega_{\mathrm{TO},\perp(\|),n}^2, \tag{7.81}
$$

so that

$$
\left|\sqrt{n_n \mu_n} \mathbf{u}_{\perp,n}(\mathbf{r})\right|^2 + \left|\sqrt{n_n \mu_n} \mathbf{u}_{\|,n}(\mathbf{r})\right|^2
$$
$$
= \frac{1}{4\pi} \frac{1}{2\omega} \frac{\partial \epsilon(\omega)_{\perp,n}}{\partial \omega} \left|\mathbf{E}_{\perp,n}\right|^2 + \frac{1}{4\pi} \frac{1}{2\omega} \frac{\partial \epsilon(\omega)_{\|,n}}{\partial \omega} \left|\mathbf{E}_{\|,n}\right|^2 . \tag{7.82}
$$

Using this identity, the normalization condition becomes

$$
\int \left(\frac{1}{4\pi} \frac{1}{2\omega} \frac{\partial \epsilon(\omega)_{\perp,n}}{\partial \omega} \left|\mathbf{E}_{\perp,n}\right|^2 + \frac{1}{4\pi} \frac{1}{2\omega} \frac{\partial \epsilon(\omega)_{\|,n}}{\partial \omega} \left|\mathbf{E}_{\|,n}\right|^2 \right) d\mathbf{r} = \frac{\hbar}{2\omega} . \tag{7.83}
$$

As discussed in Section 7.3, the normalization condition for the case where there is dimensional confinement in only the z-direction is then

$$
\int L^2 \left(\frac{1}{4\pi} \frac{1}{2\omega} \frac{\partial \epsilon(\omega)_{\perp,n}}{\partial \omega} \left|\mathbf{E}_{\perp,n}\right|^2 + \frac{1}{4\pi} \frac{1}{2\omega} \frac{\partial \epsilon(\omega)_{\|,n}}{\partial \omega} \left|\mathbf{E}_{\|,n}\right|^2 \right) dz = \frac{\hbar}{2\omega} , \tag{7.84}
$$

where $\omega = \omega_q$ and $L^2 = L_x L_y$. It should be noted that this integral is of the same form as (7.44). As for the zincblende case, the electron–optical-phonon Hamiltonian is then given by

$$
H_{\mathrm{IF}} = -e \sum_q \Phi(\mathbf{q}, z) e^{i\mathbf{q}\cdot\rho} (a_q + a^{\dagger}_{-q}), \tag{7.85}
$$

where the potential, $\Phi(\mathbf{q}, z)$, is associated with $\mathbf{E}_{\perp,n}$ and $\mathbf{E}_{\|,n}$. The form of normalization (7.84) is particularly convenient for optical modes in the dielectric continuum model since the phonons may be described in terms of the associated electric fields and potentials (Kim and Stroscio, 1990; Lee *et al.*, 1998; Komirenko *et al.*, 2000a).

In this subsection, the quantization condition for a uniaxial crystal will be applied to determine the electron–optical-phonon interaction Hamiltonian for interface phonons in a single-heterointerface structure. Appendix D summarizes the electron–optical-phonon interaction Hamiltonians for all the optical phonon modes in single- and double-heterointerface uniaxial crystals for the case where the c-axis is perpendicular to the heterointerface(s).

As an illustration, consider the interface optical mode in a würtzite structure composed of two semi-infinite regions separated by a single heterointerface situated at $z = 0$. The c-axis is taken to be normal to the heterointerface. In the region $z < 0$ the dielectric functions are $\epsilon(\omega)_{\perp(\parallel),2}$ and in the region $z > 0$ they are $\epsilon(\omega)_{\perp(\parallel),1}$. Each of the four functions, $\epsilon(\omega)_{\perp(\parallel),1(2)}$ obeys the generalized Lyddane–Sachs–Teller relation; in particular, each is less than zero for frequencies selected from one range of interest from among the four ranges $\omega_{TO,\perp(\parallel),1(2)} < \omega < \omega_{LO,\perp(\parallel),1(2)}$. In addition, for $\omega_{LO,\perp(\parallel),1(2)} < \omega < \omega_{TO,\perp(\parallel),1(2)}$, it is clear that $\epsilon(\omega)_{\perp(\parallel),1(2)} > 0$ for the range corresponding to the one dielectric function of interest. For each of these four ranges, the positivity or negativity properties are as for zincblende crystals.

However, for uniaxial crystals the products $\epsilon(\omega)_{\perp,1}\epsilon(\omega)_{\parallel,1}$ and $\epsilon(\omega)_{\perp,2}\epsilon(\omega)_{\parallel,2}$ may be either positive or negative since, for a material n, $\epsilon(\omega)_{\perp,n}$ and $\epsilon(\omega)_{\parallel,n}$ may have different signs depending on the overlap of the two regions $\omega_{TO,\perp,n} < \omega < \omega_{LO,\perp,n}$ and $\omega_{TO,\parallel,n} < \omega < \omega_{LO,\parallel,n}$.

For binary zincblende heterostructures such as GaAs/AlAs, the IF modes exist for the frequencies in the two ranges $\omega_{TO,AlAs} < \omega < \omega_{LO,AlAs}$ and $\omega_{TO,GaAs} < \omega < \omega_{LO,GaAs}$. Since these two ranges do not overlap, the frequency condition for the existence of IF modes in such zincblende heterostructures is typified by $\epsilon_{GaAs}(\omega)\epsilon_{AlAs}(\omega) < 0$ for all allowed IF mode frequencies ω. Such a simple characterization is not possible for uniaxial crystals.

As will become obvious, this situation leads to significant differences in the optical phonon modes in würtzite and zincblende structures. For the zincblende case

$$\phi(r) = e^{i\mathbf{q}\cdot\rho} \times \begin{cases} Ae^{\kappa_2 z} & z < 0 \\ Be^{-\kappa_1 z} & z > 0 \end{cases} \tag{7.86}$$

and in the absence of free charge $\nabla \cdot \mathbf{D} = 0$ so that

$$\begin{aligned} \epsilon(\omega)_{\parallel,1}\kappa_1^2 - \epsilon(\omega)_{\perp,1}q^2 = 0 & \quad \epsilon(\omega)_{\parallel,1}\epsilon(\omega)_{\perp,1} > 0, \\ \epsilon(\omega)_{\parallel,2}\kappa_2^2 - \epsilon(\omega)_{\perp,2}q^2 = 0 & \quad \epsilon(\omega)_{\parallel,2}\epsilon(\omega)_{\perp,2} > 0. \end{aligned} \tag{7.87}$$

From the continuity of the tangential component of the electric field at $z = 0$, it then follows that $A = B$. From the continuity of the normal component of the electric displacement at $z = 0$, $\epsilon_{\parallel,2}\kappa_2 A = -\epsilon_{\parallel,1}\kappa_1 B$; thus $\epsilon_{\parallel,1}\kappa_1 + \epsilon_{\parallel,2}\kappa_2 = 0$ must be satisfied in the allowed range of IF modes, which corresponds to the frequencies satisfying $\epsilon_{\parallel,1}(\omega)\epsilon_{\parallel,2}(\omega) < 0$. However,

$$\kappa_1^2 = \frac{\epsilon(\omega)_{\perp,1}}{\epsilon(\omega)_{\parallel,1}}q^2 \qquad \epsilon(\omega)_{\parallel,1}\epsilon(\omega)_{\perp,1} > 0,$$

$$\kappa_2^2 = \frac{\epsilon(\omega)_{\perp,2}}{\epsilon(\omega)_{\parallel,2}}q^2 \qquad \epsilon(\omega)_{\parallel,2}\epsilon(\omega)_{\perp,2} > 0,$$

$$(7.88)$$

so that

$$\epsilon_{\parallel,1}\kappa_1 + \epsilon_{\parallel,2}\kappa_2 = \epsilon(\omega)_{\parallel,1}\sqrt{\frac{\epsilon(\omega)_{\perp,1}}{\epsilon(\omega)_{\parallel,1}}}\,q + \epsilon_{\parallel,2}(\omega)\sqrt{\frac{\epsilon(\omega)_{\perp,2}}{\epsilon(\omega)_{\parallel,2}}} = 0. \qquad (7.89)$$

Accordingly,

$$-\sqrt{\epsilon(\omega)_{\perp,1}\epsilon(\omega)_{\parallel,1}} + \sqrt{\epsilon(\omega)_{\perp,2}\epsilon(\omega)_{\parallel,2}} = 0 \qquad \epsilon(\omega)_{\parallel,1} < 0 \text{ and } \epsilon(\omega)_{\parallel,2} > 0,$$

$$+\sqrt{\epsilon(\omega)_{\perp,1}\epsilon(\omega)_{\parallel,1}} - \sqrt{\epsilon(\omega)_{\perp,2}\epsilon(\omega)_{\parallel,2}} = 0 \qquad \epsilon(\omega)_{\parallel,1} > 0 \text{ and } \epsilon(\omega)_{\parallel,2} < 0,$$

$$(7.90)$$

and it follows that $\sqrt{\epsilon(\omega)_{\perp,1}\epsilon(\omega)_{\parallel,1}} = \sqrt{\epsilon(\omega)_{\perp,2}\epsilon(\omega)_{\parallel,2}}$. Thus

$$\phi(r) = \phi_0 e^{i\mathbf{q}\cdot\boldsymbol{\rho}} \times \begin{cases} Ae^{\kappa_2 z} & z < 0 \\ Be^{-\kappa_1 z} & z > 0 \end{cases}$$

$$= \phi_0 e^{i\mathbf{q}\cdot\boldsymbol{\rho}} \times \begin{cases} \exp(\sqrt{\epsilon(\omega)_{\perp,2}/\epsilon(\omega)_{\parallel,2}}\,qz) & z < 0 \\ \exp(-\sqrt{\epsilon(\omega)_{\perp,1}/\epsilon(\omega)_{\parallel,1}}\,qz) & z > 0. \end{cases} \qquad (7.91)$$

Now $\mathbf{E}_{\perp}$ and $\mathbf{E}_{\parallel}$ are given by the appropriate gradients of ϕ_0, and the integrals needed to calculate the normalization condition are related to $|\mathbf{E}_{\perp,n}|^2$ and $|\mathbf{E}_{\parallel,n}|^2$ through

$$\int_{-\infty}^0 |\mathbf{E}_{\perp,2}|^2 dz = \phi_0^2 q^2 \int_{-\infty}^0 e^{2\kappa_2 z} dz = \frac{q^2}{2\kappa_2}\phi_0^2 = \frac{1}{2}\sqrt{\frac{\epsilon(\omega)_{\perp,2}}{\epsilon(\omega)_{\parallel,2}}}\,q\phi_0^2,$$

$$\int_0^\infty |\mathbf{E}_{\perp,1}|^2 dz = \phi_0^2 q^2 \int_0^\infty e^{-2\kappa_1 z} dz = \frac{q^2}{2\kappa_1}\phi_0^2 = \frac{1}{2}\sqrt{\frac{\epsilon(\omega)_{\perp,1}}{\epsilon(\omega)_{\parallel,1}}}\,q\phi_0^2,$$

$$\int_{-\infty}^0 |\mathbf{E}_{\parallel,2}|^2 dz = \phi_0^2 \kappa_2^2 \int_{-\infty}^0 e^{2\kappa_2 z} dz = \frac{\kappa_2}{2}\phi_0^2 = \frac{1}{2}\sqrt{\frac{\epsilon(\omega)_{\perp,2}}{\epsilon(\omega)_{\parallel,2}}}\,q\phi_0^2,$$

$$\int_0^\infty |\mathbf{E}_{\parallel,1}|^2 dz = \phi_0^2 \kappa_1^2 \int_0^\infty e^{-2\kappa_1 z} dz = \frac{\kappa_1}{2}\phi_0^2 = \frac{1}{2}\sqrt{\frac{\epsilon(\omega)_{\perp,1}}{\epsilon(\omega)_{\parallel,1}}}\,q\phi_0^2.$$

$$(7.92)$$

Then we have

$$
\frac{1}{L^2}\frac{\hbar}{2\omega}
$$

$$
= \int \left(\frac{1}{4\pi}\frac{1}{2\omega}\frac{\partial\epsilon(\omega)_{\perp,n}}{\partial\omega}\left|\mathbf{E}_{\perp,n}\right|^2 + \frac{1}{4\pi}\frac{1}{2\omega}\frac{\partial\epsilon(\omega)_{\|,n}}{\partial\omega}\left|\mathbf{E}_{\|,n}\right|^2 \right)dz
$$

$$
= \frac{1}{4\pi}\frac{1}{2\omega}\int_{-\infty}^{0}\frac{\partial\epsilon(\omega)_{\perp,2}}{\partial\omega}\left|\mathbf{E}_{\perp,2}\right|^2 dz + \frac{1}{4\pi}\frac{1}{2\omega}\int_{0}^{\infty}\frac{\partial\epsilon(\omega)_{\perp,1}}{\partial\omega}\left|\mathbf{E}_{\perp,1}\right|^2 dz
$$

$$
+ \frac{1}{4\pi}\frac{1}{2\omega}\int_{-\infty}^{0}\frac{\partial\epsilon(\omega)_{\|,2}}{\partial\omega}\left|\mathbf{E}_{\|,2}\right|^2 dz + \frac{1}{4\pi}\frac{1}{2\omega}\int_{0}^{\infty}\frac{\partial\epsilon(\omega)_{\|,1}}{\partial\omega}\left|\mathbf{E}_{\|,1}\right|^2 dz
$$

$$
= \frac{1}{4\pi}\frac{1}{2\omega}\frac{q}{2}\left[\left(\frac{\partial\epsilon_{\perp,1}}{\partial\omega}\frac{q}{\kappa_1} + \frac{\partial\epsilon_{\|,1}}{\partial\omega}\frac{\kappa_1}{q} \right) + \left(\frac{\partial\epsilon_{\perp,2}}{\partial\omega}\frac{q}{\kappa_2} + \frac{\partial\epsilon_{\|,2}}{\partial\omega}\frac{\kappa_2}{q} \right) \right]\phi_0^2,
$$

$$(7.93)$$

and it follows that

$$
\phi_0^2 = \frac{4\pi\hbar}{L^2}\frac{2}{q}\left[\left(\frac{\partial\epsilon_{\perp,1}}{\partial\omega}\frac{q}{\kappa_1} + \frac{\partial\epsilon_{\|,1}}{\partial\omega}\frac{\kappa_1}{q} \right) + \left(\frac{\partial\epsilon_{\perp,2}}{\partial\omega}\frac{q}{\kappa_2} + \frac{\partial\epsilon_{\|,2}}{\partial\omega}\frac{\kappa_2}{q} \right) \right]^{-1}
$$

$$(7.94)$$

and

$$
H_{\mathrm{IF}} = \sum_q -(e\Phi(\mathbf{q},z))\, e^{i\mathbf{q}\cdot\boldsymbol{\rho}}(a_{\mathbf{q}} + a_{-\mathbf{q}}^{\dagger})
$$

$$
= \sum_q \sqrt{\frac{4\pi e^2\hbar}{L^2}}(2q^{-1})^{1/2}
$$

$$
\times \left[\left(\frac{\partial\epsilon_{\perp,1}}{\partial\omega}\frac{q}{\kappa_1} + \frac{\partial\epsilon_{\|,1}}{\partial\omega}\frac{\kappa_1}{q} \right) + \left(\frac{\partial\epsilon_{\perp,2}}{\partial\omega}\frac{q}{\kappa_2} + \frac{\partial\epsilon_{\|,2}}{\partial\omega}\frac{\kappa_2}{q} \right) \right]^{-1/2}
$$

$$
\times e^{i\mathbf{q}\cdot\boldsymbol{\rho}}(a_{\mathbf{q}} + a_{-\mathbf{q}}^{\dagger})
$$

$$
\times \begin{cases} e^{-\sqrt{\epsilon(\omega)_{\perp,1}/\epsilon(\omega)_{\|,1}}\,qz} & z > 0, \\ e^{\sqrt{\epsilon(\omega)_{\perp,2}/\epsilon(\omega)_{\|,2}}\,qz} & z < 0, \end{cases}
$$

$$(7.95)$$

which may be written in an alternative form by use of the relation

$$
2\left| \frac{\partial}{\partial\omega}\left(\sqrt{\epsilon(\omega)_{\perp,1}\epsilon(\omega)_{\|,1}} - \sqrt{\epsilon(\omega)_{\perp,2}\epsilon(\omega)_{\|,2}} \right) \right|
$$

$$
= \left[\left(\frac{\partial\epsilon_{\perp,1}}{\partial\omega}\frac{q}{\kappa_1} + \frac{\partial\epsilon_{\|,1}}{\partial\omega}\frac{\kappa_1}{q} \right) + \left(\frac{\partial\epsilon_{\perp,2}}{\partial\omega}\frac{q}{\kappa_2} + \frac{\partial\epsilon_{\|,2}}{\partial\omega}\frac{\kappa_2}{q} \right) \right].
$$

$$(7.96)$$

The dispersion relation for the modes described by this Hamiltonian is given by the condition, $\sqrt{\epsilon(\omega)_{\perp,1}\epsilon(\omega)_{\|,1}} = \sqrt{\epsilon(\omega)_{\perp,2}\epsilon(\omega)_{\|,2}}$ resulting from the requirements that the tangential component of the electric field as well as the normal component of the displacement field be continuous at the heterointerface. A similar analysis for the case of a structure with two heterointerfaces is given in Appendix D (Komirenko et al., 1999), where the dispersion of polar optical phonons in würtzite quantum wells is considered at length. Gleize et al. (1999) have extended Komirenko's results to the case of würtzite superlattices. For the specific case of a GaN/AlN superlattice with the c-axis normal to the heterointerfaces and with a superlattice period, d, as shown in Figure 7.2, the requirement of periodicity and the boundary conditions imposed on the fields for the quantum-well case lead to dispersion relations of the form

$$a_1(\omega)\cosh[\gamma_1(\omega)d/2] + a_2(\omega)\sinh[\gamma_1(\omega)d/2] = 0, \tag{7.97}$$

the antisymmetric modes and

$$a_1(\omega)\sinh[\gamma_1(\omega)d/2] + a_2(\omega)\cosh[\gamma_1(\omega)d/2] = 0, \tag{7.98}$$

for the symmetric modes. In these dispersion relations we have

$$\gamma_1(\omega) = q_\perp\sqrt{\epsilon(\omega)_{\perp,1}\big/\epsilon(\omega)_{\|,1}}\,,$$

$$a_1(\omega) = \text{sign}[\epsilon(\omega)_{\|,1}]\sqrt{\epsilon(\omega)_{\perp,1}\epsilon(\omega)_{\|,1}}\,, \tag{7.99}$$

$$a_2(\omega) = \text{sign}[\epsilon(\omega)_{\|,2}]\sqrt{\epsilon(\omega)_{\perp,2}\epsilon(\omega)_{\|,2}}\,.$$

The dispersion relations (7.97), (7.98) are depicted in Figures 7.3 and 7.4 for the AlN(5 nm)/GaN(5 nm) superlattice along with quasi-confined modes. In the limit $q_\perp d \to \infty$, these dispersion relations reduce to the condition $\sqrt{\epsilon(\omega)_{\perp,1}\epsilon(\omega)_{\|,1}} = \sqrt{\epsilon(\omega)_{\perp,2}\epsilon(\omega)_{\|,2}}$, as they must since in the short-wavelength limit the frequencies of these modes cannot depend on d and, in fact, should be given by the dispersion relation for a single heterointerface between GaN and AlN.

Let us consider in more detail the Hamiltonian for the single-heterointerface structure. For the case when $\epsilon(\omega)_{\perp,1} = \epsilon(\omega)_{\|,1}$ and $\epsilon(\omega)_{\perp,2} = \epsilon(\omega)_{\|,2}$, $q^2 = \kappa_1^2 = \kappa_2^2$ and the Hamiltonian reduces to

$$H_{\text{IF}} = \sum_q \sqrt{\frac{4\pi e^2 \hbar L^2}{\partial\epsilon_1(\omega)/\partial\omega + \partial\epsilon_2(\omega)/\partial\omega}}\sqrt{\frac{1}{q}}e^{-q|z|}e^{i\mathbf{q}\cdot\boldsymbol{\rho}}(a_q + a_{-q}^\dagger),$$

$$\tag{7.100}$$

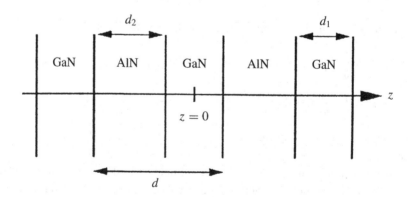

Figure 7.2. Würtzite superlattice considered by Gleize *et al.* (1999). From Gleize *et al.* (1999). American Physical Society, with permission.

which can be recognized as the Hamiltonian for the IF optical phonon modes in a zincblende single-heterointerface system.

In this subsection the interface optical phonon modes in a single-heterointerface würtzite structure have been normalized to construct the Fröhlich-like electron–

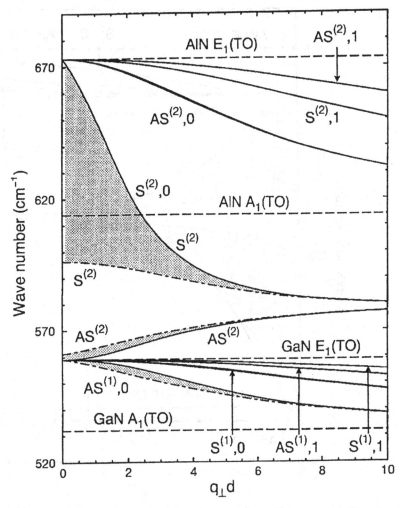

Figure 7.3. Dispersion of interface and quasi-confined modes for an infinite and unstrained AlN(5 nm)/GaN(5 nm) superlattice in the transverse optical (TO) frequency range. The shaded areas depict the bands for all values of Q_z, $-\pi/d < Q_z < \pi/d$, lying in the first Brillouin zone of the superlattice: $-\!\cdot\!-\!\cdot\!-$, $Q_z = 0$; ———, $Q_z = \pi/d$. The $S^{(j)}$ and the $AS^{(j)}$ are the symmetric and antisymmetric modes with respect to the middle plane of any layer. $j = 1$ for GaN and $j = 2$ for AlN. The quasi-confined modes are identified by their order m, an integer following a comma. From Gleize *et al.* (1999), American Physical Society, with permission.

optical-phonon interaction Hamiltonian H_{IF}. Appendix D provides a summary of all the optical phonon modes in single- and double-heterointerface würtzite structures based on the macroscopic dielectric continuum model and Loudon's model for uniaxial crystals.

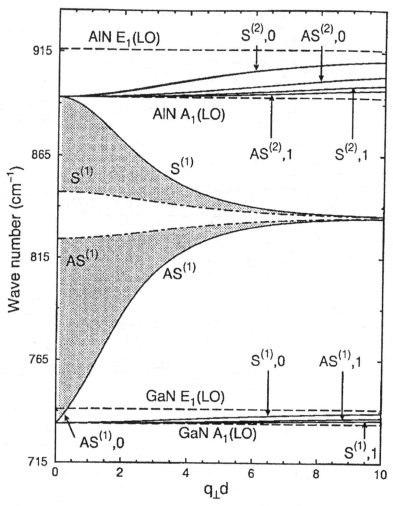

Figure 7.4. Dispersion of interface and quasi-confined modes for an infinite and unstrained AlN(5 nm)/GaN(5 nm) superlattice in the longitudinal optical (LO) frequency range. The shaded areas depict the bands for all values of Q_z, $-\pi/d < Q_z < \pi/d$, lying in the first Brillouin zone of the superlattice: - — -, $Q_z = 0$; ——, $Q_z = \pi/d$. The $S^{(j)}$ and the $AS^{(j)}$ are the symmetric and antisymmetric modes with respect to the middle plane of any layer. $j = 1$ for GaN and $j = 2$ for AlN. The quasi-confined modes are identified by their order m, an integer following a comma. From Gleize *et al.* (1999), American Physical Society, with permission.

7.3.4 Transfer matrix model for multi-heterointerface structures

Yu *et al.* (1997) derived a very useful set of normalization conditions for heterostructures containing multiple parallel heterointerfaces separating different semiconductor layers. These normalization conditions are essential for examining the optical phonon bandstructure in superlattices and they provide the basis for relatively straightforward calculations of the normalization factors for heterostructures containing just a few heterointerfaces. Since translational invariance holds in the two-dimensional planes parallel to the heterointerfaces, the electrostatic potential describing the carrier–optical-phonon interaction in each region $\mathbf{R}_i = (z_i, z_{i+1})$ is denoted by $\Phi_i(\mathbf{r})$ and is taken to be of the form

$$\Phi_i(\mathbf{r}) = \sum_{\mathbf{q}} e^{-i\mathbf{q}\cdot\rho} \Phi_i(\mathbf{q}, z) \tag{7.101}$$

with

$$\Phi_i(\mathbf{q}, z) = c_{i-}e^{-qz} + c_{i+}e^{+qz} \equiv c_{i-}\phi_{i-} + c_{i+}\phi_{i+}, \tag{7.102}$$

where the z-axis is taken to be normal to the heterointerfaces and where, as usual, $\rho = (x, y)$ and $\mathbf{q}$ denote the position and wavevector in two dimensions. c_{i-} and c_{i+} are the relative amplitudes of the exponentially decaying and growing potentials, respectively, in layer i; as will become clear, these relative amplitudes are related through a transfer matrix. Figure 7.5 depicts a generic potential $\mathbf{\Phi}_i(z)$ for regions $\mathbf{R}_0, \mathbf{R}_1, \ldots, \mathbf{R}_n$.

According to the electrostatic boundary conditions the electrostatic potential $\Phi_i(\mathbf{q}, z)$ and the normal component of the electric displacement, $\epsilon_i \mathbf{E}_i =$

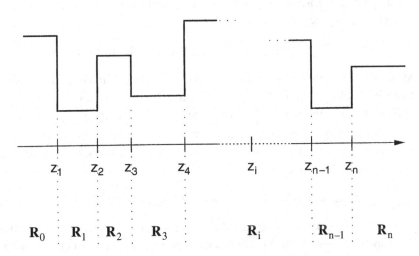

Figure 7.5. A possible generic potential for regions $\mathbf{R}_0, \mathbf{R}_1, \ldots, \mathbf{R}_n$. From Yu *et al.* (1997), American Institute Physics, with permission.

$-\epsilon_i \partial \Phi_i(\mathbf{q}, z)/\partial z$, must be continuous at each heterointerface; thus, at the hetero-interface located at z_i,

$$\Phi_i(\mathbf{q}, z_i) = \Phi_{i-1}(\mathbf{q}, z_i) \quad \text{and} \quad \epsilon_i \frac{\partial \Phi_i(\mathbf{q}, z_i)}{\partial z} = \epsilon_{i-1} \frac{\partial \Phi_{i-1}(\mathbf{q}, z_i)}{\partial z}$$

(7.103)

are the boundary conditions at the heterointerface separating regions $\mathbf{R}_{i-1} = (z_{i-1}, z_i)$ and $\mathbf{R}_i = (z_i, z_{i+1})$. Yu et al. (1997) wrote these results in matrix form by defining

$$\mathbf{C}_i = \begin{pmatrix} c_{i-} \\ c_{i+} \end{pmatrix} \quad \text{and} \quad \mathbf{M}_i(z) = \begin{pmatrix} \phi_{i-}(z) & \phi_{i+}(z) \\ \epsilon_i \phi'_{i-}(z) & \epsilon_i \phi'_{i+}(z) \end{pmatrix},$$

(7.104)

so that

$$\mathbf{M}_i(z_i)\mathbf{C}_i = \mathbf{M}_{i-1}(z_i)\mathbf{C}_{i-1}.$$

(7.105)

Then with the matrix $\mathbf{C}_0$ for region $\mathbf{R}_0$, the column vector $\mathbf{C}_i$ and therefore the electrostatic potential $\Phi_i(\mathbf{q}, z_i)$ can be determined in any region through the sequence

$$\mathbf{C}_i = \mathbf{Q}_i(z_i)\mathbf{C}_{i-1} = \mathbf{Q}_i(z_i)\mathbf{Q}_{i-1}(z_{i-1}) \cdots \mathbf{Q}_1(z_1)\mathbf{C}_0,$$

(7.106)

where the transfer matrix relating $\mathbf{R}_{i-1}$ and $\mathbf{R}_i$ is given by

$$\mathbf{Q}_i(z_i) = \mathbf{M}_i(z_i)^{-1}\mathbf{M}_{i-1}(z_i).$$

Clearly, the electrostatic potential for an n-interface heterostructure, $\Phi(\mathbf{q}, z_i)$, is given by joining the solutions $\Phi_i(\mathbf{q}, z)$ for each region:

$$\Phi_0(\mathbf{q}, z), \ z \in \mathbf{R}_0; \cdots \Phi_i(\mathbf{q}, z), \ z \in \mathbf{R}_i; \cdots \Phi_n(\mathbf{q}, z), \ z \in \mathbf{R}_n.$$

Moreover, in each region $\epsilon(\omega)$ and $\epsilon(\infty)$ are related through the generalized Lyddane–Sachs–Teller relations, subsection 2.3.3 and Section 7.1. For interface optical phonons, the potentials must decrease exponentially as $z \to \pm\infty$ so that, for an n-region heterostructure, $c_{n+} = 0$ and $c_{0-} = 0$. Thus, the dispersion relation for this interface mode is obtained by setting the $(2, 2)$ component of the transfer matrix equal to zero; that is,

$$[\mathbf{Q}_n(z_n, q, \omega)\mathbf{Q}_{n-1}(z_{n-1}, q, \omega) \cdots \mathbf{Q}_1(z_1, q, \omega)]_{2,2} = 0.$$

(7.107)

The number of interface optical phonon modes may be determined by examining the dispersion relation obtained in this way; indeed, since a given $\mathbf{Q}_i$ is proportional to $\epsilon_i(\omega)\epsilon_{i-1}(\omega)$, and since for a heterostructure with n interfaces $\mathbf{C}_n$ is the product of n $\mathbf{Q}$-matrices and $\mathbf{C}_0$, it follows that for a system with binary layers only the dispersion relation goes as $(\omega^2)^{2n}$. Thus, for such an n-interface structure with only binary layers, there are $2n$ interface optical phonons. By extending this argument, it follows that such a heterostructure with alternating layers of binary and ternary semiconductors has $3n$ interface optical phonon modes, since each ternary layer has two binary-like optical phonon modes.

The normalization condition for these modes is a straightforward generalization of the normalization condition for optical phonon modes in simple heterostructures, namely,

$$\sum_i L^2 \int_{\mathbf{R}_i} dz \, |\mu_i n_i \mathbf{u}_i(q, z)|^2 = \frac{\hbar}{2\omega}. \tag{7.108}$$

To normalize the optical phonon modes of such multiple heterointerface structures, it is convenient to take

$$\mathbf{E}_i(\mathbf{q}, z) = -\nabla \Phi_i(\mathbf{q}, z) = -iq\Phi_i(\mathbf{q}, z)\hat{\mathbf{q}} - \frac{\partial \Phi_i(\mathbf{q}, z)}{\partial z}\hat{\mathbf{z}},$$

$$\mathbf{P}_i(\mathbf{q}, z) = -\chi_i(\omega)\left[iq\Phi_i(\mathbf{q}, z)\hat{\mathbf{q}} + \frac{\partial \Phi_i(\mathbf{q}, z)}{\partial z}\hat{\mathbf{z}}\right], \tag{7.109}$$

where $\hat{\mathbf{q}}$ and $\hat{\mathbf{z}}$ are the unit vectors for $\mathbf{q}$ and the z-direction respectively. Then, for $\mathbf{R}_i$ it follows from our previous expression relating the displacement to the electric polarization that

$$\mathbf{u}_i(q, z) = \frac{\mathbf{P}_i(\mathbf{q}, z)}{n_i e_i^*\left[1 + \alpha_i \mu_i e_i^{*-2}(\omega_{0i}^2 - \omega^2)\right]}$$

$$= \frac{-\chi_i(\omega)\left[iq\Phi_i(\mathbf{q}, z)\hat{\mathbf{q}} + \frac{\partial \Phi_i(\mathbf{q}, z)}{\partial z}\hat{\mathbf{z}}\right]}{n_i e_i^*[1 + \alpha_i \mu_i e_i^{*-2}(\omega_{0i}^2 - \omega^2)]}. \tag{7.110}$$

Then, it follows straightforwardly that

$$|\mu_i n_i \mathbf{u}_i(q, z)|^2 = \mu_i n_i \frac{\chi_i^2(\omega)\left[q^2 |\Phi_i(\mathbf{q}, z)|^2 + \left|\frac{\partial \Phi_i(\mathbf{q}, z)}{\partial z}\right|^2\right]}{\left\{n_i e_i^*[1 + \alpha_i \mu_i e_i^{*-2}(\omega_{0i}^2 - \omega^2)]\right\}^2}. \tag{7.111}$$

Generalizing our previous expression for $\mathbf{P}_i(\mathbf{q}, z)$ to the corresponding result for layer i,

$$\mathbf{P}_i(\mathbf{q}, z) = n_i e_i^* \mathbf{u}_i(q, z) + n_i \alpha_i \mathbf{E}_{\text{local},i}(\mathbf{q}, z)$$

$$= \left[n_i e_i^* \frac{e_i^*}{\mu_i(\omega_{0i}^2 - \omega^2)} + n_i \alpha_i\right] \mathbf{E}_{\text{local},i}(\mathbf{q}, z)$$

$$= \left[n_i e_i^* \frac{e_i^*}{\mu_i(\omega_{0i}^2 - \omega^2)} + n_i \alpha_i\right]\left[\mathbf{E}_i(\mathbf{q}, z) + \frac{4\pi}{3}\mathbf{P}_i(\mathbf{q}, z)\right]. \tag{7.112}$$

Thus

$$\mathbf{P}_i(\mathbf{q}, z) = \chi_i(\omega)\mathbf{E}_i(\mathbf{q}, z)$$

$$= \frac{n_i e_i^* \dfrac{e_i^*}{\mu_i(\omega_{0i}^2 - \omega^2)} + n_i \alpha_i}{1 - \frac{4}{3}\pi \left[n_i e_i^* \dfrac{e_i^*}{\mu_i(\omega_{0i}^2 - \omega^2)} + n_i \alpha_i \right]} \mathbf{E}_i(\mathbf{q}, z), \qquad (7.113)$$

and it follows that

$$\frac{|\mu_i n_i \mathbf{u}_i(q, z)|^2}{q^2 |\Phi_i(\mathbf{q}, z)|^2 + |\partial \Phi_i(\mathbf{q}, z)/\partial z|^2}$$

$$= \mu_i n_i \left\{ n_i e_i^* \left[1 + \frac{\alpha_i \mu_i (\omega_{0i}^2 - \omega^2)}{e_i^{*2}} \right] \right\}^{-2}$$

$$\times \frac{\left[n_i e_i^* \dfrac{e_i^*}{\mu_i(\omega_{0i}^2 - \omega^2)} + n_i \alpha_i \right]^2}{\left\{ 1 - \frac{4}{3}\pi \left[n_i e_i^* \dfrac{e_i^*}{\mu_i(\omega_{0i}^2 - \omega^2)} + n_i \alpha_i \right] \right\}^2}$$

$$= n_i \frac{e_i^{*2}}{(\omega_{0i}^2 - \omega^2)^2} \left\{ -1 + \frac{4}{3}\pi n_n \left[\frac{e_i^{*2}}{\mu_i(\omega_{0i}^2 - \omega^2)} + \alpha_i \right] \right\}^{-2}$$

$$= \frac{1}{4\pi} \frac{\omega_{pi}^2}{(1 - \frac{4\pi}{3}n_n\alpha_n)^2} \left[-(\omega_{0i}^2 - \omega^2) + \frac{\frac{1}{3}\omega_{pi}^2}{1 - \frac{4\pi}{3}n_n\alpha_n} \right]^{-2}, \qquad (7.114)$$

where as defined previously $\omega_{pi}^2 = 4\pi n_i e_i^{*2}/\mu_i$ is the plasma frequency squared. Finally, using the expressions for $\omega_{LO,i}^2$, $\omega_{TO,i}^2$, and $\epsilon_i(\infty)$ given in Appendix B, equation (B.15), we have

$$\frac{|\mu_i n_i \mathbf{u}_i(q, z)|^2}{\left(q^2 |\Phi_i(\mathbf{q}, z)|^2 + \left| \dfrac{\partial \Phi_i(\mathbf{q}, z)}{\partial z} \right|^2 \right)} = \frac{1}{4\pi} \epsilon_i(\infty) \frac{\omega_{LO,i}^2 - \omega_{TO,i}^2}{(\omega^2 - \omega_{TO,i}^2)^2}$$

$$= \frac{1}{4\pi} \frac{1}{2\omega} \frac{\partial \epsilon_i(\omega)}{\partial \omega} \qquad (7.115)$$

and

$$\sum_i \frac{1}{4\pi} \frac{1}{2\omega} \frac{\partial \epsilon_i(\omega)}{\partial \omega} \int_{\mathbf{R}_i} dz \left\{ q^2 |\Phi_i(\mathbf{q}, z)|^2 + \left| \frac{\partial \Phi_i(\mathbf{q}, z)}{\partial z} \right|^2 \right\} = \frac{\hbar}{2\omega L^2}. \qquad (7.116)$$

Thus, for a phonon potential of the form

$$\Phi_i(\mathbf{q}, z) = A(c_{i-}' e^{-qz} + c_{i+}' e^{+qz}) = A\Psi_i(\mathbf{q}, z), \qquad (7.117)$$

the normalization constant A is given by

$$A = \left(\frac{\hbar}{2\omega L^2}\right)^{1/2} \left(\sum_i \frac{1}{4\pi} \frac{1}{2\omega} \frac{\partial \epsilon_i(\omega)}{\partial \omega}\right.$$

$$\left. \times \int_{R_i} dz \left\{ q^2 |\Psi_i(\mathbf{q}, z)|^2 + \left|\frac{\partial \Psi_i(\mathbf{q}, z)}{\partial z}\right|^2 \right\} \right)^{-1/2},$$

$$(7.118)$$

and so the Hamiltonian is

$$H_{IF} = e\Phi_i(\mathbf{r}) = e \sum_{\mathbf{q}} e^{-i\mathbf{q}\cdot\boldsymbol{\rho}} \Phi_i(\mathbf{q}, z)(a^\dagger_{-q} + a_q)$$

$$= e \sum_{\mathbf{q}} e^{-i\mathbf{q}\cdot\boldsymbol{\rho}} A\Psi_i(\mathbf{q}, z)(a^\dagger_{-q} + a_q). \qquad (7.119)$$

The utility of this formulation is illustrated by its application for heterostructures with known phonon modes. Consider the interface optical phonons for the case of two semi-infinite semiconductor regions joined at a single interface at $z = 0$. Clearly, $\Psi_0(\mathbf{q}, z) = c_{0-}e^{+qz} = e^{+qz}$ for $z \leq 0$ and $\Psi_1(\mathbf{q}, z) = c_{1-}e^{-qz} = e^{-qz}$ for $z \geq 0$. Then

$$A = \left(\frac{4\pi\hbar}{2\omega L^2}\right)^{1/2} \left[\frac{1}{2\omega} \frac{\partial \epsilon_0(\omega)}{\partial \omega} \int_{-\infty}^0 dz\, 2q^2 e^{+2qz}\right.$$

$$\left. + \frac{1}{2\omega} \frac{\partial \epsilon_1(\omega)}{\partial \omega} \int_0^\infty dz\, 2q^2 e^{-2qz}\right]^{-1/2}$$

$$= \left(\frac{4\pi\hbar}{2\omega L^2}\right)^{1/2} \left\{\frac{1}{2\omega} q \left[\frac{\partial \epsilon_0(\omega)}{\partial \omega} + \frac{\partial \epsilon_1(\omega)}{\partial \omega}\right]\right\}^{-1/2}, \qquad (7.120)$$

so that

$$H_{IF} = \sum_{\mathbf{q}} \sqrt{\frac{4\pi e^2 \hbar L^{-2}}{\partial\epsilon_0(\omega)/\partial\omega + \partial\epsilon_1(\omega)/\partial\omega}} \sqrt{\frac{1}{q}} e^{-i\mathbf{q}\cdot\boldsymbol{\rho}} e^{-q|z|}(a^\dagger_{-q} + a_q),$$

$$(7.121)$$

which is identical to the result obtained previously.

A second illustrative example is given by the case of a layer of one material situated in the region from $z = -d/2$ to $z = +d/2$ and bounded by two semi-infinite regions of another material; for example, we might consider a GaAs quantum well of thickness d embedded in AlAs barriers; the center of the quantum well is at $z = 0$. For this case, the phonon potential must decrease exponentially for $z \to \pm\infty$ and the phonon potential in the quantum well must be a combination of increasing and decreasing exponentials. Consider the case where the phonon potential in the quantum well is even. Let the dielectric constant in the quantum well be $\epsilon_1(\omega)$ and that of the barriers be $\epsilon_0(\omega)$. Since the barriers are taken to be the same material, $\epsilon_2(\omega) = \epsilon_0(\omega)$. Then, it is clear that $\Psi_0(\mathbf{q}, z) = e^{+q(z+d/2)}$ for $z \leq -d/2$,

$\Psi_1(\mathbf{q}, z) = (\cosh qz)/(\cosh qd/2)$ for $|z| < d/2$, and $\Psi_2(\mathbf{q}, z) = e^{-q(z-d/2)}$ for $z \geq d/2$ define an admissible envelope for the phonon potential. It follows that

$$\int_{\mathbf{R}_0} dz \left(q^2 |\Psi_0(\mathbf{q}, z)|^2 + \left| \frac{\partial \Psi_0(\mathbf{q}, z)}{\partial z} \right|^2 \right) = \int_{-\infty}^{-d/2} dz\, 2q^2 e^{2qz} e^{qd} = q,$$

$$\int_{\mathbf{R}_1} dz \left\{ q^2 |\Psi_1(\mathbf{q}, z)|^2 + \left| \frac{\partial \Psi_1(\mathbf{q}, z)}{\partial z} \right|^2 \right\}$$
$$= \int_{-d/2}^{+d/2} dz\, q^2 \frac{\cosh^2 qz + \sinh^2 qz}{\cosh^2 qd/2} = 2q \tanh qd/2,$$

$$\int_{\mathbf{R}_2} dz \left\{ q^2 |\Psi_2(\mathbf{q}, z)|^2 + \left| \frac{\partial \Psi_2(\mathbf{q}, z)}{\partial z} \right|^2 \right\} = \int_{d/2}^{\infty} dz\, 2q^2 e^{-2qz} e^{qd} = q,$$

$$(7.122)$$

and, accordingly,

$$A = \left(\frac{4\pi\hbar}{2\omega L^2} \right)^{1/2} \left[\frac{1}{2\omega} \frac{\partial \epsilon_0(\omega)}{\partial \omega} 2q + \frac{1}{2\omega} \frac{\partial \epsilon_1(\omega)}{\partial \omega} 2q \tanh \frac{qd}{2} \right]^{-1/2} \quad (7.123)$$

so that for the symmetric case

$$H_{\mathrm{IF},S} = \sum_{\mathbf{q}} \sqrt{\frac{4\pi e^2 \hbar L^{-2}}{\dfrac{\partial \epsilon_0(\omega)}{\partial \omega} + \dfrac{\partial \epsilon_1(\omega)}{\partial \omega} \tanh qd/2}} \sqrt{\frac{1}{2q}} e^{-i\mathbf{q}\cdot\boldsymbol{\rho}} f_S(\mathbf{q}, z)(a_{-q}^{\dagger} + a_q),$$

$$(7.124)$$

where $f_S(\mathbf{q}, z) = \Psi_i(\mathbf{q}, z)$. The dispersion relation for this optical phonon mode is given from the requirement that the normal components of the electric displacement field be continuous at the heterointerfaces. At $z = -d/2$, $\epsilon_0(\omega)E_{0,z}\big|_{z=-d/2} = \epsilon_1(\omega)E_{1,z}\big|_{z=-d/2}$.

From this condition it follows immediately that the frequencies of the IF optical phonons must satisfy $\epsilon_0(\omega) + \epsilon_1(\omega) \tanh qd/2 = 0$. This same dispersion relation is obtained from the continuity of the normal component of the electric displacement field at $z = d/2$. Recall that in a bulk semiconductor the optical phonon frequencies must satisfy $\epsilon(\omega) = 0$; indeed, since this is the condition necessary for the propagation of any longitudinal electromagnetic disturbance, it is expected that the frequencies of longitudinal optical phonons should be given by this dispersion relation.

In the case of a two-material, double-heterointerface structure, the IF longitudinal optical phonon frequencies depend on both $\epsilon_0(\omega)$ and $\epsilon_1(\omega)$. The expression for $H_{\mathrm{IF},S}$ is identical to that of Kim and Stroscio (1990) and can be rewritten to be in the form given by Mori and Ando (1989). The mode described by this Hamiltonian is the symmetric IF optical phonon for the quantum well being considered. As a final example, if $\Psi_0(\mathbf{q}, z) = -e^{+q(z+d/2)}$ for $z \le -d/2$, $\Psi_1(\mathbf{q}, z) = (\sinh qz)/(\sinh qd/2)$ for $|z| < d/2$, and $\Psi_2(\mathbf{q}, z) = e^{-q(z-d/2)}$ for $z \ge d/2$, it follows that for the antisymmetric case

$$H_{\text{IF},A} = \sum_q \sqrt{\frac{4\pi e^2 \hbar L^{-2}}{\dfrac{\partial \epsilon_0(\omega)}{\partial \omega} + \dfrac{\partial \epsilon_1(\omega)}{\partial \omega} \coth \dfrac{qd}{2}}} \sqrt{\frac{1}{2q}} e^{-i\mathbf{q}\cdot\boldsymbol{\rho}} f_A(\mathbf{q}, z)(a^{\dagger}_{-q} + a_q),$$

(7.125)

where $f_A(\mathbf{q}, z) = \Psi_i(\mathbf{q}, z)$. As before, the dispersion relation for this mode follows from the requirement that the normal component of the electric displacement field be continuous at the heterointerface. In this case, $\epsilon_0(\omega) + \epsilon_1(\omega) \coth qd/2 = 0$. This result reproduces the Hamiltonian derived by Kim and Stroscio (1990) for the antisymmetric IF optical phonon of the quantum-well system in question.

The transfer-matrix approach of Yu et al. (1997) may be used to gain insights into the nature of phonons in superlattices. Indeed, application of the transfer matrix method to a multiple-barrier AlAs/GaAs structure, Figure 7.6(a), leads to the dispersion relations such as those depicted in Figure 7.7 for various AlAs/GaAs heterostructures.

As will become evident in Chapter 10, the five-interface heterostructure of Figure 7.6(b) is of importance in narrow-well semiconductor lasers. The transfer-matrix method of Yu et al. (1997) may be applied to determine the IF phonon dispersion relations and the associated IF phonon potentials. For the case where the two barriers in Figure 7.6(b) are $Al_{0.6}Ga_{0.4}As$ and the shallow barrier to the far left is $Al_{0.25}Ga_{0.75}As$, the dispersion relations are as in Figure 7.8 and the five AlAs-like interface modes are as shown in Figure 7.9. As discussed previously, there are in total 15 IF modes in such a five-interface binary–ternary heterostructure. Indeed, as

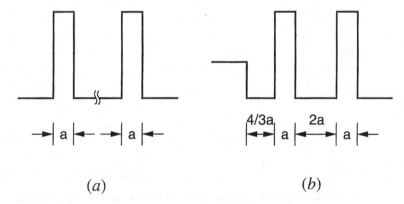

(a)   (b)

Figure 7.6. Potential profiles for (a) a multiple-barrier AlAs/GaAs heterostructure and for (b) a five-interface asymmetric heterostructure. From Yu et al. (1997), American Institute of Physics, with permission.

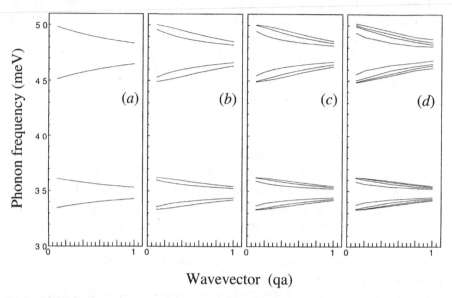

Figure 7.7. Longitudinal optical IF phonon dispersion relations for (*a*) one-barrier, (*b*) two-barrier, (*c*) three-barrier, (*d*) four-barrier structures of the type shown in Figure 7.6(*a*). From Yu *et al.* (1997), American Institute of Physics, with permission.

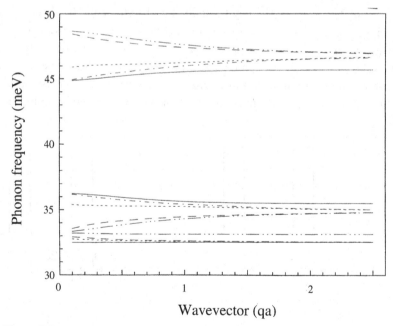

Figure 7.8. Dispersion relations for the 15 interface modes in a five-interface AlAs/GaAs heterostructure. From Yu *et al.* (1997), American Institute of Physics, with permission.

deduced previously, such an n-interface structure with alternating layers of binary and ternary semiconductors has $3n$ IF optical phonon modes.

The transfer-matrix approach of Yu *et al.* (1997) is also useful in understanding the IF phonons in the double-barrier heterostructures discussed widely in connection with new types of electronic diodes and transistors. As will be discussed in Chapter 10, the so-called valley current in such double-barrier structures is due in large measure to phonon-assisted tunneling. Consider Figure 7.10, which depicts the potential profile for a typical double-barrier heterostructure. The Fröhlich interaction Hamiltonian and the dispersion for a heterostructure of the generic type shown in Figure 7.10 was derived by Mori *et al.* (1992) and by Kim *et al.* (1992) without the benefit of the transfer-matrix method of Yu *et al.* These results may be summarized as follows:

$$
H_{\mathrm{IF},S} = \sum_{\mathbf{q}} (4\pi e^2 \hbar L^{-2})^{1/2}
$$

$$
\times \left[\frac{\partial \epsilon_1(\omega)}{\partial \omega} \left(\tanh \frac{qd_1}{2} + a^2 \right) + \frac{\partial \epsilon_2(\omega)}{\partial \omega} (b^2 + c^2)(1 - e^{-2qd_2}) \right]^{-1/2}
$$

$$
\times \sqrt{\frac{1}{2q}} e^{-i\mathbf{q}\cdot\boldsymbol{\rho}} (a_{-q}^\dagger + a_q)
$$

$$
\times \begin{cases}
ae^{-q(z-d_1/2-d_2)} & d_1/2 + d_2 \leq z, \\
be^{-q(z-d_1/2)} + ce^{q(z-d_1/2-d_2)} & d_1/2 \leq z \leq 2 + d_2, \\
(\cosh qz)/(\cosh qd_1/2) & |z| \leq d_1/2, \\
be^{q(z+d_1/2)} + ce^{-q(z+d_1/2+d_2)} & -d_1/2 - d_2 \leq z \leq -d_1/2, \\
ae^{q(z+d_1/2+d_2)} & z \leq -d_1/2 - d_2
\end{cases}
$$

$$(7.126)$$

is the Fröhlich interaction Hamiltonian for the symmetric LO phonon interface modes, and

$$
H_{\mathrm{IF},A} = \sum_{\mathbf{q}} (4\pi e^2 \hbar L^{-2})^{1/2}
$$

$$
\times \left[\frac{\partial \epsilon_1(\omega)}{\partial \omega} \left(\coth \frac{qd_1}{2} + a^2 \right) + \frac{\partial \epsilon_2(\omega)}{\partial \omega} (b^2 + c^2)(1 - e^{-2qd_2}) \right]^{-1/2}
$$

$$
\times \sqrt{\frac{1}{2q}} e^{-i\mathbf{q}\cdot\boldsymbol{\rho}} (a_{-q}^\dagger + a_q)
$$

$$
\times \begin{cases}
ae^{-q(z-d_1/2-d_2)} & d_1/2 + d_2 \leq z, \\
be^{-q(z-d_1/2)} + ce^{q(z-d_1/2-d_2)} & d_1/2 \leq z \leq 2 + d_2, \\
(\sinh qz)/(\sinh qd_1/2) & |z| \leq d_1/2, \\
-be^{q(z+d_1/2)} - ce^{-q(z+d_1/2+d_2)} & -d_1/2 - d_2 \leq z \leq -d_1/2, \\
-ae^{q(z+d_1/2+d_2)} & z \leq -d_1/2 - d_2
\end{cases}
$$

$$(7.127)$$

is the Fröhlich interaction Hamiltonian for the antisymmetric modes.

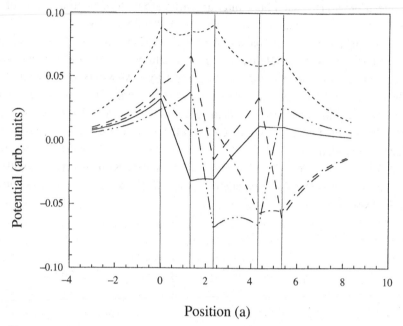

Position (a)

Figure 7.9. The phonon potentials for the five AlAs-like interface phonon modes of Figure 7.8. The line codes used in Figure 7.8 are employed here also to indicate which phonon potentials correspond to which. From Yu *et al.* (1997), American Institute of Physics, with permission.

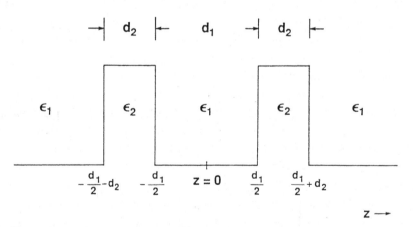

Figure 7.10. Potential profile for a double-barrier heterostructure with a quantum well of thickness d_1 and dielectric constant ϵ_1, and barriers of thickness d_2 and dielectric constant ϵ_2. The growth direction is taken to be the z-direction. From Kim *et al.* (1992), American Institute of Physics, with permission.

For the symmetric modes,

$$a = \cosh qd_2 + \frac{\epsilon_1}{\epsilon_2} \tanh \frac{qd_1}{2} \sinh qd_2,$$

$$b = \frac{1}{2}\left(1 - \frac{\epsilon_1}{\epsilon_2} \tanh \frac{qd_1}{2}\right), \qquad\qquad (7.128)$$

$$c = \frac{1}{2}e^{qd_2}\left(1 + \frac{\epsilon_1}{\epsilon_2} \tanh \frac{qd_1}{2}\right),$$

and the dispersion relation is given by

$$0 = \left(1 \pm \left\{1 - [(2\tanh qd_2)/(1+\tanh qd_1/2)]^2 \tanh qd_1/2\right\}^{1/2}\right)$$
$$\times \epsilon_1(1 + \tanh qd_1/2)/(2\tanh qd_2) + \epsilon_2. \qquad (7.129)$$

The antisymmetric modes a, b, c and the dispersion relation are obtained from these results by substituting $\coth qd_1/2$ for $\tanh qd_1/2$.

The dispersion relations determined by Kim *et al.* (1992) are displayed in Figures 7.11 and 7.12 for the case of a 60-ångstrom-wide GaAs quantum well with 60-ångstrom-wide AlAs barriers. There are four symmetric (S) and four antisymmetric (A) IF optical phonon modes for this heterostructure. The AlAs-like modes are denoted by the subscript 2 and the GaAs-like modes by the subscript 1.

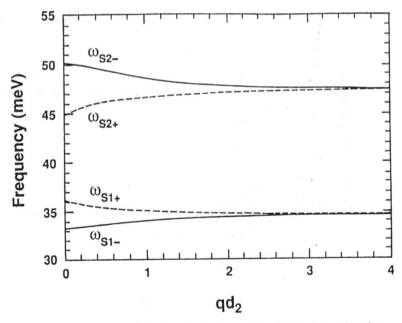

Figure 7.11. Dispersion relation for the four symmetric (S) LO phonon interface modes for the heterostructure shown in Figure 7.10 for the case of a 60-ångstrom-wide GaAs quantum well with 60-ångstrom-wide AlAs barriers. From Kim *et al.* (1992), American Institute of Physics, with permission.

The ± subscripts are used to distinguish the two different roots of the dispersion relation for each of the modes.

As will become clear in Chapter 10 the interface LO phonon modes contribute significantly to the valley current in such a double-barrier quantum-well structure. There it will be explained how the mode labeled by ω_{S2-} in Figure 7.11 makes a major contribution to the valley current in certain double-barrier quantum-well structures, through phonon-assisted tunneling of carriers into the quantum well.

7.4   Comparison of continuum and microscopic models for phonons

The dielectric continuum model of optical phonons has been compared with the microscopic theories by many authors, including Rücker *et al.* (1992), Molinari *et al.* (1992), Molinari *et al.* (1993), Bhatt *et al.* (1993a), and Lee *et al.* (1995). In view of the variety of continuum models of dimensionally confined phonons, as discussed in Appendix C, there has been a clear motivation to perform microscopic calculations in order to understand the properties of phonons in nanostructures. These microscopic models have included both *ab initio* models (Molinari *et al.*, 1992) and simplified microscopic models based on empirical lattice force constants

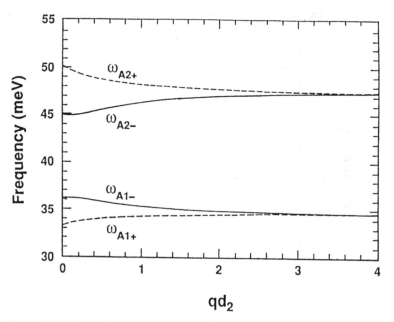

Figure 7.12. Dispersion for the four antisymmetric (A) LO phonon interface modes for the heterostructure shown in Figure 7.10 for the case of a 60-ångstrom-wide GaAs quantum well with 60-ångstrom-wide AlAs barriers. From Kim *et al.* (1992), American Institute of Physics, with permission.

(Bhatt *et al.*, 1995). Molinari *et al.* (1993) applied a microscopic model to calculate the atomic displacement amplitudes and the Fröhlich potentials for quantum wells and quantum wires. These calculations show that none of the macroscopic models – including the so-called 'slab' model – gives a completely accurate representation of the microscopic situation. However, as discussed in Appendix C the intersubband and intrasubband scattering rates computed with the phonon modes of all these models are in good agreement as long as the macroscopic model selected is based on a complete set of orthogonal modes. Further discussions concerning the comparisons of these macroscopic models are given in Appendix C. In this section, the intersubband and intrasubband scattering rates calculated with slab modes will be compared with the results of the microscopic models (Bhatt *et al.*, 1993a). Figure 7.13 presents a comparison of the scattering rates for intersubband and intrasubband electron–phonon scattering for a GaAs quantum well embedded in AlAs barriers. For well widths in the 2 to 10 nanometer range and for a temperature of 300 K, the scattering rates calculated from the microscopic model (Bhatt *et al.*, 1993a) and from the slab modes are in excellent agreement for an electron energy of 50 meV.

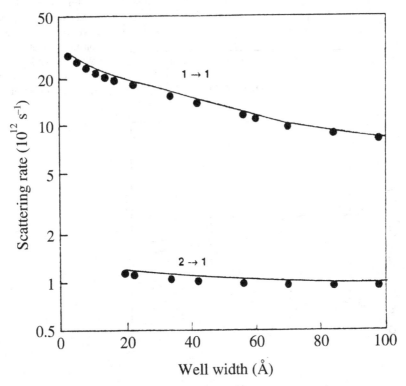

Figure 7.13. Intrasubband $(1 \rightarrow 1)$ and intersubband $(2 \rightarrow 1)$ scattering rates as a function of quantum-well width at a temperature of 300 K and for an electron energy of 50 meV: solid line, microscopic theory; dots, macroscopic theory. From Bhatt *et al.* (1993a), American Physical Society, with permission.

For the specific case of a 20-monolayer-wide GaAs quantum well embedded
in AlAs, Figure 7.14 presents a comparison of the scattering rates at 300 K for
(*a*) intrasubband (1 → 1) and (*b*) intersubband (2 → 1) electron–optical-phonon

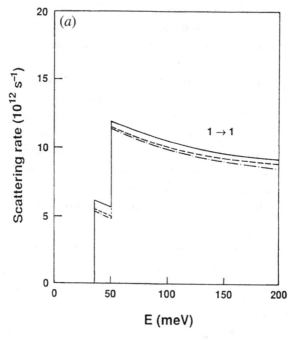

Figure 7.14. Comparison of
the scattering rates at 300 K
for (*a*) intrasubband (1 → 1)
and (*b*) intersubband
(2 → 1)
electron–optical-phonon
transitions in a
20-monolayer-wide GaAs
quantum well embedded in
AlAs. The result from three
different phonon models are
given: solid lines,
macroscopic models with
slab modes; broken and
dotted lines, *ab initio* model
of Molinari *et al.* (1992);
broken lines, microscopic
model with empirical force
constants (Bhatt *et al.*,
1993a). From Bhatt *et al.*
(1993a), American Physical
Society, with permission.

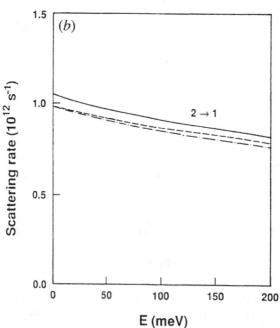

transitions for three different types of phonon model: macroscopic models with slab modes (solid lines), the *ab initio* model (broken and dotted lines) of Molinari *et al.* (1992), and a microscopic model (Bhatt *et al.*, 1993a) with empirical force constants (broken lines). These results indicated that the slab model and the simplified microscopic model provide good approximations to the scattering rates predicted by the fully microscopic model.

7.5   Comparison of dielectric continuum model predictions with results of Raman measurements

Chapters 8 and 10 will survey numerous examples on the applications of dimensionally confined phonon modes of the dielectric continuum model. In the present section, selected Raman measurements of optical phonon modes in dimensionally confined polar semiconductors are discussed, since these measurements illuminate the properties of such modes.

Figure 7.15 depicts Raman spectra taken at 15 K for $GaAs_N/AlAs_N$ quantum-well heterostructures with thicknesses of $N = 10 \pm 1$ monolayers, for three different laser energies: (*a*) $E_L = 1.933$ eV, (*b*) $E_L = 1.973$ eV, and (*c*) $E_L = 2.410$ eV (Fasol *et al.*, 1988). In cases (*a*) and (*b*) the measurements were made in the $z(xx)\bar{z}$ polarization configuration and for case (*c*) the polarization configuration was $z(xy)\bar{z}$. The peaks on the various spectra are denoted by $10_N, 8_N, 6_N, 6_{N-1}, 5_N, 5_{N-1}, 4_N,$ $4_{N-1}, 3_N, 2_N,$ and 1_N. This notation is used to identify the LO_4 mode – confined mode with $n = 4$ – in a quantum well with N monolayers as the 4_N peak. Likewise, the peak associated with the LO_4 mode in a quantum well with $N - 1$ monolayers is identified by 4_{N-1}. A remarkable feature of Figure 7.15 is that the confined-phonon wavevectors, $q_z = n\pi/L_z$, for $n = 4, 5,$ and 6 are sensitive to even a one-monolayer fluctuation in the thickness of the GaAs quantum well. Moreover, Fasol *et al.* (1988) showed that the change in q_z corresponds to changing L_z from 10 monolayers to 9 monolayers. These observations indicate that the phonons in AlAs/GaAs/AlAs quantum-well heterostructures are confined very strongly at the heterointerfaces. This result is expected on the basis of the dielectric continuum model, since the LO phonon energies differ substantially for the two materials; the LO phonon modes in AlAs at zone center have energies of about 50 meV while those of GaAs are about 36 meV.

One of the major predictions of the dielectric continuum model of optical phonon modes is the existence of IF optical phonons which are joint modes of the two or more heterostructure materials and which have frequencies characteristic of their individual optical phonon modes. A recent example of the Raman analysis of such an interface mode – and the first known for a würtzite heterostructure – is provided by Dutta *et al.* (2000). In these measurements, the IF modes were observed for GaN/AlN würtzite superlattices, as shown in Figure 7.16 for a 20-period GaN(9 nm)/AlN(8.5 nm) superlattice (lower curve) and a 40-period GaN(3 nm)/AlN(3 nm) superlattice (upper curve). In both cases, the wavelength of the incident laser radiation is 244 nm.

The second case, with the narrower wells, is characterized by a broader Raman peak as expected (Komirenko *et al.*, 1999; Gleize *et al.*, 1999). Moreover, the energies and asymmetries of the IF modes are as expected from the dielectric continuum model, as is clear by comparison with Figure D.2 of Appendix D.

The superperiodicity of the multilayered structures lowers the crystal symmetry and increases the size of the unit cell so that it includes more atoms per cell. The effect on the Raman properties is to allow more modes to exist and perhaps to allow optically inactive modes to become active. For instance, in the zincblende structure, for heterostructures grown in the [001] direction, the crystal symmetry is lowered from T_d to D_{2d} and the A_1 mode, previously inactive, becomes active. Similar behavior has been predicted by group-theory analysis (Kitaev *et al.*, 1998) in würtzite $(GaN)_m(AlN)_n$ superlattices, in cases where the Raman-active modes

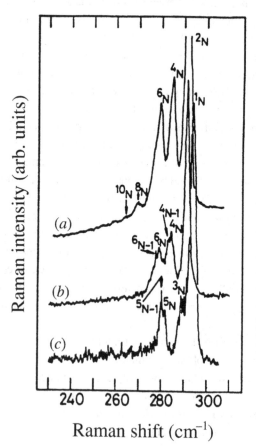

Figure 7.15. Raman spectra taken at 15 K for $GaAs_N/AlAs_N$ quantum-well heterostructures with thicknesses of $N = 10 \pm 1$ monolayers for three different laser energies: (*a*) $E_L = 1.933$ eV, (*b*) $E_L = 1.973$ eV, and (*c*) $E_L = 2.410$ eV. The peaks in the spectra are denoted by n_N and n_{N-1}, where n represents the nth confined mode for the quantum well. In this case, there are only two quantum-well thicknesses, $N = 10$ monolayers and $N - 1 = 9$ monolayers. From Fasol *et al.* (1988), American Physical Society, with permission.

vary non-monotonically with the increase in supercell size. The material quality in these structures, however, is not good enough yet for these effects to be observed experimentally.

Resonant Raman studies on (AlGa)N/GaN quantum wells have allowed the observation of the A_1 (LO) phonons in the quantum wells (Behr *et al.*, 1997; Gleize *et al.*, 2000), but the observation of a series of confined phonons similar to those observed in the zincblende structures has yet to be made in the würtzite nitride system.

The first really significant observation of the impact of heterostructures on light scattering was the observation of doublets in the acoustic phonons in superlattices by Colvard *et al.* (1980). A number of other studies on different materials have since been done. These have been reviewed extensively by Jusserand and Cardona (1991). As yet no similar observations have been reported for the würtzite nitride system, although recently Göppert *et al.* (1998) have reported that confined optical and folded acoustic phonons have been observed in the würtzite CdSe/CdS superlattices.

As a final example of where Raman measurements provide insights into the properties of confined optical phonon modes in polar semiconductor heterostructures,

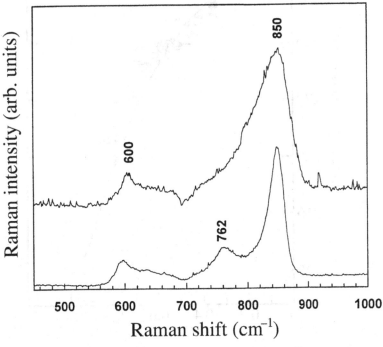

Figure 7.16. Raman spectra for a 20-period GaN(9 nm)/AlN(8.5 nm) würtzite superlattice (lower line) and a 40-period GaN(3 nm)/AlN(3 nm) würtzite superlattice (upper line). The incident laser wavelength was 244 nm. From Dutta *et al.* (2000), to be published.

Figure 7.17 (Fasol *et al.*, 1988) illustrates that the confined phonon energies and wavevectors for superlattices and quantum wells fall on the dispersion curve (solid line) measured by neutron scattering at 10 K for bulk GaAs (Richter and Strauch, 1987). The neutron scattering data are in agreement with the earlier results of Waugh and Dolling (1963).

The Raman data in Figure 7.17 are from works as follows: circles, Klein (1986); diagonal crosses, Worlock (1985); squares, Castro and Cardona (1987); diamonds, Colvard *et al.* (1980); and upright crosses, Sood *et al.* (1985), Jusserand and Paguet (1986), and Sood *et al.* (1986).

The agreement between the Raman and neutron scattering measurements is extremely enlightening. These results tell us – at least for the GaAs/AlAs system –

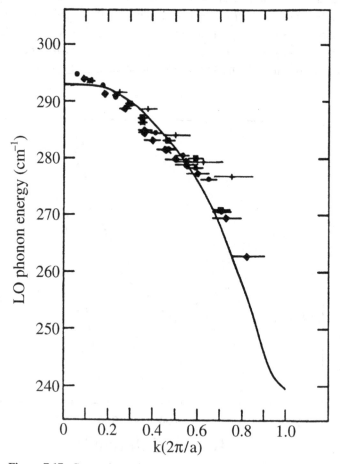

Figure 7.17. Comparison of LO phonon energies and wavevectors determined by Raman scattering with those of neutron scattering (solid line). See text for references to the works summarized in this figure. From Fasol *et al.* (1988), American Physical Society, with permission.

that phonon confinement effects serve to restrict the phase space of the phonons but not to alter substantially the basic energy–wavevector relationship for the phonons. This may be understood by considering the dominant role of the nearest-neighbor coupling of adjacent ions in a polar semiconductor. Indeed, the presence of a hetero-interface will not influence the local LO phonon frequency in the region between two ions located about two or more monolayers away from the heterointerface. Thus, the basic energy–wavevector relationship – the dispersion relationship – is expected to be approximately that of the bulk phonons. The phase space, however, is altered dramatically by the effects of dimensional confinement: only wavevectors corre-sponding to multiples of half-wavelengths, as discussed previously, are allowed.

7.6 Continuum model for acoustic modes in dimensionally confined structures

In this section, the elastic continuum model of Section 7.2 is applied to the analysis of acoustic phonon modes in a variety of nanostructures. The acoustic phonon amplitudes and dispersion relations are determined for free-standing layers as well as for double-interface structures. In addition, the acoustic phonon modes are analyzed for both rectangular and cylindrical wires. For the case of hollow wires, the acoustic phonon modes are applied to the analysis of acoustic disturbances in the microtubulin structures found in many biological systems. Finally, the elastic continuum model is applied to describe acoustic phonon modes in quantum wires.

7.6.1 Acoustic phonons in a free-standing and unconstrained layer

The acoustic phonons in a free-standing and unconstrained layer illustrate key fea-tures of the confined modes in dimensionally confined structures. Such free-standing layers are referred to occasionally as free-standing quantum wells in the case when the layer is thin enough for quantum confinement to modify the properties of the electron de Broglie waves in the layer. From Section 7.2 the three force equations (7.33) describing the amplitudes of the phonon displacements may be written as the vector equation

$$\partial^2 \mathbf{u}/\partial t^2 = c_t^2 \nabla^2 \mathbf{u} + (c_l^2 - c_t^2)\,\mathrm{grad}\,\Delta, \qquad (7.130)$$

where

$$\Delta = \nabla \cdot \mathbf{u} = \mathrm{div}\,\mathbf{u}, \qquad c_l^2 = (\lambda + 2\mu)/\rho, \qquad \text{and} \qquad c_t^2 = \mu/\rho. \tag{7.131}$$

Let us consider a slab of width a and free standing in space with unconstrained surfaces at $z = \pm a/2$. At such surfaces the material displacements are unrestricted but the normal components of the stress tensor – the traction force – must vanish; with the expressions for stress given in Section 7.2 it follows that at $z = \pm a/2$ (Bannov *et al.*, 1994a, b, 1995),

$$T_{x,z} = \mu \left(\frac{\partial u_x}{\partial z} + \frac{\partial u_z}{\partial x} \right) = 0,$$

$$T_{y,z} = \mu \left(\frac{\partial u_y}{\partial z} + \frac{\partial u_z}{\partial y} \right) = 0, \qquad (7.132)$$

$$T_{z,z} = \lambda \Delta + 2\mu \frac{\partial u_z}{\partial z} = 0.$$

Taking the displacement eigenmodes to be of the form

$$\mathbf{u}(\mathbf{r}, t) = \left(\frac{1}{2\pi} \right)^2 \sum_n \int_{-\infty}^{\infty} d^2 \mathbf{q}_{\|} \, e^{i\mathbf{q}_{\|} \cdot \mathbf{r}_{\|}} \mathbf{u}_n(\mathbf{q}_{\|}, z), \qquad (7.133)$$

the force equations reduce to an eigenvalue problem

$$\mathcal{D}\mathbf{u}_n(\mathbf{q}_{\|}, z) = -\omega_n^2 \mathbf{u}_n(\mathbf{q}_{\|}, z), \qquad (7.134)$$

where

$$\mathcal{D} = \begin{pmatrix} c_t^2 \dfrac{d^2}{dz^2} - c_l^2 q_x^2 & 0 & (c_l^2 - c_t^2) i q_x \dfrac{d}{dz} \\ 0 & c_t^2 \dfrac{d^2}{dz^2} - c_t^2 q_x^2 & 0 \\ (c_l^2 - c_t^2) i q_x \dfrac{d}{dz} & 0 & c_l^2 \dfrac{d^2}{dz^2} - c_t^2 q_x^2 \end{pmatrix}, \qquad (7.135)$$

with boundary conditions corresponding to zero traction force at $z = \pm a/2$

$$\frac{du_x}{dz} = -iq_x u_z, \qquad \frac{du_y}{dz} = 0, \qquad \text{and} \qquad \frac{du_z}{dz} = -iq_x \frac{c_l^2 - 2c_t^2}{c_l^2} u_x. \qquad (7.136)$$

As discussed in Appendix A, it is convenient to consider the quantity $\mathbf{w} = \sqrt{\rho}\mathbf{u}$; for acoustic modes the quantity ρ replaces $\sqrt{\mu N/V}$, which applies for optical modes, μ being the reduced mass. Accordingly, let us take the orthogonal eigenvectors for these modes to satisfy

$$\int \mathbf{w}\dagger_n(\mathbf{q}_{\|}, z) \cdot \mathbf{w}_m(\mathbf{q}_{\|}, z) dz = 0, \qquad n \neq m. \qquad (7.137)$$

From classical acoustics (Auld, 1973) it is known that the problem at hand admits to three types of solution: shear waves, dilatational waves, and flexural waves. First, consider the shear waves. For these modes, the only component of the displacement is parallel to the surfaces $z = \pm a/2$; taking this non-zero component to be in the y-direction, we have $\mathbf{u}_n(\mathbf{q}_{\parallel}, z) = (0, u_y, 0)$ with

$$u_y = \begin{cases} \cos q_{z,n} z & \text{for } n = 0, 2, 4, \ldots \\ \sin q_{z,n} z & \text{for } n = 1, 3, 5, \ldots \end{cases} \tag{7.138}$$

and $q_{z,n} = n\pi/a$. These transverse modes are designated as rotational modes. The frequency–wavevector relation for these shear waves is

$$\omega_n = c_t \sqrt{q_{z,n}^2 + q_x^2}. \tag{7.139}$$

Clearly, these shear modes have wavelengths such that an integral number n of half wavelengths fits into the confinement region of length a. In nanoscale crystalline layers the number of half-wavelengths n is limited to the number of unit cells in the thickness a. Thus, for an elemental semiconductor layer with N_m monolayers in the thickness a, n takes on integer values from 0 to N_m.

The second class of solutions is associated with so-called dilatational modes. These dilatational modes are irrotational modes and they are associated with compressional distortions of the medium. The compressional character of these modes leads to local changes in the volume of the medium. They have two non-zero components: $\mathbf{u}_n(\mathbf{q}_{\parallel}, z) = (u_x, 0, u_z)$ with

$$u_x = iq_x(q_x^2 - q_t^2)\left(\sin\frac{q_t a}{2}\cos q_l z + 2q_l q_t \sin\frac{q_l a}{2}\cos q_t z\right),$$
$$u_z = q_l\left[-(q_x^2 - q_t^2)\sin\frac{q_t a}{2}\sin q_l z + 2q_x^2 \sin\frac{q_l a}{2}\sin q_t z\right], \tag{7.140}$$

where q_l and q_t are solutions of

$$\frac{\tan q_t a/2}{\tan q_l a/2} = -\frac{4q_x^2 q_l q_t}{(q_x^2 - q_t^2)^2},$$
$$c_l^2(q_x^2 + q_l^2) = c_t^2(q_x^2 + q_t^2). \tag{7.141}$$

For each value of q_x this pair of equations has either pure imaginary or real solutions, denoted by $q_{l,n}(q_x)$ and $q_{t,n}(q_x)$; here, the label n is used to denote the different branches of the solutions $q_{l,n}(q_x)$ and $q_{t,n}(q_x)$. The dilatational modes have frequencies ω_n satisfying

$$\omega_n = c_l\sqrt{q_{l,n}^2 + q_x^2} = c_t\sqrt{q_{t,n}^2 + q_x^2}. \tag{7.142}$$

Numerical solutions of these dispersion relations were given by Bannov et al. (1995) for a 100-ångstrom-wide GaAs slab under the assumption that $c_l = 5.7 \times 10^5$ cm s^{-1} and $c_t = 3.35 \times 10^5$ cm s^{-1}.

Finally, the third class of solutions is referred to as the flexural modes. These flexural modes are of the form $\mathbf{u}_n(\mathbf{q}_\parallel, z) = (u_x, 0, u_z)$ with

$$
u_x = iq_x \left[(q_x^2 - q_t^2) \cos\frac{q_l a}{2} \sin q_l z + 2q_l q_t \cos\frac{q_l a}{2} \sin q_t z \right],
$$
$$
u_z = q_l \left[(q_x^2 - q_t^2) \cos\frac{q_l a}{2} \cos q_l z - 2q_x^2 \cos\frac{q_l a}{2} \cos q_t z \right],
$$

(7.143)

q_l and q_t being determined as solutions of the pair of equations

$$
\frac{\tan q_l a/2}{\tan q_t a/2} = -\frac{4q_x^2 q_l q_t}{(q_x^2 - q_t^2)^2},
$$
$$
c_l^2(q_x^2 + q_l^2) = c_t^2(q_x^2 + q_t^2).
$$

(7.144)

Just as for the dilatational modes, this pair of equations for the flexural modes admits solutions of the form $q_{l,n}(q_x)$ and $q_{t,n}(q_x)$, where n labels the different branches of the solutions. These modes are normalized, according to the procedures of Section 5.1, in terms of $\mathbf{w}_n(\mathbf{q}_\parallel, z)$ instead of $\mathbf{u}_n(\mathbf{q}_\parallel, z)$ since, as mentioned above, the considerations of Appendix A make it clear that it is convenient to use $\mathbf{w}_n = \sqrt{\rho}\mathbf{u}_n$:

$$
\mathbf{u}(\mathbf{r}) = \sum_{\mathbf{q}_\parallel, n} \sqrt{\frac{\hbar}{2L^2 \rho \omega_n(\mathbf{q}_\parallel)}} (a_{n,q} + a_{n,-q}^\dagger) \mathbf{w}_n(\mathbf{q}_\parallel, z) \, e^{i\mathbf{q}_\parallel \cdot \mathbf{r}_\parallel}.
$$

(7.145)

This last result is, of course, consistent with the normalization condition of subsection 7.3.1,

$$
\int L^2 \, dz \{\sqrt{\mu n}\mathbf{u}(\mathbf{q}, z)\}^* \cdot \{\sqrt{\mu n}\mathbf{u}(\mathbf{q}, z)\} = \frac{\hbar}{2\omega(q)};
$$

(7.146)

this equivalence follows straightforwardly by noting that $\mathbf{q}_\parallel = \mathbf{q}$ and by taking $\rho = \mu n$, as is appropriate since the mass density ρ is clearly the appropriate quantity for acoustic modes.

7.6.2 Acoustic phonons in double-interface heterostructures

Wendler and Grigoryan (1988) considered the acoustic phonon modes supported by an isotropic slab of density ρ_1 and width a bounded at $z = \pm a/2$ by semi-infinite embedding materials of density ρ_2. The slab of width a is taken to have transverse and longitudinal sound speeds c_{t1} and c_{l1} respectively. The sound speeds of the two identical embedding materials are taken to be c_{t2} and c_{l2} respectively. According to the cardinal boundary conditions of classical acoustics, the material displacements $\mathbf{u}$ and the normal components of the stress tensor T_{i3}, also known as the traction force, must be continuous at $z = \pm a/2$. The Young's modulus, E_α, and the Poisson ratio, ν_α, for medium α may be expressed in terms of the Lamé constants as

$$E_\alpha = \frac{\mu_\alpha(3\lambda_\alpha + 2\mu_\alpha)}{\lambda_\alpha + \mu_\alpha} \quad \text{and} \quad \nu_\alpha = \frac{\lambda_\alpha}{2(\lambda_\alpha + \mu_\alpha)} \tag{7.147}$$

where $\alpha = 1$ for the slab and $\alpha = 2$ for the embedding materials. Then, the inverse relations are

$$\lambda_\alpha = \frac{E_\alpha \nu_\alpha}{(1 - 2\nu_\alpha)(1 + \nu_\alpha)} \quad \text{and} \quad \mu_\alpha = \frac{E_\alpha}{2(1 + \nu_\alpha)}, \tag{7.148}$$

and it follows that

$$\lambda_\alpha + \mu_\alpha = \frac{E_\alpha}{2(1 - 2\nu_\alpha)(1 + \nu_\alpha)} \quad \text{and} \quad \nu_\alpha = \frac{\lambda_\alpha}{2(\lambda_\alpha + \mu_\alpha)}. \tag{7.149}$$

Writing the displacement field $\mathbf{u}$ as the sum of its longitudinal part, $\mathbf{u}_l$, satisfying $\nabla \times \mathbf{u}_l = 0$, and its transverse part, $\mathbf{u}_t$, satisfying $\nabla \cdot \mathbf{u}_t = 0$, it follows from the wave equation (7.130) of subsection 7.6.1 that

$$\frac{\partial^2}{\partial t^2}\mathbf{u}_l - c_{l\alpha}^2 \Delta \mathbf{u}_l = 0 \quad \text{and} \quad \frac{\partial^2}{\partial t^2}\mathbf{u}_t - c_{t\alpha}^2 \Delta \mathbf{u}_t = 0, \tag{7.150}$$

where

$$c_{l\alpha}^2 = \frac{E_\alpha(1 - \nu_\alpha)}{\rho_\alpha(1 + \nu_\alpha)(1 - 2\nu_\alpha)} \quad \text{and} \quad c_{t\alpha}^2 = \frac{E_\alpha}{2\rho_\alpha(1 + \nu_\alpha)}. \tag{7.151}$$

Following Wendler and Grigoryan (1988), the media are assumed to be isotropic and, without loss of generality, we may consider acoustic modes propagating in the x-direction with wavevector $q_\parallel$. Wendler and Grigoryan (1988) classified the acoustic modes for such an embedded quantum well as symmetric shear vertical waves, antisymmetric shear vertical waves, symmetric shear horizontal waves, and antisymmetric shear horizontal waves. Defining $\mathbf{u}(z)$ through the relationship

$$\mathbf{u}(x, y, z) = \mathbf{u}(z) \cdot \exp\left[i(q_\parallel x - \omega t)\right], \tag{7.152}$$

it is possible to consider all the acoustic phonons as belonging to two classes of waves: shear vertical (SV) modes with two non-zero components,

$$\mathbf{u}(z) = \left(u_1(z), 0, u_3(z)\right), \tag{7.153}$$

and shear horizontal (SH) modes with

$$\mathbf{u}(z) = \left(0, u_2(z), 0\right). \tag{7.154}$$

As in subsection 7.6.1, the symmetric modes satisfy

$$u_1(z) = u_1(-z), \quad u_2(z) = u_2(-z), \quad u_3(z) = -u_3(-z), \tag{7.155}$$

and the antisymmetric modes satisfy

$$u_1(z) = -u_1(-z), \qquad u_2(z) = -u_2(-z), \qquad u_3(z) = u_3(-z). \quad (7.156)$$

The localized modes for the embedded slab under consideration must satisfy the boundary conditions

$$\mathbf{u}(z)|_{z=\pm\infty} = 0. \quad (7.157)$$

Then from the wave equations for the displacements, the symmetric shear vertical (SSV) modes must have the form

$$u_1^S(z) = \begin{cases} A_2^S \exp\left(-\eta_{l2}z\right) + B_2^S \exp\left(-\eta_{t2}z\right) & z > a/2, \\ A_1^S \cosh \eta_{l1}z + B_1^S \cosh \eta_{t1}z & a/2 > z > -a/2, \\ A_2^S \exp \eta_{l2}z + B_2^S \exp \eta_{t2}z & z < -a/2, \end{cases}$$

$$u_3^S(z) = \begin{cases} i\left[\dfrac{\eta_{l2}}{q_\parallel} A_2^S \exp\left(-\eta_{l2}z\right) + \dfrac{q_\parallel}{\eta_{t2}} B_2^S \exp\left(-\eta_{t2}z\right)\right] & z > a/2, \\ i\left(\dfrac{\eta_{l1}}{q_\parallel} A_1^S \sinh \eta_{l1}z - \dfrac{q_\parallel}{\eta_{t1}} B_1^S \sinh \eta_{t1}z\right) & a/2 > z > -a/2, \\ i\left(\dfrac{\eta_{l2}}{q_\parallel} A_2^S \exp\left(\eta_{l2}z\right) - \dfrac{q_\parallel}{\eta_{t2}} B_2^S \exp\left(\eta_{t2}z\right)\right) & z < -a/2. \end{cases}$$

$$(7.158)$$

and the antisymmetric shear vertical (ASV) modes have the form

$$u_1^A(z) = \begin{cases} A_2^A \exp\left(-\eta_{l2}z\right) + B_2^A \exp\left(-\eta_{t2}z\right) & z > a/2, \\ A_1^A \sinh \eta_{l1}z + B_1^A \sinh \eta_{t1}z & a/2 > z > -a/2, \\ -A_2^A \exp \eta_{l2}z - B_2^A \exp \eta_{t2}z & z < -a/2, \end{cases}$$

$$u_3^A(z) = \begin{cases} i\left[\dfrac{\eta_{l2}}{q_\parallel} A_2^A \exp\left(-\eta_{l2}z\right) + \dfrac{q_\parallel}{\eta_{t2}} B_2^A \exp\left(-\eta_{t2}z\right)\right] & z > a/2, \\ i\left(-\dfrac{\eta_{l1}}{q_\parallel} A_1^A \cosh \eta_{l1}z - \dfrac{q_\parallel}{\eta_{t1}} B_1^A \cosh \eta_{t1}z\right) & a/2 > z > -a/2, \\ i\left(\dfrac{\eta_{l2}}{q_\parallel} A_2^A \exp \eta_{l2}z - \dfrac{q_\parallel}{\eta_{t2}} B_2^A \exp \eta_{t2}z\right) & z < -a/2. \end{cases}$$

$$(7.159)$$

The functions $\eta_{l\alpha}$ and $\eta_{t\alpha}$ are defined by

$$\eta_{l\alpha} = (q_\parallel^2 - \omega^2/c_{l\alpha}^2)^{1/2} \quad \text{and} \quad \eta_{t\alpha} = (q_\parallel^2 - \omega^2/c_{t\alpha}^2)^{1/2}, \quad (7.160)$$

where $\alpha = 1$ for the slab and $\alpha = 2$ for the embedding materials. The conditions $\eta_{l\alpha} = 0$ and $\eta_{t\alpha} = 0$ are again recognized as the bulk dispersion relations for medium α for the longitudinal and transverse acoustic modes, respectively. From the definitions of the stress, T_{ij}, and strain, S_{ij}, of Section 7.2, it follows that for medium α

$$T_{ij}^{(\alpha)} = \frac{E_\alpha}{1 + v_\alpha} \left[S_{ij}^{(\alpha)} + \frac{v_\alpha}{1 - 2v_\alpha} \left(S_{11}^{(\alpha)} + S_{22}^{(\alpha)} + S_{33}^{(\alpha)} \right) \delta_{ij} \right]. \tag{7.161}$$

From the expressions (7.158) and (7.159) for $u_1^S(z)$, $u_3^S(z)$, $u_1^A(z)$, and $u_3^A(z)$ it then follows that

$$\begin{aligned}
T_{13}^{(\alpha)} &= 2\rho_a c_{t\alpha}^2 S_{13}^{(\alpha)}, \\
T_{23}^{(\alpha)} &= 0, \\
T_{33}^{(\alpha)} &= \rho_a (c_{l\alpha}^2 - c_{t\alpha}^2) S_{11}^{(\alpha)} + \rho_a c_{l\alpha}^2 S_{33}^{(\alpha)}.
\end{aligned} \tag{7.162}$$

Requiring that $T_{i3}^{(\alpha)}$ and $\mathbf{u}$ be continuous at $z \pm a/2$ yields four equations. The determinant of these equations, $\det d_{mn}$, then yields the dispersion relations for the SSV modes. In d_{mn}, m and n take on the the values 1, 2, 3, and 4, and the d_{mn} are given by

$$d_{11} = \exp\left(-\frac{\eta_{l2}a}{2}\right), \qquad\qquad d_{12} = \exp\left(-\frac{\eta_{t2}a}{2}\right),$$

$$d_{13} = -\cosh\frac{\eta_{l1}a}{2}, \qquad\qquad d_{14} = -\cosh\frac{\eta_{t1}a}{2},$$

$$d_{21} = -\eta_{l2}\exp\left(-\frac{\eta_{l2}a}{2}\right), \qquad\qquad d_{22} = -\frac{q_\parallel^2}{\eta_{t2}}\exp\left(-\frac{\eta_{t2}a}{2}\right),$$

$$d_{23} = -\eta_{l1}\sinh\frac{\eta_{l1}a}{2}, \qquad\qquad d_{24} = -\frac{q_\parallel^2}{\eta_{t1}}\sinh\frac{\eta_{t1}a}{2},$$

$$d_{31} = -2\rho_2 c_{t2}^2 \eta_{l2}\exp\left(-\frac{\eta_{l2}a}{2}\right), \qquad d_{32} = -\rho_2 c_{t2}^2 \frac{\eta_{t2}^2 + q_\parallel^2}{\eta_{t2}}\exp\left(-\frac{\eta_{t2}a}{2}\right),$$

$$d_{33} = -2\rho_1 c_{t1}^2 \eta_{l1}\sinh\frac{\eta_{l1}a}{2}, \qquad d_{34} = -\rho_1 c_{t1}^2 \frac{\eta_{t1}^2 + q_\parallel^2}{\eta_{t1}}\sinh\frac{\eta_{t1}a}{2},$$

$$d_{41} = \rho_2 c_{t2}^2 \left(\eta_{t2}^2 + q_\parallel^2\right)\exp\left(-\frac{\eta_{l2}a}{2}\right), \qquad d_{42} = 2\rho_2 c_{t2}^2 q_\parallel^2 \exp\left(-\frac{\eta_{t2}a}{2}\right),$$

$$d_{43} = -\rho_1 c_{t1}^2 \left(\eta_{t1}^2 + q_\parallel^2\right)\cosh\left(\frac{\eta_{l1}a}{2}\right), \qquad d_{44} = -2\rho_1 c_{t1}^2 q_\parallel^2 \cosh\left(\frac{\eta_{t1}a}{2}\right).$$

$$\tag{7.163}$$

For the ASV modes, the corresponding elements d_{mn} are given in terms of these results by making the replacements $\cosh \leftrightarrows \sinh$, $d_{14} \to -d_{14}$, $d_{24} \to -d_{24}$, $d_{31} \to -d_{31}$, $d_{32} \to -d_{32}$, $d_{33} \to -d_{33}$, and $d_{44} \to -d_{44}$. As in Wendler and Grigoryan

(1988), $A_1^{S,A}$, $A_2^{S,A}$, $B_1^{S,A}$, and $B_2^{S,A}$ are then given in terms of the new elements d_{mn} by

$$A_1^{S,A} = \frac{1}{\det d_{mn}} \begin{vmatrix} -d_{14} & d_{12} & d_{13} \\ -d_{24} & d_{22} & d_{23} \\ -d_{34} & d_{32} & d_{33} \end{vmatrix},$$

$$A_2^{S,A} = \frac{1}{\det d_{mn}} \begin{vmatrix} d_{11} & d_{12} & -d_{14} \\ d_{21} & d_{22} & -d_{24} \\ d_{31} & d_{32} & -d_{34} \end{vmatrix}, \qquad (7.164)$$

$$B_1^{S,A} = \frac{1}{\det d_{mn}} \begin{vmatrix} d_{11} & -d_{14} & d_{13} \\ d_{21} & -d_{24} & d_{23} \\ d_{31} & -d_{34} & d_{33} \end{vmatrix},$$

and

$$B_2^{S,A} = 1, \qquad (7.165)$$

where in $\det d_{mn}$ m and n take on the values 1, 2, and 3.

The SSV modes are of special interest because they possess a non-zero deformation potential. They will be considered further in Chapter 10 in connection with the radiation of coherent acoustic phonons from an electron current in an embedded layer. The shear horizontal (SH) modes may be determined by a derivation analogous to that for the SV modes (Wendler and Grigoryan, 1988) and take the form

$$u_2^S(z) = \begin{cases} D_2^S \exp(-\eta_{t2}z) & z > a/2, \\ D_1^S \cos\theta_{t1}z & a/2 > z > -a/2, \\ D_2^S \exp\eta_{t2}z & z < -a/2 \end{cases} \qquad (7.166)$$

for the SSH modes and

$$u_2^A(z) = \begin{cases} D_2^A \exp(-\eta_{t2}z) & z > a/2, \\ D_1^A \sin\theta_{t1}z & a/2 > z > -a/2, \\ -D_2^A \exp\eta_{t2}z & z < -a/2, \end{cases} \qquad (7.167)$$

for the ASH modes, where

$$\theta_{t1} = \left(\omega^2/c_{t1}^2 - q_\parallel^2\right) = i\eta_{t1}. \qquad (7.168)$$

By procedures analogous to those for the SSV modes, it follows that the dispersion relation for the SSH modes is given by

$$\frac{\rho_2 c_{t2}^2 \eta_{t2}}{\rho_1 c_{t1}^2 \theta_{t1}} - \tan\frac{\theta_{t1}a}{2} = 0. \qquad (7.169)$$

The dispersion relation for the ASH modes is given by

$$\frac{\rho_1 c_{t1}^2 \theta_{t1}}{\rho_2 c_{t1}^2 \eta_{t2}} + \tan \frac{\theta_{t1} a}{2} = 0, \tag{7.170}$$

and the amplitude coefficients for the displacement fields are given by

$$D_1^S = D_1^A = 1, \qquad D_2^S = \exp \frac{\eta_{t2} a}{2} \cos \frac{\theta_{t1} a}{2}, \qquad D_2^A = \exp \frac{\eta_{t2} a}{2} \sin \frac{\theta_{t1} a}{2}. \tag{7.171}$$

The SH modes are seen to exist only in the region $\omega^2/c_{t2}^2 < q_\parallel^2 < \omega^2/c_{t1}^2$.

Wendler and Grigoryan (1988), see also Mitin et al. (1999), classified the localized acoustic modes in a symmetrical embedded layer in terms of the regions in the $\omega q_\parallel$-plane. They found that localized acoustic modes exist in this structure provided that $c_{t1} < c_{t2}, c_{l1}, c_{l2}$. These localized modes propagate along the layer and decay outside the layer. For such a symmetrical embedded layer, Wendler and Grigoryan (1988) give many numerical results for the dispersion relations and mode amplitudes for the localized symmetric shear vertical waves (SSVWs), antisymmetric shear vertical waves (ASVWs), symmetric shear horizontal waves (SSHWs), and antisymmetric shear horizontal waves (ASHWs). As mentioned previously, the SV modes will be considered further in Chapter 10 in connection with the radiation of coherent acoustic phonons from an electron current in an embedded layer.

7.6.3   Acoustic phonons in rectangular quantum wires

The classical compressional acoustic modes in free-standing rods with rectangular cross sections have been examined experimentally (Morse, 1948) and theoretically (Morse, 1949, 1950). The solutions obtained by Morse are based on the elastic continuum model as well as on the approximation method of separation of variables. As illustrated previously, these classical elastic continuum solutions provide the basis for describing the compressional – that is, the longitudinal – phonon modes in a nanoscale quantum wire with a rectangular cross section. For cross-sectional dimensions with aspect ratios of approximately two or greater, Morse (1948, 1949, 1950) found that these solutions provide simple and accurate analytical expressions in agreement with the experimentally observed modes over a wide range of conditions. Consider a free-standing rectangular rod of infinite length in the z-direction with an x-directed height $2a$ and a width $2d$ in the y-direction, as shown in Figure 7.18.

Taking the origin of the coordinates in the geometric center of the of the xy-plane, the acoustic mode displacements determined by Morse are given by $\mathbf{u}(x, y, z) = (u_1, v_1, w_1)$, where

$$u_1 = u(x, y) \, e^{i\gamma(z-ct)}, \qquad v_1 = v(x, y) \, e^{i\gamma(z-ct)}, \qquad w_1 = w(x, y) \, e^{i\gamma(z-ct)}. \tag{7.172}$$

Here c is the phase velocity and γ is the wavevector in the z-direction. Morse considered the wave equations

$$\left(\frac{\partial^2}{\partial x^2} + \frac{\partial^2}{\partial y^2} + q_l^2\right)\phi = 0, \qquad \left(\frac{\partial^2}{\partial x^2} + \frac{\partial^2}{\partial y^2} + q_t^2\right)\psi_i = 0, \qquad (7.173)$$

by writing solutions in the form of a product of trigonometric functions:

$$\phi \propto \begin{Bmatrix} \sin q_1 x \\ \cos q_1 x \end{Bmatrix} \begin{Bmatrix} \sin h_1 y \\ \cos h_1 y \end{Bmatrix}, \qquad \psi \propto \begin{Bmatrix} \sin q_2 x \\ \cos q_2 x \end{Bmatrix} \begin{Bmatrix} \sin h_2 y \\ \cos h_2 y \end{Bmatrix} \qquad (7.174)$$

where

$$q_l^2 = q_1^2 + h_1^2 \qquad \text{and} \qquad q_t^2 = q_2^2 + h_2^2. \qquad (7.175)$$

The boundary conditions at $x = \pm a$ and $y = \pm d$ are taken as

$$\begin{aligned} T_{xx} = 0, \quad T_{yx} = 0, \quad T_{zx} = 0 \qquad &\text{at} \quad x = \pm a, \\ T_{yy} = 0, \quad T_{xy} = 0, \quad T_{zy} = 0 \qquad &\text{at} \quad y = \pm d. \end{aligned} \qquad (7.176)$$

These boundary conditions cannot be satisfied completely. Morse adopted a simple approach to satisfying them. Specifically, he noted that the simplest way to attempt to meet such conditions is to make the stress components factor into products of functions of x and y. This is possible if either $q_1 = q_2$ or $h_1 = h_2$.

The approximate separation-of-variables solution given by Morse for the compressional modes in the case where $h = h_1 = h_2$ is

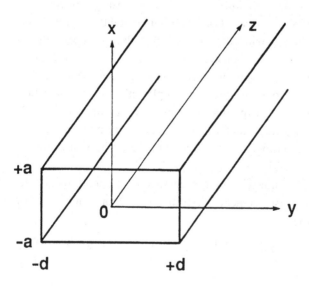

Figure 7.18. Rectangular quantum wire of width $2d$ in the y-direction and height $2a$ in the x-direction. From Yu *et al.* (1994), American Physical Society.

$$u = (A \sin q_1 x + B \sin q_2 x) \cos hy,$$

$$v = \left(\frac{h}{q_1} A \cos q_1 x + C \cos q_2 x\right) \sin hy, \tag{7.177}$$

$$w = i\left[-\frac{\gamma}{q_1} A \cos q_1 x + \frac{1}{\gamma}(q_2 B + hC) \cos q_2 x\right] \cos hy,$$

where

$$q_1^2 + h_1^2 = \gamma^2[(c/c_l)^2 - 1], \qquad q_2^2 + h_2^2 = \gamma^2[(c/c_t)^2 - 1], \tag{7.178}$$

ρ is the density of the elastic medium, and the longitudinal and transverse sound speeds are given by

$$c_l = \sqrt{(\lambda + 2\mu)/\rho} \qquad \text{and} \qquad c_t = \sqrt{\mu/\rho}. \tag{7.179}$$

The modes associated with this case are known as the 'thickness modes', as designated by Morse, who showed that $h = h_1 = h_2$ leads to an adequate description of the experimentally determined modes when $d \geq 2a$ (Morse, 1948). Using the expressions (7.177) for u, v, and w to evaluate $T_{xx} = T_{yx} = T_{zx} = 0$ at $x = \pm a$, it follows that

$$\begin{pmatrix} 2h \sin q_1 a & h \sin q_2 a & q_2 \sin q_2 a \\ -(\gamma^2 + h^2 - q_2^2) \cos q_1 a & 2q_1 q_2 \cos q_2 a & 0 \\ 2(h^2 + \gamma^2) \sin q_1 a & (\gamma^2 + h^2 - q_2^2) \sin q_2 a & 0 \end{pmatrix} \begin{pmatrix} A \\ B \\ C \end{pmatrix}$$
$$= 0. \tag{7.180}$$

The dispersion relation for $q_2 \neq 0$ is given by the expression resulting from the condition that the determinant of the coefficients vanishes, that is,

$$\frac{\tan q_2 a}{\tan q_1 a} = -\frac{4q_1 q_2(h^2 + \gamma^2)}{(h^2 + \gamma^2 - q_2^2)^2}, \tag{7.181}$$

which is similar to the dispersion relation for a free-standing layer discussed in subsection 7.6.1. For calculating the acoustic phonon frequencies as functions of the wavevector, γ, it is convenient to rewrite this dispersion relation as

$$\frac{\tan\left(\pi\sqrt{\chi^2 - \psi^2}\right)}{\tan\left(\pi\sqrt{\epsilon\chi^2 - \psi^2}\right)} = -\frac{4\psi^2\sqrt{\chi^2 - \psi^2}\sqrt{\epsilon\chi^2 - \psi^2}}{(2\psi^2 - \chi^2)^2}, \tag{7.182}$$

where χ^2 and ψ^2 are related to q_1 and q_2 through $q_1 = (\pi/a)\sqrt{\epsilon\chi^2 - \psi^2}$, $q_2 = (\pi/a)\sqrt{\chi^2 - \psi^2}$ and in accordance with $\omega_\gamma^2 = c_l(\gamma^2 + h^2 + q_1^2)$ and $\omega_\gamma^2 = c_t(\gamma^2 +$

$h^2 + q_2^2$), where $\epsilon = (c_t/c_l)^2 = (1 - 2\sigma)/(2 - 2\sigma)$. Defining $s = a\gamma/\pi$ and recalling that $\omega_\gamma = c\gamma$, it follows that

$$
\begin{aligned}
\chi^2 &= s^2 \left(\frac{c}{c_t}\right)^2 = \left(\frac{a}{\pi}\right)^2 \gamma^2 \left(\frac{c}{c_t}\right)^2 = \left(\frac{a}{\pi}\right)^2 \left(\frac{\omega}{c_t}\right)^2, \\
\psi^2 &= s^2 + \left(\frac{ah}{\pi}\right)^2 = \left(\frac{a}{\pi}\right)^2 (\gamma^2 + h^2), \\
\gamma^2 &= \left(\frac{\pi}{a}\right)^2 \psi^2 - h^2 = \left(\frac{\pi}{a}\right) \left[\psi^2 - \left(\frac{\tilde{r}}{2}\right)^2\right]
\end{aligned}
\tag{7.183}
$$

with $\tilde{r} = a/d$.

Solving (7.180) for B and C in terms of A, it follows that

$$
\begin{aligned}
u_1 &= A(\sin q_1 x + \alpha \sin q_2 x) \cos hy\, e^{i\gamma(z-ct)}, \\
v_1 &= A \left(\frac{h}{q_1} \cos q_1 x + \beta \cos q_2 x\right) \sin hy\, e^{i\gamma(z-ct)}, \\
w_1 &= iA \left[-\frac{\gamma}{q_1} \cos q_1 x + \frac{1}{\gamma}(q_2\alpha + h\beta) \cos q_2 x\right] \cos hy\, e^{i\gamma(z-ct)},
\end{aligned}
\tag{7.184}
$$

where

$$
\begin{aligned}
B &= \alpha A = -\frac{\sin q_1 a}{\sin q_2 a} \frac{2(h^2 + \gamma^2)}{(\gamma^2 + h^2 - q_2^2)} A, \\
C &= -\left(\frac{q_2 h}{h^2 + \gamma^2}\right) B = \frac{\sin q_1 a}{\sin q_2 a} \frac{2q_2 h}{(\gamma^2 + h^2 - q_2^2)} A = \beta A.
\end{aligned}
\tag{7.185}
$$

The remaining constant A is determined by quantizing the phonon amplitude according to

$$
\frac{1}{4ad} \int_{-a}^{a} dx \int_{-d}^{d} dy\, (u^*u + v^*v + w^*w) = \frac{\hbar}{2M\omega_\gamma},
\tag{7.186}
$$

where ω_γ is the angular frequency of the mode with wavevector γ. Performing the indicated integrations, it follows that

$$
\frac{A^2}{4ad} \left\{ f_1(h, d) \left[f_2(q_1, a) + 2\alpha g_1(q_1, q_2, a) + \alpha^2 f_2(q_2, a) \right] \right.
$$

$$
- f_1(h, d) \left[\frac{h^2}{q_1^2} f_1(q_1, a) + \frac{2\beta h}{q_1} g_2(q_1, q_2, a) + \beta^2 f_1(q_2, a) \right]
$$

$$
+ f_1(h, d) \left[\frac{\gamma^2}{q_1^2} f_1(q_1, a) - \frac{2}{q_1}(q_2\alpha + h\beta) g_2(q_1, q_2, a) \right.
$$

$$
\left. + \frac{q_1\alpha + h\beta^2}{\gamma^2} f_1(q_2, a) \right]
$$

$$
+ 2d \left[\frac{h^2}{q_1^2} f_1(q_1, a) + \frac{2\beta h}{q_1} g_2(q_1, q_2, a) + \beta^2 f_1(q_2, a) \right] \right\}
$$

$$
= \frac{\hbar}{2M\omega_\gamma}, \tag{7.187}
$$

with

$$
f_1(h, d) = d \left(1 + \frac{\sin 2hd}{2hd} \right), \qquad f_2(h, d) = 2d - f_1(h, d),
$$

$$
g_1(q_1, q_2, a) = \frac{\sin(q_1 - q_2)a}{q_1 - q_2} - \frac{\sin(q_1 + q_2)a}{q_1 + q_2}, \tag{7.188}
$$

$$
g_2(q_1, q_2, a) = \frac{\sin(q_1 - q_2)a}{q_1 - q_2} + \frac{\sin(q_1 + q_2)a}{q_1 + q_2}.
$$

It is convenient to define a new normalization constant B_γ, through

$$
A^2 = \frac{2\hbar}{M\omega_\gamma B_\gamma} \equiv A_\gamma^2. \tag{7.189}
$$

As discussed by Morse, h must be chosen to satisfy the boundary condition on the stress components at $y = \pm d$, that is, $T_{yy} = T_{xy} = T_{zy} = 0$. This can be accomplished for $d \gtrsim 2a$ since in this case T_{xy} and T_{zy} become negligible; with $T_{yy} = 0$ this implies that

$$
hd = \left(n + \tfrac{1}{2} \right)\pi, \qquad n = 0, 1, 2, \dots . \tag{7.190}
$$

The principal propagation mode has no nodal surfaces parallel to the length of the quantum wire; this corresponds to the case $n = 0$. Morse found close agreement between theory and experiment for $a/d = 1/8$ and as expected less agreement for $a/d = 1/2$. In addition to the 'thickness modes' another set of modes was observed experimentally by Morse (1948, 1950). These modes are known as 'width modes' and are determined by a procedure used to analyze the 'thickness modes'. Specifically, Morse took $q_1 = q_2 = q$ and obtained a set of equations similar to those for u_1, v_1, and w_1 but with x and y interchanged. By imposing the boundary conditions at $y = \pm d$, the 'width modes' were found to have a dispersion relation identical in form to that for the 'thickness modes.' The dispersion curves for selected acoustic modes are shown in Figure 7.19 for a 28.3 Å × 56.6 Å GaAs quantum wire and in Figure 7.20 for a 50 Å × 200 Å GaAs quantum wire.

For carriers at the non-degenerate Γ point in band α, $E_\alpha(\mathbf{k})$, the deformation-potential interaction Hamiltonian H_{def}^α is given in terms of the displacement operator $\hat{\mathbf{u}}(\mathbf{r})$ by

$$H_{\text{def}}^\alpha = E_1^\alpha \nabla \cdot \hat{\mathbf{u}}(\mathbf{r}). \tag{7.191}$$

At such a symmetry point, only the irrotational – that is, the longitudinal – components of $\mathbf{u}(\mathbf{r})$ contribute to H_{def}^α. Accordingly, only the the potential ϕ contributes to H_{def}^α. Since there are multiple modes for a given value of n, another index, m, is needed to describe the phonon spectrum at each value of γ. For the case of a quantum wire, the quantization of the acoustic phonons may be performed by taking

$$\hat{\mathbf{u}}(\mathbf{r}) = \frac{1}{\sqrt{L}} \sum_{n,m,\gamma} [\mathbf{u}(\gamma, x, y)a_\gamma + \text{c.c.}]\, e^{i\gamma z}, \tag{7.192}$$

where the components of $\mathbf{u}(\gamma, x, y) = (u, v, w)$ were normalized previously over the area $4ad$. The deformation potential is then given by

$$H_{\text{def}}^\alpha = \frac{E_1^\alpha}{\sqrt{L}} \sum_{n,m,\gamma} [a_{n,m}(\gamma) + a_{n,m}(-\gamma)]\left(\frac{\partial u}{\partial x} + \frac{\partial v}{\partial y} + i\gamma w\right) e^{i\gamma z}, \tag{7.193}$$

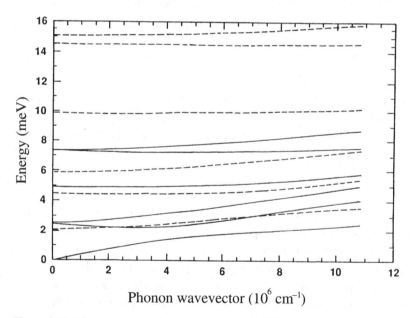

Figure 7.19. The six lowest-order ($m = 1, 2, \ldots, 6$) width modes (solid lines) and thickness modes (broken lines) for a 28.3 Å × 56.6 Å GaAs quantum wire. From Yu *et al.* (1994), American Physical Society, with permission.

and, upon applying the Fermi golden rule, these combinations lead to conditions enforcing the conservation of energy.

The Hamiltonian is independent of time. In Chapters 8 and 9, such time-independent carrier–phonon Hamiltonians will be used in applying the Fermi golden rule to calculate carrier–phonon scattering rates. The carrier–phonon interaction also has a time dependence of the form $e^{i\omega_\gamma t}$, where ω_γ is the phonon frequency. As will become obvious in Chapters 8 and 9, such time-dependent factors are combined with the time-independent factors of carrier wavefunctions. Since

$$\frac{\partial u}{\partial x} = A(q_1 \cos q_1 x + \alpha q_2 \cos q_2 x) \cos hy,$$

$$\frac{\partial v}{\partial y} = A\left(\frac{h^2}{q_1} \cos q_1 x + \beta \cos q_2 x\right) \cos hy, \tag{7.194}$$

$$-i\gamma w = A\left[-\frac{\gamma^2}{q_1} \cos q_1 x + (q_2\alpha + h\beta) \cos q_2 x\right] \cos hy,$$

it follows that

$$\left(\frac{\partial u}{\partial x} + \frac{\partial v}{\partial y} + i\gamma w\right) = A\left(q_1 + \frac{h^2}{q_1} + \frac{\gamma^2}{q_1}\right) \cos q_1 x \cos hy$$

$$= A\frac{1}{q_1}\left(q_1^2 + h^2 + \gamma^2\right) \cos q_1 x \cos hy$$

$$= A\frac{\omega_\gamma}{c_l^2 q_1} \cos q_1 x \cos hy \tag{7.195}$$

and, accordingly,

$$H_{\text{def}}^\alpha = \frac{E_1^\alpha}{\sqrt{L}} \sum_{n,m,\gamma} A\frac{\omega_\gamma}{c_l^2 q_1} \cos q_1 x \cos hy[a_{n,m}(\gamma) + a_{n,m}(-\gamma)]\, e^{i\gamma z}. \tag{7.196}$$

In Chapter 8 this expression for the deformation-potential interaction will be used to calculate electron–acoustic-phonon scattering rates for a rectangular quantum wire.

7.6.4 Acoustic phonons in cylindrical structures

The acoustic phonons in a cylindrical waveguide and in a cylindrical shell illustrate key features of the confined modes in dimensionally confined structures. The cylindrical waveguide is of obvious practical importance. Furthermore, the cylindrical shell is of interest because it approximates a single-walled buckytube and also because it resembles the microtubuline structure found in many parts of the human body. As discussed previously in this section and in Section 7.3, the elastic continuum model provides an approximate description of the acoustic phonon modes in such dimensionally confined nanostructures. The force equations for a cylindrical elastic medium may be written as (Auld, 1973; Sirenko *et al.*, 1995)

$$\rho \frac{\partial^2 u_r}{\partial t^2} = \frac{\partial T_{rr}}{\partial r} + \frac{1}{r}\frac{\partial T_{r\varphi}}{\partial \varphi} + \frac{\partial T_{rz}}{\partial z} + \frac{T_{rr} - T_{\varphi\varphi}}{r},$$

$$\rho \frac{\partial^2 u_\varphi}{\partial t^2} = \frac{\partial T_{r\varphi}}{\partial r} + \frac{1}{r}\frac{\partial T_{\varphi\varphi}}{\partial \varphi} + \frac{\partial T_{\varphi z}}{\partial z} + \frac{2 T_{r\varphi}}{r}, \tag{7.197}$$

$$\rho \frac{\partial^2 u_{z\varphi}}{\partial t^2} = \frac{\partial T_{rz}}{\partial r} + \frac{1}{r}\frac{\partial T_{\varphi z}}{\partial y} + \frac{\partial T_{zz}}{\partial z} + \frac{T_{rz}}{r},$$

where the axis of the cylinder is oriented along the z-direction, φ is the azimuthal angle, and r is the radial coordinate of the cylindrical structure. As before, the stress tensor T is related to the strain tensor S through the Hooke's law relationship

$$T_{\mu\nu} = \lambda S_{\mu\mu}\delta_{\mu\nu} + 2\mu S_{\mu\nu}; \tag{7.198}$$

in this stress–strain relation, λ and μ are the Lamé constants. Alternatively, these force equations are frequently written in the form

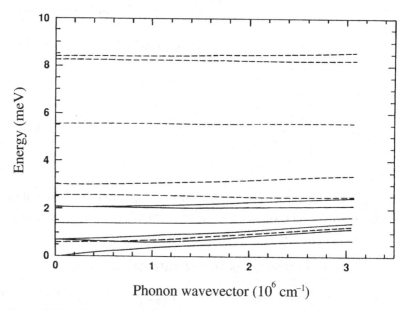

Figure 7.20. The six lowest-order ($m = 1, 2, \ldots, 6$) width modes (solid lines) and thickness modes (broken lines) for a 50 Å × 200 Å GaAs quantum wire. From Yu *et al.* (1994), American Physical Society, with permission.

$$\rho(\mathbf{r})\ddot{u}_i(\mathbf{r}, t) = \frac{\partial}{\partial x_j}\left[\lambda_{ijkl}(\mathbf{r})\frac{\partial u_k(\mathbf{r}, t)}{\partial x_l}\right], \tag{7.199}$$

where the elastic stiffness tensor for a particular isotropic medium is expressed as

$$\lambda_{ijkl} = \lambda\delta_{ij}\delta_{kl} + 2\mu\delta_{ik}\delta_{jl}. \tag{7.200}$$

These equations are more complicated than their counterparts in rectilinear co-ordinates, Section 7.2. Indeed, the additional complexity of the force equations in cylindrical coordinates is a direct consequence of the fact that in curvilinear coordinates the basis vectors are coordinate dependent.

Consider the acoustic phonon modes in a cylindrical waveguide of radius a embedded in an elastic medium. Both of these media are taken to be isotropic. From the normalization procedures of Section 5.1, the modes are normalized in terms of $\mathbf{w}$ instead of $\mathbf{u}$ since the considerations of Appendix A make it clear that it is convenient to use $\mathbf{w} = \sqrt{\rho}\mathbf{u}$; the displacement operator $\hat{\mathbf{u}}(\mathbf{r})$ is then given by

$$\hat{\mathbf{u}}(\mathbf{r}) = \sum_{q,mn}\sqrt{\frac{\hbar}{2L\rho\omega_{mn}(q)}}[\mathbf{w}_{mn,q}(r)a_{mn,q} + \mathbf{w}_{mn,-q}^*(r)a_{mn,-q}^\dagger]e^{im\varphi+iqz/a}. \tag{7.201}$$

The quantum number n labels modes with the same m and q in the set $\mathbf{w}_{mn,q}(r)$, where q represents the z-component of the wavevector q_z. In determining the normalization constants for the normal modes $\mathbf{w}_{mn,q}(r)$, it is convenient to write

$$\mathbf{w}_{mn,q}(\mathbf{r}) = \mathbf{w}_{mn,q}(r)\,e^{im\varphi+iqz/a} = \frac{1}{\sqrt{\pi a^2 \mathcal{N}}}\mathbf{u}(r)\,e^{im\varphi+iqz/a}, \tag{7.202}$$

where $\mathbf{u}(r)$ is the classical displacement given by the elastic continuum model and the normalization constant $\mathcal{N}$ is then determined by the normalization condition

$$\int d^2r\,\rho(\mathbf{r})\mathbf{w}_{n,m,q}^*(\mathbf{r}) \cdot \mathbf{w}_{n',m',q'}(\mathbf{r}) = \delta_{n,m,q;n',m',q'}, \tag{7.203}$$

and $q \equiv aq_z$.

Let the density and Lamé constants of the cylindrical waveguide be ρ_1, λ_1 and μ_1 respectively, and those of the surrounding material be ρ_2, λ_2 and μ_2. The general solution of the classical elastic continuum equations for such a cylindrical structure may be written (Beltzer, 1988; Stroscio *et al.*, 1996) in terms of three scalar potentials ϕ, ψ, and χ as

$$\mathbf{u} = \nabla\phi + \nabla \times \hat{\mathbf{e}}_z\psi + a\nabla \times \nabla \times \hat{\mathbf{e}}_z\chi, \tag{7.204}$$

where $\hat{\mathbf{e}}_z$ is a unit vector along the z-direction. The second and third terms in this last result correspond to the usual irrotational contribution to $\mathbf{u}$, expressed as a sum of two mutually normal vectors. The potentials ϕ, ψ, and χ satisfy scalar wave equations with longitudinal and transverse sound speeds given by

$$c_{l\xi} = \sqrt{(\lambda_\xi + 2\mu_\xi)/\rho_\xi} \quad \text{and} \quad c_{t\xi} = \sqrt{\mu_\xi/\rho_\xi}; \tag{7.205}$$

the subscript ξ takes on the value 1 to designate the material parameters of the cylinder and the subscript 2 to designate those of the surrounding material. Solutions of the classical elastic continuum equations are sought with vibration frequency ω, wavevector $q_z = q/a$, and azimuthal quantum number m. Seeking acoustic modes confined near the cylindrical waveguide, the scalar potentials for $r < a$ are taken to be

$$\begin{pmatrix} \phi \\ \psi \\ \chi \end{pmatrix} = \frac{1}{a} \begin{pmatrix} i b_{l1} J_m(k_l r/a) \\ B_{t1} J_m(k_t r/a) \\ b_{t1} J_m(k_t r/a) \end{pmatrix} e^{im\varphi + iqz/a - i\omega t}. \tag{7.206}$$

Outside the cylindrical waveguide, where $r > a$, the solutions are taken to be

$$\begin{pmatrix} \phi \\ \psi \\ \chi \end{pmatrix} = \frac{1}{a} \begin{pmatrix} i b_{l2} K_m(\kappa_l r/a) \\ B_{t2} K_m(\kappa_t r/a) \\ b_{t2} K_m(\kappa_t r/a) \end{pmatrix} e^{im\varphi + iqz/a - i\omega t}, \tag{7.207}$$

where $k_{l,t}$ and κ_l are defined by

$$k_{l,t}^2 = q^2 - \omega^2 a^2/c_{(l,t)1}^2 \quad \text{and} \quad \kappa_{l,t}^2 = \omega^2 a^2/c_{(l,t)2}^2 - q^2 \tag{7.208}$$

and the b_{l1} etc. are normalization constants, to be determined. In the expressions for ϕ, ψ, and χ, it is assumed that $k_{l,t}^2 > 0$ and $\kappa_{l,t}^2 > 0$, since confined acoustic modes are desired. Substituting these potentials into the general expression for **u** it follows that

$$-i u_r(r) = b_{l1} k_l J_m'\left(\frac{k_l r}{a}\right) + B_{t1} m \frac{a}{r} J_m\left(\frac{k_t r}{a}\right) + b_{t1} q k_t J_m'\left(\frac{k_t r}{a}\right),$$

$$-u_\varphi(r) = b_{l1} m \frac{a}{r} J_m\left(\frac{k_l r}{a}\right) + B_{t1} k_t J_m'\left(\frac{k_t r}{a}\right) + b_{t1} m q \frac{a}{r} J_m\left(\frac{k_t r}{a}\right),$$

$$-u_z(r) = b_{l1} q J_m\left(\frac{k_l r}{a}\right) - b_{t1} k_t^2 J_m\left(\frac{k_t r}{a}\right) \tag{7.209}$$

for $r < a$ and

$$-i u_r(r) = b_{l2} k_l K_m'(\kappa_l r/a) + B_{t2} m \frac{a}{r} K_m(\kappa_t r/a) + b_{t2} q \kappa_t K_m'(\kappa_t r/a),$$

$$-u_\varphi(r) = b_{l2} m \frac{a}{r} K_m\left(\frac{\kappa_l r}{a}\right) + B_{t1} \kappa_t K_m'\left(\frac{\kappa_t r}{a}\right) + b_{t1} m q \frac{a}{r} K_m\left(\frac{\kappa_t r}{a}\right), \tag{7.210}$$

$$-u_z(r) = b_{l2}q\,J\,K_m\left(\frac{\kappa_l r}{a}\right) - b_{t1}\kappa_t^2 K_m\left(\frac{\kappa_t r}{a}\right)$$

for $r > a$. By applying the boundary conditions of continuity of displacement and continuity of the normal components of the stress tensor at $r = a$, it follows that

$$\begin{pmatrix} \mathbf{U}_1 & -\mathbf{U}_2 \\ \mu_1\mathbf{F}_1 & -\mu_2\mathbf{F}_2 \end{pmatrix}\begin{pmatrix} \mathbf{B}_1 \\ \mathbf{B}_2 \end{pmatrix} = 0, \tag{7.211}$$

where $\mathbf{B}_\xi = [b_{l\xi}, B_{t\xi}, b_{t\xi}]^T$ and where the 'displacement' matrices $\mathbf{U}_\xi$ are evaluated at $r = a$ and are given by

$$\begin{pmatrix} G_\xi & my_\xi & qY_\xi \\ -mg_\xi & -Y_\xi & -mqy_\xi \\ -qg_\xi & 0 & k_{t\xi}y_\xi \end{pmatrix}; \tag{7.212}$$

the matrices $\mathbf{F}_\xi$ are given by

$$\begin{pmatrix} -2qG_\xi & -mqy_\xi & (k_{t\xi}^2 - q^2)Y_\xi \\ 2m(g_\xi - G_\xi) & (k_{t\xi}^2 - 2m^2)y_\xi + 2Y_\xi & 2mq(y_\xi - Y_\xi) \\ (2m^2 + q^2 - k_{l\xi}^2)g_\xi - 2G_\xi & 2mq(y_\xi - Y_\xi) & 2q[(m^2 - k_{t\xi}^2)y_\xi - Y_\xi] \end{pmatrix}, \tag{7.213}$$

where $k_{(l,t)\xi=1}^2 = k_{(l,t)}^2$, $k_{(l,t)\xi=2}^2 = -\kappa_{(l,t)}^2$, $g_1 = J_m(k_l)$, $G_1 = k_l J_m'(k_l)$, $y_1 = J_m(k_t)$, $Y_1 = k_t J_m'(\kappa_t)$, $g_2 = K_m(k_l)$, $G_2 = \kappa_l K_m'(\kappa_l)$, $y_2 = K_m(\kappa_t)$, and $Y_2 = \kappa_t K_m'(\kappa_t)$. From the components of $\mathbf{u}(r)$, $u_r(r)$, $u_\varphi(r)$, and $u_z(r)$, it is then straightforward to construct $\mathbf{w}_{mn,q}(\mathbf{r})$ and $\hat{\mathbf{u}}(\mathbf{r})$. The case where $m = 0$ is of practical interest and leads to relatively simple results. Specifically, for $m = 0$, submatrices of the 6×6 matrix decouple into matrices representing distinct axisymmetric torsional and radial–axial modes. For the axisymmetric torsional modes the dispersion relation is given by

$$\mu_1 k_t \frac{J_2(k_t)}{J_1(k_t)} = \mu_2 \kappa_t \frac{K_2(\kappa_t)}{K_1(\kappa_t)} \tag{7.214}$$

and the components of the normal modes are given by

$$w_r = 0, \qquad w_\varphi = \frac{1}{\sqrt{\pi a^2 \mathcal{N}_\varphi}}\begin{cases} K_1(\kappa_t)J_1(k_t r/a) & r < a, \\ J_1(k_t)K_1(\kappa_t r/a) & r > a, \end{cases} \qquad w_z = 0, \tag{7.215}$$

where the normalization condition gives

$$\mathcal{N}_\varphi = \rho_1 K_1^2(\kappa_t)[J_1^2(k_t) - J_0(k_t)J_2(k_t)] \\ + \rho_2 J_1^2(k_t)[(K_0(\kappa_t)K_2(\kappa_t) - K_1^2(\kappa_t)]. \tag{7.216}$$

For the axisymmetric radial–axial modes, the dispersion relation is defined by

$$
\begin{pmatrix} \mathbf{Q}_{11} & \mathbf{Q}_{12} \\ \mathbf{Q}_{21} & \mathbf{Q}_{22} \end{pmatrix}
\begin{pmatrix} b_{1l} \\ b_{1t} \\ b_{2l} \\ b_{2t} \end{pmatrix} = 0,
\tag{7.217}
$$

with $A = q^2 - k_l^2$, $B = q^2 + k_l^2$, and

$$
\mathbf{Q}_{11} = \begin{pmatrix} -k_l J_1(k_l) & -q k_t J_1(k_t) \\ -q J_0(k_l) & k_t^2 J_0(k_t) \end{pmatrix},
$$

$$
\mathbf{Q}_{12} = \begin{pmatrix} \kappa_1 K_1(\kappa_l) & q \kappa_t K_1(\kappa_t) \\ q K_0(\kappa_l) & \kappa_t^2 K_0(\kappa_t) \end{pmatrix},
$$

$$
\mathbf{Q}_{21} = \begin{pmatrix} 2\mu_1 q k_l J_1(k_l) & \mu_2 k_t (q^2 - k_t^2) J_1(k_t) \\ \mu_1[A J_0(k_l) + 2 k_l J_1(k_l)] & 2\mu_1 q k_t[J_1(k_t) - k_t J_0(k_t)] \end{pmatrix},
$$

$$
\mathbf{Q}_{22} = \begin{pmatrix} -2\mu_2 q \kappa_l K_1(\kappa_l) & -2\mu_2 \kappa_t(\kappa_t^2 + q^2) K_1(\kappa_l) \\ -\mu_2[B K_0(k_l) + 2\kappa_l K_1(\kappa_l)] & -\mu_2 q \kappa_t[J(k_t) + \kappa_t J_0(k_t)] \end{pmatrix}.
\tag{7.218}
$$

The components of the normal modes are given by

$$
i w_r = \frac{1}{\sqrt{\pi a^2 \mathcal{N}_{rz}}}
\begin{cases}
b_{l1} k_l J_1(k_l r/a) + b_{t1} q k_t J_1(k_t r/a) & r < a, \\
b_{l2} \kappa_l K_1(\kappa_l r/a) + b_{t2} q \kappa_t K_1(\kappa_t r/a) & r > a,
\end{cases}
$$

$$
w_\varphi = 0,
$$

$$
-w_z = \frac{1}{\sqrt{\pi a^2 \mathcal{N}_{rz}}}
\begin{cases}
b_{l1} q J_0(k_l r/a) - b_{t1} k_t^2 J_0(k_t r/a) & r < a, \\
b_{l2} q K_0(\kappa_l r/a) + b_{t2} \kappa_t^2 K_0(\kappa_t r/a) & r > a.
\end{cases}
\tag{7.219}
$$

For the axisymmetric radial–axial modes, the normalization condition requires that

$$
\begin{aligned}
\mathcal{N}_{rz} = & \rho_1 \big(b_{1t}^2 \{ q^2 [J_0^2(k_l) + J_1^2(k_l)] + k_t^2 [J_1^2(k_l) - J_0(k_l) J_1(k_l)] \} \\
& + b_{1t}^2 k_t^2 \{ k_t^2 [J_0^2(k_t) + J_1^2(k_t)] + q^2 [J_1^2(k_t) - J_0(k_t) J_2(k_t)] \} \\
& - 4 b_{1l} b_{1t} q k_t J_0(k_l) J_1(k_t) \big) \\
& + \rho_2 \big(b_{2l}^2 \{ q^2 [K_1^2(\kappa_l) - K_0^2(\kappa_l)] + \kappa_t^2 [K_0(\kappa_l) K_2(\kappa_l) - K_1^2(\kappa_l)] \} \\
& + b_{2t}^2 \kappa_t^2 \{ \kappa_t^2 [K_1^2(\kappa_t) - K_0^2(\kappa_t)] + q^2 [K_0(\kappa_t) K_2(\kappa_t) - K_1^2(\kappa_t)] \} \\
& - 4 b_{2l} b_{2t} q \kappa_t K_0(\kappa_t) K_1(\kappa_t) \big).
\end{aligned}
\tag{7.220}
$$

As for the case of rectangular quantum wires, subsection 7.6.3, for carriers at the non-degenerate Γ point in band α, $E_\alpha(\mathbf{k})$, the deformation-potential interaction Hamiltonian H_{def}^α is given in terms of the displacement operator $\hat{\mathbf{u}}(\mathbf{r})$ by

$$H_{\text{def}}^{\alpha} = E_1^{\alpha} \nabla \cdot \hat{\mathbf{u}}(\mathbf{r}). \tag{7.221}$$

Again, at such a symmetry point only the irrotational – that is, the longitudinal – components of $\mathbf{u}(\mathbf{r})$ contribute to H_{def}^{α}. Accordingly, only the the potential ϕ, (7.204), contributes to H_{def}^{α}. Indeed, from the normalized components $\mathbf{w}_{mn,q}(\mathbf{r})$ (7.202), $\hat{\mathbf{u}}(\mathbf{r})$ is obtained readily and by using the relation $\nabla^2 \phi = -(\omega/c_l)^2 \phi$ it follows that

$$H_{\text{def}}^{\alpha} = -E_1^{\alpha} \sum_{\mathbf{q},mn} \left[\frac{\omega_{mn}(q)}{ac_l} \right]^2 \sqrt{\frac{\hbar}{2\pi L \rho \mathcal{N} \omega_{mn}(q)}}$$

$$\times \left[\Phi_{mn,q}(r) a_{mn,q} + \Phi_{mn,-q}^*(r) a_{mn,-q}^{\dagger} \right] e^{im\varphi + iqz/a}, \tag{7.222}$$

the potential $\Phi_{mn,q}(r)$ being given by

$$\Phi_{mn,q}(r) = \begin{cases} i b_{l1} J_m(k_l r/a) & r < a, \\ i b_{l2} K_m(\kappa_l r/a) & r > a. \end{cases} \tag{7.223}$$

Let us consider the case of a thin cylindrical shell. For a cylindrical shell, the boundary conditions on the inner and outer surfaces are

$$\mathcal{P}_{\mu} = T_{\mu\nu} n_{\nu}, \tag{7.224}$$

where $\mathcal{P}$ represents an external pressure that would be present, for example, in the case where the cylindrical shell is in contact with a liquid, $T_{\mu\nu}$ is the stress tensor, and n_{ν} is the normal to the surface of the shell. In particular, $\mathcal{P} = \pm \mathcal{P}_{\substack{\text{in} \\ \text{out}}} \hat{\mathbf{e}}_r$ and $\mathbf{n} = \mp \hat{\mathbf{e}}_r$ where $\hat{\mathbf{e}}_r$ is the unit vector in the r-direction. The subscripts 'in' and 'out' are alternatives. For a cylindrical shell of infinite length in the z-direction and of thickness h and radius R such that $h \ll R$, the boundary conditions are

$$T_{r\varphi}\big|_{R \mp h/2} = T_{rz}\big|_{R \mp h/2} = 0, \qquad T_{rr}\big|_{R \mp h/2} = \mathcal{P}_{\substack{\text{in} \\ \text{out}}}. \tag{7.225}$$

Assuming that all quantities except T_{rr} are nearly constant with respect to r over the interval from $R - h/2$ to $R + h/2$, it is possible to show that

$$-\frac{T_{\varphi\varphi}}{R} + \frac{\mathcal{P}_{\text{in}} - \mathcal{P}_{\text{out}}}{h} = \rho \ddot{u}_r,$$

$$\frac{1}{R} \frac{\partial T_{\varphi\varphi}}{\partial \varphi} + \frac{\partial T_{\varphi z}}{\partial z} = \rho \ddot{u}_\varphi, \tag{7.226}$$

$$\frac{1}{R} \frac{\partial T_{\varphi z}}{\partial \varphi} + \frac{\partial T_{zz}}{\partial z} = \rho \ddot{u}_z.$$

These results follow straightforwardly by integrating the right- and left-hand sides of each force equation over the interval from $R - h/2$ to $R + h/2$, invoking the boundary conditions in the radial force equation, and cancelling factors of h. From the stress–strain relation (7.198), it follows that

$$T_{\varphi\varphi} = (\lambda + 2\mu)S_{\varphi\varphi} + \lambda(S_{zz} + S_{rr}),$$

$$T_{\varphi z} = 2\mu S_{\varphi z},$$

$$T_{zz} = (\lambda + 2\mu)S_{zz} + \lambda(S_{\varphi\varphi} + S_{rr}), \tag{7.227}$$

$$T_{rr} = (\lambda + 2\mu)S_{zz} + \lambda(S_{\varphi\varphi} + S_{zz})$$

$$= 0.$$

These stress–strain relations then imply that

$$S_{rr} = -\frac{\lambda}{\lambda + 2\mu}(S_{\varphi\varphi} + S_{zz}),$$

$$T_{\varphi\varphi} = \frac{4\mu(\lambda + \mu)}{\lambda + 2\mu}S_{\varphi\varphi} + \frac{2\lambda\mu}{\lambda + 2\mu}S_{zz}, \tag{7.228}$$

$$T_{zz} = \frac{4\mu(\lambda + \mu)}{\lambda + 2\mu}S_{zz} + \frac{2\lambda\mu}{\lambda + 2\mu}S_{\varphi\varphi}.$$

The Young's modulus E and the Poisson ratio ν may be expressed in terms of the Lamé constants as

$$E = \frac{\mu(3\lambda + 2\mu)}{\lambda + \mu} \quad \text{and} \quad \nu = \frac{\lambda}{2(\lambda + \mu)}. \tag{7.229}$$

Then the inverse relations are

$$\lambda = \frac{E\nu}{(1 - 2\nu)(1 + \nu)} \quad \text{and} \quad \mu = \frac{E}{2(1 + \nu)} \tag{7.230}$$

and it follows that

$$\lambda + \mu = \frac{E}{2(1 - 2\nu)(1 + \nu)} \quad \text{and} \quad \nu = \frac{\lambda}{2(\lambda + \mu)}. \tag{7.231}$$

Using these stress–strain relations and eliminating the Lamé constants in favor of Young's modulus and the Poisson ratio, the force equations may be written as

$$-\frac{E}{R(1 - \nu^2)}(S_{\varphi\varphi} + \nu S_{zz}) + \frac{\mathcal{P}_{\text{in}} - \mathcal{P}_{\text{out}}}{h} = \rho \ddot{u}_r,$$

$$\frac{E}{R(1 - \nu^2)}\frac{\partial}{\partial\varphi}(S_{\varphi\varphi} + \nu S_{zz}) + \frac{E}{1 + \nu}\frac{\partial}{\partial\varphi}S_{\varphi z} = \rho \ddot{u}_\varphi, \tag{7.232}$$

$$\frac{E}{R(1 + \nu)}\frac{\partial}{\partial\varphi}S_{\varphi z} + \frac{E}{1 - \nu^2}\frac{\partial}{\partial z}(S_{zz} + \nu S_{\varphi\phi}) = \rho \ddot{u}_z.$$

Then using the relations (Auld, 1973)

$$S_{\varphi\varphi} = \frac{1}{r}\left(\frac{\partial u_\varphi}{\partial \varphi} + u_r\right), \qquad S_{zz} = \frac{\partial u_z}{\partial z}, \qquad \text{and} \qquad S_{\varphi z} = \frac{1}{r}\frac{\partial u_z}{\partial \varphi} + \frac{\partial u_\varphi}{\partial z}, \tag{7.233}$$

the force equations may be written as

$$-\frac{u_r + u_{\varphi,\varphi}}{R^2} - \frac{vu_{z,z}}{R} + \frac{p}{R} = \frac{\ddot{u}_r}{c^2},$$

$$\frac{u_{r,\varphi} + u_{\varphi,\varphi\varphi}}{R^2} + \frac{1-v}{2R}u_{\varphi,zz} + \frac{1+v}{2R}u_{z,\varphi z} = \frac{\ddot{u}_\varphi}{c^2}, \tag{7.234}$$

$$\frac{vu_{r,z}}{R} + \frac{1+v}{2R}u_{\varphi,\varphi z} + \frac{1-v}{2R^2}u_{z,\varphi\varphi} + u_{z,zz} = \frac{\ddot{u}_z}{c^2},$$

the thin-plate longitudinal sound speed squared, c^2, and the dimensionless pressure p are defined by

$$c^2 = \frac{E}{\rho(1-v^2)} \qquad \text{and} \qquad p = \frac{R}{h}\left(\frac{1-v^2}{E}\right)(\mathcal{P}_{in} - \mathcal{P}_{out}). \tag{7.235}$$

These equations have been used (Sirenko *et al.*, 1995) to describe the free vibrations of a thin shell and the vibrations of such a shell immersed in liquid. For the case of a free shell, $p = 0$ and solutions are taken to have the form

$$\begin{pmatrix} u_r \\ u_\varphi \\ u_z \end{pmatrix} = \begin{pmatrix} -ic_r \\ c_\varphi \\ c_z \end{pmatrix} \exp\left(im\varphi + i\frac{q}{R}z - i\omega t\right), \tag{7.236}$$

where the dimensionless wavevector in the z-direction, q, and the dimensionless frequency Ω are defined by

$$q = q_z R \qquad \text{and} \qquad \Omega = \frac{\omega R}{c}. \tag{7.237}$$

Then the force equations yield

$$ic_r - imc_\varphi - ivqc_z = i\Omega^2 c_r,$$

$$mc_r - m^2 c_\varphi - v_- q^2 c_\varphi - v_+ mqc_z = -\Omega^2 c_\varphi, \tag{7.238}$$

$$vqc_r - v_+ mqc_\varphi - v_- m^2 c_z = -\Omega^2 c_z$$

or

$$\begin{pmatrix} c_r \\ c_\varphi \\ c_z \end{pmatrix} \begin{pmatrix} \Omega^2 - 1 & m & vq \\ m & \Omega^2 - m^2 - v_- q^2 & -v_+ mq \\ vq & -v_+ mq & \Omega^2 - v_- m^2 - vq^2 \end{pmatrix} = 0. \tag{7.239}$$

The dispersion relation is then given by

$$\det \mathcal{D} = \begin{vmatrix} \Omega^2 - 1 & m & vq \\ m & \Omega^2 - m^2 - v_-q^2 & -v_+mq \\ vq & -v_+mq & \Omega^2 - v_-m^2 - vq^2 \end{vmatrix} = 0.$$

(7.240)

Hence

$$\Omega^6 - [(v_- + 1)(m^2 + q^2) + 1]\Omega^4$$
$$+ [(m^2 + v_-q^2)(v_-m^2 + q^2) + (1 + v_-)(m^2 + q^2)$$
$$- v^2q^2 - m^2 - v_+^2m^2q^2]\Omega^2 - v_-(1 - v^2)q^4 = 0 \quad (7.241)$$

or

$$Aq^4 - Bq^2 + C = 0, \tag{7.242}$$

where

$$A = v_-(1 + v^2),$$

$$B = v_-(\Omega^2 - 1)(\Omega^2 - v_-m^2) + (\Omega^2 - 1)(\Omega^2 - m^2)$$
$$\quad + 2vv_+m^2 + v^2(\Omega^2 - m^2) + v_+^2m^2(\Omega^2 - 1) - v_-m^4, \tag{7.243}$$

$$C = (\Omega^2 - m^2 - 1)(\Omega^2 - v_-m^2)\Omega^2.$$

Let us examine these solutions for three several special cases: symmetric azimuthal modes with $m = 0$, small q, and large q.

For the modes with $m = 0$,

$$\begin{pmatrix} c_r \\ c_\varphi \\ c_z \end{pmatrix} \begin{pmatrix} \Omega^2 - 1 & 0 & vq \\ 0 & \Omega^2 - v_-q^2 & 0 \\ vq & 0 & \Omega^2 - q^2 \end{pmatrix} = 0. \tag{7.244}$$

For $c_r = c_z = 0$, the pure torsional mode has frequency

$$\Omega_{\text{torsional}} \equiv \Omega^{\text{II}}_{m=0} = \sqrt{v_-}q. \tag{7.245a}$$

In addition, for $c_\varphi = 0$ it follows that two coupled radial–longitudinal modes have frequencies

$$\Omega^2_{\text{rad–long}} \equiv \left(\Omega^{\text{I,III}}_{m=0} \right)^2 = \tfrac{1}{2}\left[q^2 + 1 \pm \sqrt{(q^2 - 1)^2 + 4v^2} \right]. \tag{7.245b}$$

For small q, the $q \to 0$ limit yields

$$\begin{pmatrix} c_r \\ c_\varphi \\ c_z \end{pmatrix} \begin{pmatrix} \Omega^2 - 1 & m & 0 \\ m & \Omega^2 - m^2 & 0 \\ 0 & 0 & \Omega^2 - v_-m^2 \end{pmatrix} = 0. \tag{7.246}$$

For $c_z = 0$, the two torsional–radial modes have frequencies $\Omega_{\text{torsional–radial},a} = \Omega^{\text{I}}_m = \sqrt{m^2 + 1}$ and $\Omega_{\text{torsional–radial},b} \equiv \Omega^{\text{III}}_m = 0$. In addition, for $c_r = c_z = 0$, it follows that the longitudinal mode has frequency $\Omega_{\text{longitudinal}} \equiv \Omega_m^{II} = \sqrt{v_- m}$. Since $\Omega_m^{III} \approx 0$ was obtained in the lowest order in q it is necessary to find the first non-vanishing term. For $m = 0$, by making the assumption that $\left(\Omega_{m=0}^{III}\right)^2 = \alpha q^2$ and collecting terms up to order q^4 in the dispersion equation, it follows that

$$-\alpha^2 q^4 + \alpha q^2(1 + v_- - v^2)q^2 - v_-(1 - v^2)q^4 = 0. \tag{7.247}$$

One of the roots of this equation, $\alpha = v_-$, corresponds to the mode $\Omega_{m=0}^{II}$, and another root, $\alpha = 1 - v^2$, implies that $\Omega_{m=0}^{III} \approx \sqrt{1 - v^2}q$. For the case where $m \neq 0$, by making the assumption $\left(\Omega_m^{III}\right)^2 = \alpha q^4$ and collecting terms up to q^4 in the dispersion relation, it follows that

$$\alpha[v_- m^4 + (1 + v_-)m^2 - m^2]q^4 - v_-(1 - v^2)q^4 = 0, \tag{7.248}$$

so that $\alpha = (1 - v^2)/(m^4 + m^2)$ and $\Omega_m^{III} \approx \sqrt{(1 - v^2)/(m^2 + 1)}(q^2/m)$. In the limit of large q the leading terms of the dispersion relation imply that

$$\Omega^6 - (1 + v_-)q^2\Omega^4 + v_- q^4\Omega^2 - v_-(1 - v^2)q^4 = 0 \tag{7.249}$$

or making the guess $\Omega^2 \to \alpha q^2$, it follows that

$$\alpha^3 - (1 + v_-)\alpha^2 + v_-\alpha = 0, \tag{7.250}$$

so that $\alpha_1 = 1$, $\alpha_2 = v_-$, and $\alpha_3 = 0$. The result $\alpha_3 = 0$ is inconsistent with the initial guess, so solutions are of the form $\Omega^2 \to$ constant. Then from the last two terms of the dispersion relation, with leading fourth power of q, it follows that $v_- q^4\Omega^2 - v_-(1 - v^2)q^4 \approx 0$, and $\Omega^2 \to 1 - v^2$. Convenient interpolation formulae between the small-q and large-q solutions are given by

$$\Omega_m^I(q) \simeq \sqrt{q^2 + m^2 + 1},$$

$$\Omega_m^{II}(q) \simeq \sqrt{v_-(q^2 + m^2)},$$

$$\Omega_{m=0}^{III}(q) \simeq \sqrt{1 - v^2}\,\frac{q}{1 + q}, \tag{7.251}$$

$$\Omega_m^{III}(q) \simeq \sqrt{1 - v^2}\,\frac{q^2}{m\sqrt{m^2 + 1} + q^2}.$$

In the axisymmetric case with $m = 0$ the I, II, and III modes correspond to pure radial, torsional, and longitudinal modes respectively. When $m \neq 0$, the radial and torsional modes are coupled. In the limit where $q \gg m + 1$ the asymptotic expressions do not depend on m; indeed, in this limit $\Omega_m^I(q) \simeq q$, $\Omega_m^{II}(q) \simeq \sqrt{v_-}q$, and $\Omega_m^{III}(q) \simeq \sqrt{1 - v^2}$. Analysis of the coefficients c_r, c_φ, and c_z (Sirenko et al., 1996a, b) reveals that in the limit of large q the I, II, and III modes correspond to pure longitudinal, torsional, and radial vibrations, respectively.

The case of a cylindrical shell immersed in fluid was considered by Markus (1988) and solved numerically by Sirenko et al. (1996b). In this case, $p \neq 0$ and the dispersion relation takes the form

$$\det \mathcal{D} = \begin{vmatrix} \Omega^2(1 + W_{mQ}) - 1 & m & vq \\ m & \Omega^2 - m^2 - v_-q^2 & -v_+mq \\ vq & -v_+mq & \Omega^2 - v_-m^2 - vq^2 \end{vmatrix} = 0, \tag{7.252}$$

where the term describing the coupling between the shell and the fluid is given by

$$W_{mQ} = \frac{\alpha}{Q} \left[\frac{I_m(Q)}{I'_m(Q)} - \frac{K_m(Q)}{K'_m(Q)} \right]. \tag{7.253}$$

Here

$$Q^2 = R^2 \left(q_z^2 - \frac{\omega^2}{s_f^2} \right) \equiv q^2 - \left(\frac{c}{s_f} \right)^2 \Omega^2, \tag{7.254}$$

s_f being the sound speed in the fluid and

$$I_m(-i|Q|) = i^{-m} J_m(|Q|). \tag{7.255}$$

In the case of a thin cylindrical shell immersed in fluid, interface modes in the fluid are localized near the cylindrical surface and correspond to the region $Q^2 > 0$. In the case where $\omega > s_f q_z$ the acoustic disturbances are radiated from the cylindrical shell. Indeed, the ωq_z-plane is divided into two regions by the curve $\omega > s_f q_z$. As just indicated, the region defined by $Q^2 > 0$, or $\omega < s_f q_z$, is the region where interface modes are localized on the scale of R/Q from the cylindrical shell. When $\omega > s_f q_z$, $Q^2 < 0$ and the relation

$$K_m(-i|Q|) = i^{m+1} \frac{\pi}{2} H_m^{(1)}(|Q|),$$

makes manifest the radiation of cylindrical waves from the cylindrical shell into the region surrounding it, since the Hankel function of the first kind, $H_m^{(1)}$, represents outgoing cylindrical waves.

The case of a cylindrical shell immersed in fluid was solved numerically by Sirenko et al. (1996b) in order to model the vibrational behavior of microtubules (MTs) immersed in water. These microtubules are of great interest in biology as they are present in many biological structures including cytoskeleton and eukariotic cells. In these numerical calculations, the inner and outer radii of the MTs are taken to be 11.5 nm and 14.2 nm respectively, the length of the MTs is taken to be 8 nm, their mass is taken to be 1.83×10^{-19} g, the Poisson ratio, v, is taken to be 0.3, and the Young's modulus is taken to be 0.5 ± 0.1 GPa. These parameters give a thin-plate longitudinal sound speed c of 610 m s^{-1} and a density of 1.47 g cm^{-3}. The sound speed and density of water are taken to be 1.50 km s^{-1} and 1 g cm^{-3}. The calculated dispersion relations, $\Omega_m^i(q)$, for $m = 1, 2$, and 3 are shown in Figures 7.21, 7.22, and 7.23, respectively.

From Figures 7.21–7.23, it is apparent that for $q_z \gg m/R$ the mode frequencies of the immersed MTs tend to those of the free-standing MTs and do not depend on m. These modes are seen to have maximum frequencies of the order of tens of GHz. Moreover, the sound speeds of the axisymmetric acoustic modes are in the range $200\text{–}600 \text{ m s}^{-1}$.

Sirenko et al. (1996b) also considered the dynamical behavior of cytoskeletal filaments, by using the elastic continuum model to determine the mode structure for the vibrations of a solid cylinder. Particular attention was given to (a) the axisymmetric torsional mode, (b) the axisymmetric radial–longitudinal mode, and (c) the flexural mode, as depicted in Figure 7.24.

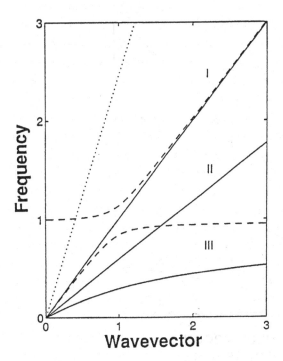

Figure 7.21. Dispersion relations Ω_m^i for $m = 1$ modes of the MT sample discussed in the text. I, pure longitudinal; II, torsional; III, radial. The frequency is in units of $\Omega = \omega R/c$ and $q = q_z R$. $\Omega = 1$ yields $\omega = 7.6$ GHz for the parameters given in the text. The dotted line separating the regions of interface and radiated waves corresponds to the condition $\omega = s_f q_z$, as discussed in the text. The solid and broken lines correspond to MTs immersed in water and free standing respectively. From Sirenko et al. (1996a), American Physical Society, with permission.

7.6.5   Acoustic phonons in quantum dots

In quantum dots, phonons and carriers alike are modified as a result of abrupt changes in the material properties at the interface between the quantum dot and the surrounding material. Indeed, carrier wavefunctions are modified as a result of the variations in the electron and hole band energies near the boundaries of the quantum dot. In the case of acoustic phonons, the changes in elastic properties near the quantum-dot boundaries lead to modifications in the displacement amplitudes. The acoustic phonon modes for spherical quantum dots and for quantum dots with rectangular-face confinement have been considered previously (Stroscio *et al.*, 1994). For the case of a free-standing spherical quantum dot, the quantization of the acoustic phonons may be performed by taking

$$\mathbf{u}(\mathbf{r}) = \frac{1}{\sqrt{N}} \sum_{\mathbf{q}} [\mathbf{u}(\mathbf{q}, \mathbf{r}) a_{\mathbf{q}} + \text{c.c.}], \tag{7.256}$$

and normalizing the acoustic phonon Fourier amplitude, $\mathbf{u}(\mathbf{q}, r, \varphi, z)$, according to

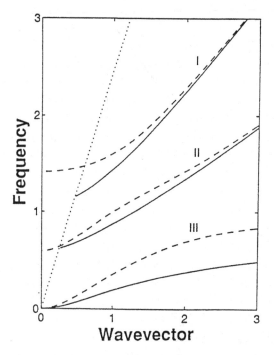

Figure 7.22. Dispersion relations Ω_m^i for $m = 2$ modes of the MT sample discussed in the text. I, pure longitudinal; II, torsional; III, radial. The frequency Ω is in units of $\omega R/c$ and $q = q_z R$. $\Omega = 1$ yields $\omega = 7.6$ GHz for the parameters given in the text. The dotted line separating the regions of interface and radiated waves corresponds to the condition $\omega = s_f q_z$, as discussed in the text. The solid and broken lines correspond to MTs immersed in water and free standing respectively. From Sirenko *et al.* (1996a), American Physical Society, with permission.

$$\frac{3}{4\pi a^3} \int_0^{2\pi} d\varphi \int_0^{\pi} d\theta \sin\theta \int_0^a dr \, r^2 \mathbf{u}(\mathbf{q}, r, \theta, \varphi) \cdot \mathbf{u}^*(\mathbf{q}, r, \theta, \varphi) = \frac{\hbar}{2M\omega_q}.$$

$$(7.257)$$

Here, a is the radius of the quantum dot, N is the number of unit cells in the normalization volume V, $a_{\mathbf{q}}$ is the phonon annihilation operator, $\mathbf{q}$ is the phonon wavevector, ω_q is the angular frequency of the phonon mode, M is the mass of the ions in the unit cell, and r, θ, ϕ are the usual spherical coordinates. For a quantum dot with rectangular faces the normalization condition for the acoustic phonon mode amplitude is given by

$$\frac{1}{abc} \int_{-a/2}^{a/2} dx \int_{-b/2}^{b/2} dy \int_{-c/2}^{c/2} dz \, \mathbf{u}(\mathbf{q}, x, y, z) \cdot \mathbf{u}^*(\mathbf{q}, x, y, z) = \frac{\hbar}{2M\omega_q}.$$

$$(7.258)$$

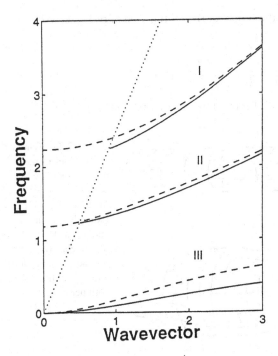

Figure 7.23. Dispersion relations Ω_m^i for $m = 3$ modes of the MT sample discussed in the text. I, pure longitudinal; II, torsional; III, radial. The frequency is in units of $\Omega = \omega R/c$ and $q = q_z R$. $\Omega = 1$ yields $\omega = 7.6$ GHz for the parameters given in the text. The dotted line separating the regions of interface and radiated waves corresponds to the condition $\omega = s_f q_z$, as discussed in the text. The solid and broken lines correspond to MT immersed in water and free standing, respectively. From Sirenko *et al.* (1996a), American Physical Society, with permission.

The classical acoustic modes in an isotropic elastic medium have been analyzed previously, and many of the most useful known results summarized, by Auld (1973). The lowest-order pure-compressional mode is referred to frequently as the breathing mode. The displacement field associated with this lowest order compressional mode of a sphere of radius a is given by

$$\mathbf{u}(q,r) = \hat{\mathbf{r}}\gamma j_1\left(\frac{\omega_q r}{c_l}\right)e^{-i\omega_q t},\tag{7.259}$$

where γ is the normalization constant, $\hat{\mathbf{r}}$ is the unit vector in the radial direction, j_1 is the spherical Bessel function of order unity, $j_1(x) = \sin x/x^2 - \cos x/x$, ω_q is the mode frequency, and the longitudinal sound speed c_l is equal to $\sqrt{(\lambda + 2\mu)/\rho}$. The frequency for a free-standing sphere is determined by the condition that the normal component of the traction force at the surface of the sphere vanishes; that is, $T_{rr} = 0$ at $r = a$:

$$\left[(\lambda + 2\mu)\frac{d^2}{dr^2}j_0\left(\frac{\omega_q r}{c_l}\right) + \frac{2\lambda}{r}\frac{d}{dr}j_0\left(\frac{\omega_q r}{c_l}\right)\right]_{r=a} = 0,\tag{7.260}$$

where $j_0(x) = \sin x/x$ is the spherical Bessel function of order zero. This last result implies that

$$\tan\left(\frac{\omega_q r}{c_l}\right) = \frac{4\mu\omega_q a/c}{4\mu - (\lambda + 2\mu)(\omega_q a/c)^2}.\tag{7.261}$$

The normalization condition for this lowest-order breathing mode is

(a)

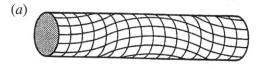

(b)

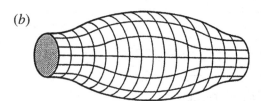

(c)

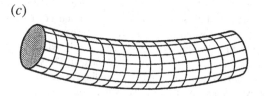

Figure 7.24. Schematic illustrations of the displacement fields of (a) the axisymmetric torsional mode, (b) the axisymmetric radial–longitudinal mode, and (c) the flexural mode for a solid elastic cylinder. From Sirenko *et al.* (1996b), American Physical Society, with permission.

$$\frac{3\gamma^2}{\tilde{\omega}_l^3} \int_0^{\tilde{\omega}_l} dr\, r^2 j_1^2(r) = \frac{\hbar}{2M\omega_q}, \tag{7.262}$$

with $\tilde{\omega}_l$ representing the quantity $\omega_q a / c_l$. This integral may be performed analytically and it follows that

$$\gamma = \frac{1}{\sqrt{j_1^2(\tilde{\omega}_l) - j_0(\tilde{\omega}_l) j_2(\tilde{\omega}_l)}} \sqrt{\frac{\hbar}{3M\omega_q}}, \tag{7.263}$$

where the second-order spherical Bessel function, $j_2(x)$, is defined by $j_2(x) = [(3/x^3) - (1/x)]\sin x - (3/x^2)\cos x$. The lowest-order breathing and torsional modes for a spherical quantum dot are shown in Figure 7.25.

The lowest-order torsional mode – a pure shear mode – of an free-standing isotropic spherical quantum dot may be determined from the known elastic-continuum solution for the lowest-order classical pure shear mode of an isotropic elastic medium (Auld, 1973):

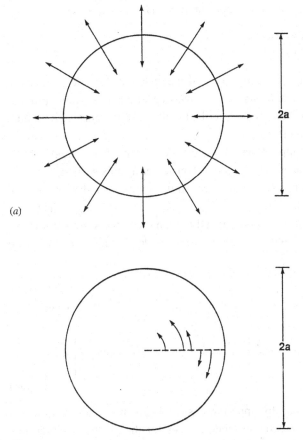

Figure 7.25. Lowest-order (a) breathing mode and (b) torsional mode in a spherical quantum dot. From Stroscio et al. (1994), American Institute of Physics, with permission.

(a)

(b)

$$\mathbf{u}(q, r, \theta) = \hat{\varphi}\tau \cos\theta \, j_1\!\left(\frac{\omega_q r}{c_t}\right) e^{-i\omega_q t}, \tag{7.264}$$

where $\hat{\varphi}$ is the unit vector in the φ-direction, τ is the normalization constant to be determined from the phonon normalization condition, and the transverse sound speed of the shear wave is given by $c_t = \sqrt{\mu/\rho}$. This mode is depicted in Figure 7.25(b). The normalization condition for this mode is

$$\frac{\tau^2}{\tilde{\omega}_t^3} \int_0^{\tilde{\omega}_t} dr \, r^2 j_1^2(r) = \frac{\hbar}{2M\omega_q}, \tag{7.265}$$

with $\tilde{\omega}_t = \omega_q a/c_t$. Thus, the normalization constant τ may be evaluated in terms of the same integral used to calculate γ; indeed,

$$\gamma = \frac{1}{\sqrt{j_1^2(\tilde{\omega}_t) - j_0(\tilde{\omega}_t) j_2(\tilde{\omega}_t)}} \sqrt{\frac{\hbar}{M\omega_q}}. \tag{7.266}$$

Following the same procedure as for the breathing mode, it follows that the dispersion relation for the lowest-order torsional mode is

$$\tan\!\left(\frac{\omega_q a}{c_t}\right) = \frac{3\omega_q a/c_t}{3 - (\omega_q a/c_t)^2}. \tag{7.267}$$

Krauss and Wise (1997) have recently observed the coherent acoustic phonons in spherical quantum dots. The damping of the lowest-order acoustic phonon modes observed by Krauss and Wise has been described in terms of the elastic continuum model by Stroscio and Dutta (1999).

McSkimin (1944) gave approximate classical flexural thickness modes for a structure with rectangular faces. The structure considered in this section has faces joining each other at right angles, and the faces in the xy-, yz-, and xz-planes are rectangles such that the width of the structure – in the x-direction – is a, the height of the structure – in the y-direction – is b, and the length of the structure – in the z-direction – is c. The approximate flexural thickness modes given by McSkimin are

$$u(x, y, z) = A \sin mx \, (\sin l_1 y + \alpha \sin l_2 y + \beta \sin l_3 y) \cos nz,$$

$$v(x, y, z) = A \cos mx \left[\frac{-l_1}{m} \cos l_1 y - \frac{\alpha l_2}{m} \cos l_2 y + \frac{\beta(m^2 + n^2)}{l_3 m} \cos l_3 y \right] \cos nz,$$

$$w(x, y, z) = A \cos mx \left[\frac{n}{m} \sin l_1 y - \frac{\alpha(m^2 + l_2^2)}{nm} \sin l_2 y + \frac{\beta n}{m} \sin l_3 y \right] \sin nz,$$

$$\tag{7.268}$$

where α and β are determined by applying desired boundary conditions on two sets of rectangular faces. The discrete mode indices for the x- and z-dimensions are labeled by m and n respectively. For the y-dimension, the mode index for the irrotational, or longitudinal, part of the mode is l_1, and l_2 and l_3 correspond to the indices for the rotational parts of the mode. As will become clear, the deformation potential depends only on l_1, as expected. Indeed, it follows that

$$\Delta(x, y, z) = \frac{\partial u}{\partial x} + \frac{\partial v}{\partial y} + \frac{\partial w}{\partial z}$$

$$= A \cos mx \left(m \sin l_1 y + \frac{l_1^2}{m} \sin l_1 y + \frac{n^2}{m} \sin l_1 y \right) \cos nz. \quad (7.269)$$

The flexural acoustic modes of McSkimin are used in this section to illustrate the quantization of modes for such a structure. Obviously, these modes do not form a complete set. Indeed, in addition to compressional modes there are flexural modes corresponding to the width and length modes. The normalization condition for McSkimin's flexural thickness modes,

$$\frac{1}{abc} \int_{-a/2}^{a/2} dx \int_{-b/2}^{b/2} dy \int_{-c/2}^{c/2} dz \left(uu^* + vv^* + ww^* \right) = \frac{\hbar}{2M\omega}, \quad (7.270)$$

reduces to

$$\frac{1}{abc} \left\{ f_2(m, a/2) f_1(n, c/2) [f_2(l_1, b/2) + 2\alpha g_1(l_1, l_2, b/2) \right.$$

$$+ 2\beta g_1(l_1, l_3, b/2) + \alpha^2 f_2(l_2, b/2) + 2\alpha\beta g_1(l_2, l_3, b/2) + \beta^2 f_2(l_3, b/2)]$$

$$+ f_1(m, a/2) f_1(n, c/2) \left[\frac{l_1^2}{m^2} f_1(l_1, b/2) + \frac{2\alpha l_1 l_2}{m^2} g_2(l_1, l_2, b/2) \right.$$

$$- \frac{2\beta l_1(m^2 + n^2)}{l_3 m^2} g_2(l_1, l_3, b/2) + \frac{\alpha^2 l_2^2}{m^2} f_1(l_2, b/2) - \frac{2\alpha\beta l_2(m^2 + n^2)}{l_3 m^2}$$

$$\times g_2(l_2, l_3, b/2) + \frac{2\beta^2(m^2 + n^2)}{l_3^2 m^2} f_1(l_3, b/2) \right]$$

$$+ f_1(m, a/2) f_2(n, c/2) \left[\frac{n^2}{m^2} f_2(l_1, b/2) - \frac{2\alpha n(m^2 + l_2^2)}{n m^2} g_1(l_1, l_2, b/2) \right.$$

$$+ \frac{2\beta n^2}{m^2} g_1(l_1, l_3, b/2) + \frac{\alpha^2(m^2 + l_2^2)}{n^2 m^2} f_2(l_2, b/2)$$

$$- \frac{2\alpha\beta n(m^2 + l_2^2)}{n m^2} g_1(l_2, l_3, b/2) + \frac{\beta^2 n^2}{m^2} f_2(l_3, b/2) \right] \right\}$$

$$= \frac{\hbar}{2M\omega}, \quad (7.271)$$

where

$$f_1(h, a/2) = \int_{-a/2}^{+a/2} dy \cos^2 hy = \frac{a}{2}\left(1 + \frac{\sin ha}{ha}\right),$$

$$f_2(h, a/2) = \int_{-a/2}^{+a/2} dy \sin^2 hy = a - f_1(h, a/2),$$

$$\begin{aligned}
g_1(l_i, l_j, b/2) &= \int_{-b/2}^{+b/2} dy \sin l_i y \sin l_j y \\
&= \frac{\sin(l_i - l_j)b/2}{l_i - l_j} - \frac{\sin(l_i + l_j)b/2}{l_i + l_j},
\end{aligned}$$

(7.272)

$$\begin{aligned}
g_2(l_i, l_j, b/2) &= \int_{-b/2}^{+b/2} dy \cos l_i y \cos l_j y \\
&= \frac{\sin(l_i - l_j)b/2}{l_i - l_j} + \frac{\sin(l_i + l_j)b/2}{l_i + l_j}.
\end{aligned}$$

The deformation potential for these flexural thickness modes has an especially simple form at a non-degenerate Γ-point, namely,

$$\begin{aligned}
H_{\mathrm{def}} &= E_1 \Delta(x, y, z) \\
&= \frac{E_1}{\sqrt{N}} \sum_q \left[\frac{A}{m}(m^2 + l_1^2 + n^2)(\cos mx \sin l_1 y \cos nz) a_q + \mathrm{c.c.} \right].
\end{aligned}$$

(7.273)

In fact, the spatially dependent terms in H_{def} do not depend on α and β.

Chapter 8

Carrier–LO-phonon scattering

Beauty is Nature's brag, and must be shown in courts, at feasts,
and high solemnities where most may wonder at the
workmanship.
John Milton, 1637

8.1 Fröhlich potential for LO phonons in bulk zincblende and würtzite structures

In this section, the Fermi golden rule and the dielectric continuum model of optical phonons are applied to determine the electron–LO-phonon scattering rates in bulk zincblende and bulk würtzite structures.

8.1.1 Scattering rates in bulk zincblende semiconductors

A transition rate for electron–LO-phonon scattering may be estimated by calculating the transition rate predicted by the Fermi golden rule for the perturbation Hamiltonian H_{Fr} of Section 5.2. According to the Fermi golden rule, the transition rate for emission or absorption, $S^{\{e,a\}}(\mathbf{k}, \mathbf{k}')$, from an initial electron state $|\mathbf{k}\rangle$ and an initial phonon state $\left|N_q + \frac{1}{2} \pm \frac{1}{2}\right\rangle$ – denoted by $\left|\mathbf{k}, N_q + \frac{1}{2} \pm \frac{1}{2}\right\rangle$ – to a final electron state $\langle\mathbf{k}'|$ and a final phonon state $\left\langle N_q + \frac{1}{2} \pm \frac{1}{2}\right|$ – denoted by $\left\langle\mathbf{k}', N_q + \frac{1}{2} \pm \frac{1}{2}\right|$ – per unit time per unit volume is given by

$$S^{\{e,a\}}(\mathbf{k}, \mathbf{k}') = \frac{2\pi}{\hbar} \sum_q \left|M^{\{e,a\}}(q)\right|^2 \delta(E(k') - E(k) \pm \hbar\omega), \qquad (8.1)$$

where

$$M^{\{e,a\}}(q) = \left\langle\mathbf{k}', N_q + \frac{1}{2} \pm \frac{1}{2}\right| H_{Fr} \left|\mathbf{k}, N_q + \frac{1}{2} \pm \frac{1}{2}\right\rangle. \qquad (8.2)$$

In these expressions, the positive and negative signs are chosen to select the desired initial and final phonon states. In the case of phonon emission, the final phonon state contains one phonon more than the initial phonon state; accordingly, the initial and final states are selected to be $|\mathbf{k}, N_q\rangle$ and $\langle \mathbf{k}', N_q + \frac{1}{2} \pm \frac{1}{2}|$ respectively. For phonon absorption processes the initial and final states are chosen in a similar manner. In the energy-conserving delta function in $S^{\{e,a\}}(\mathbf{k}, \mathbf{k}')$, the upper sign corresponds to phonon emission and the lower sign to phonon absorption. In this expression, the electron and phonon states represent different spaces so the 'product' $|\mathbf{k}\rangle |N_q\rangle \equiv |\mathbf{k}, N_q\rangle$; that is, calculating the matrix elements of electron states involves only integration over coordinate space and calculating matrix elements of phonon states involves only taking matrix elements of the phonon creation and annihilation operators as in Section 5.1. Clearly these two types of matrix element may be calculated independently of each other. For this reason, the 'product' $|\mathbf{k}\rangle |N_q\rangle \equiv |\mathbf{k}, N_q\rangle$ should not cause any confusion. The phonon states are as defined in Section 5.1 and the Fröhlich Hamiltonian is that of Section 5.2,

$$H_{\text{Fr}} = -i\sqrt{\frac{2\pi e^2 \hbar \omega_{\text{LO}}}{V}\left[\frac{1}{\epsilon(\infty)} - \frac{1}{\epsilon(0)}\right]} \sum_q \frac{1}{q}(a_q e^{i\mathbf{q}\cdot\mathbf{r}} - a_q^\dagger e^{-i\mathbf{q}\cdot\mathbf{r}})$$

$$= \sum_q \frac{C}{q}(a_q e^{i\mathbf{q}\cdot\mathbf{r}} - a_q^\dagger e^{-i\mathbf{q}\cdot\mathbf{r}}), \tag{8.3}$$

where

$$C = -i\sqrt{\frac{2\pi e^2 \hbar \omega_{\text{LO}}}{V}\left[\frac{1}{\epsilon(\infty)} - \frac{1}{\epsilon(0)}\right]}. \tag{8.4}$$

The electron states are plane-wave states normalized in the volume V,

$$|\mathbf{k}\rangle = e^{i\mathbf{k}\cdot\mathbf{r}}/\sqrt{V} \qquad \text{and} \qquad \langle\mathbf{k}'| = e^{-i\mathbf{k}'\cdot\mathbf{r}}/\sqrt{V}. \tag{8.5}$$

The matrix element, $M^{\{e,a\}}(q)$, is then given by

$$M^{\{e,a\}}(q) = \int d^3\mathbf{r} \frac{e^{-i\mathbf{k}'\cdot\mathbf{r}+i\mathbf{k}\cdot\mathbf{r}}}{V} \frac{C}{q}\langle N_q + \frac{1}{2} + \frac{1}{2}\epsilon| (a_q e^{i\mathbf{q}\cdot\mathbf{r}} - a_q^\dagger e^{-i\mathbf{q}\cdot\mathbf{r}})|N_q\rangle$$

$$= \int d^3\mathbf{r} \frac{e^{-i\mathbf{k}'\cdot\mathbf{r}+i\mathbf{k}\cdot\mathbf{r}\mp i\mathbf{q}\cdot\mathbf{r}}}{V} \frac{C}{q}(\mp)(n_q + \frac{1}{2} \pm \frac{1}{2})^{1/2}, \tag{8.6}$$

where the phonon matrix elements have been evaluated using the expressions of Section 5.1; in particular,

$$\langle N_q - 1| a_q |N_q\rangle = \sqrt{n_q} \qquad \text{and} \qquad \langle N_q + 1| a_q^\dagger |N_q\rangle = \sqrt{n_q + 1}. \tag{8.7}$$

In (8.6) the upper signs correspond to phonon emission and the lower signs to phonon absorption. The integral over the factor $e^{-i\mathbf{k}'\cdot\mathbf{r}+i\mathbf{k}\cdot\mathbf{r}\mp i\mathbf{q}\cdot\mathbf{r}}/V$ is, of course, dimensionless and contributes a delta function that depends on the quantity $\mathbf{k}' - \mathbf{k} \pm \mathbf{q}$. Hence, it follows that the final state electron wavevector, $\mathbf{k}'$, is related to the initial state electron wavevector $\mathbf{k}$ and the phonon wavevector $\mathbf{q}$ by $\mathbf{k}' = \mathbf{k} \mp \mathbf{q}$.

The case where $\mathbf{k} = \mathbf{k}' + \mathbf{q}$ corresponds to an electron of wavevector $\mathbf{k}$ in the initial state and an electron and phonon in the final state with wavevectors $\mathbf{k}'$ and $\mathbf{q}$ respectively. That is, this case represents phonon emission or phonon creation.

For the remaining case, $\mathbf{k} + \mathbf{q} = \mathbf{k}'$ and it is clear that the final state represents an electron of wavevector $\mathbf{k}'$ and that the initial state is composed of an electron and phonon with wavevectors $\mathbf{k}$ and $\mathbf{q}$ respectively. Clearly, this case represents phonon absorption or phonon annihilation. The two conditions $\mathbf{k} = \mathbf{k}' + \mathbf{q}$ and $\mathbf{k} + \mathbf{q} = \mathbf{k}'$ are equivalent to momentum conservation, since the product of each wavevector times $\hbar$ is the momentum of the particle in question. The integral over the factor $\exp(-i\mathbf{k}' \cdot \mathbf{r} + i\mathbf{k} \cdot \mathbf{r} \mp i\mathbf{q} \cdot \mathbf{r})/V$ may be written as a Kronecker delta function requiring $\mathbf{k}' - \mathbf{k} \pm \mathbf{q} = 0$ or it may be expressed as $V^{-1}\delta(\mathbf{k}' - \mathbf{k} \pm \mathbf{q})$. Then it follows that

$$\sum_q \left| M^{\{e,a\}}(q) \right|^2 = \int \frac{d^3q}{(2\pi)^3} \frac{|C|^2}{q^2} \left(n_q + \tfrac{1}{2} \pm \tfrac{1}{2} \right) \delta(\mathbf{k}' - \mathbf{k} \pm \mathbf{q})$$

$$= \frac{|C|^2}{4\pi\hbar^2} \frac{1}{|\mathbf{k} - \mathbf{k}'|^2} \left(n_q + \tfrac{1}{2} \pm \tfrac{1}{2} \right), \tag{8.8}$$

where

$$|C|^2 = \frac{2\pi}{V} e^2 \hbar\omega \left[\frac{1}{\epsilon(\infty)} - \frac{1}{\epsilon(0)} \right]. \tag{8.9}$$

In evaluating the integral over q it has been assumed that the frequency of the phonon mode, ω_q, is independent of q and is equal to the zone-center phonon frequency, ω_q. This is a good approximation for many materials. For example, in GaAs the LO phonon frequency varies by about 10% over the entire Brillouin zone. Moreover, since the largest contributions to the integral occur near the zone center – as a result of the q^{-1} dependence in the integrand – the approximation that $\omega_q = \omega_{LO}$ causes little error in the value of the integral in question.

Accordingly, the transition rate $S^{\{e,a\}}(\mathbf{k}, \mathbf{k}')$ is given by

$$S^{\{e,a\}}(\mathbf{k}, \mathbf{k}') = \frac{|C|^2}{4\pi\hbar^2} \frac{1}{|\mathbf{k} - \mathbf{k}'|^2} \left(n_q + \tfrac{1}{2} \pm \tfrac{1}{2} \right) \delta(E(\mathbf{k}') - E(\mathbf{k}) \pm \hbar\omega)$$

$$= \frac{|C|^2}{4\pi\hbar^2} \frac{1}{q^2} \left(n_q + \tfrac{1}{2} \pm \tfrac{1}{2} \right) \delta(E(\mathbf{k}') - E(\mathbf{k}) \pm \hbar\omega), \tag{8.10}$$

where $q^2 = |\mathbf{q}|^2$ and the delta function expresses the conservation of energy, $E(\mathbf{k}') - E(\mathbf{k}) = \mp\hbar\omega$, the upper sign representing phonon emission and the lower sign representing phonon absorption. The $|\mathbf{q}|^{-2}$ dependence in $S^{\{e,a\}}(\mathbf{k}, \mathbf{k}')$ is the same as that appearing in the Coulomb interaction. Accordingly, it is anticipated that finite scattering rates will be obtained only if screening is treated properly. Therefore, $S^{\{e,a\}}(\mathbf{k}, \mathbf{k}')$ is rewritten as

$$S^{\{e,a\}}(\mathbf{k}, \mathbf{k}') = \frac{|C|^2}{4\pi\hbar^2} \frac{q^2}{(q^2 + q_0^2)^2} \left(n_q + \tfrac{1}{2} \pm \tfrac{1}{2}\right) \delta(E(\mathbf{k}') - E(\mathbf{k}) \pm \hbar\omega),$$

(8.11)

where q_0 is a cut-off parameter introduced to represent the effect of screening of the Coulomb-like interaction. The final scattering rate will be calculated in the limit $q_0 \to 0$.

The total, three-dimensional, scattering rate, $1/\tau_{3D}^{\{e,a\}}(\mathbf{k})$, associated with $S_{3D}^{\{e,a\}}(\mathbf{k}, \mathbf{k}')$ results from integrating $S_{3D}^{\{e,a\}}(\mathbf{k}, \mathbf{k}')$ over $\mathbf{q}$ and multiplying by the volume of the heterostructure, V:

$$\frac{1}{\tau_{3D}^{\{e,a\}}(\mathbf{k})} = \frac{|C|^2 V}{4\pi^2\hbar} \left(n_q + \tfrac{1}{2} \pm \tfrac{1}{2}\right)$$
$$\times \int_0^{2\pi} d\varphi \int_0^{\pi} d\theta \sin\theta \int dq\, q^2 \frac{q^2}{(q^2 + q_0^2)^2} \delta(E(\mathbf{k}') - E(\mathbf{k}) \pm \hbar\omega).$$

(8.12)

The integral over φ contributes a factor of 2π since there is no φ-dependence in the problem at hand. In performing the integration over θ it is necessary to consider the fact that the argument of the delta function depends on θ. By assuming parabolic carrier bands and considering the case where the initial carrier wavevector is $\mathbf{k}$ and the carrier mass is m, the argument of the delta function may be rewritten as follows:

$$E(\mathbf{k}') - E(\mathbf{k}) \pm \hbar\omega = E(\mathbf{k} \mp \mathbf{q}) - E(\mathbf{k}) \pm \hbar\omega$$
$$= \frac{\hbar^2 (\mathbf{k} \mp \mathbf{q})^2}{2m} - \frac{\hbar^2 (\mathbf{k})^2}{2m} \pm \hbar\omega$$
$$= \frac{\hbar^2}{2m} \left[(\mathbf{k}')^2 - (\mathbf{k})^2 \right] \pm \hbar\omega$$
$$= \frac{\hbar^2}{2m} (q^2 \mp 2qk\cos\theta) \pm \hbar\omega.$$

(8.13)

The integral over θ is now performed by use of the relation

$$\int g(\theta)\delta[f(\theta) - a]dy = \frac{g(\theta)}{df/d\theta}\bigg|_{\theta=\theta_0},$$

(8.14)

where θ_0 is determined by $f(\theta_0) = a$. Then, taking

$$f(\theta) = \mp \frac{\hbar^2}{2m} (2qk\cos\theta),$$

(8.15)

it follows by letting $u = \cos\theta$ that

$$\int_0^{2\pi} d\theta \, \sin\theta \, \delta\left(\frac{\hbar^2}{2m}(q^2 \mp 2qk\cos\theta) \pm \hbar\omega\right)$$

$$= \int_{-1}^{1} du \, \delta\left(\frac{\hbar^2}{2m}(\mp 2qku) + \frac{\hbar^2}{2m}q^2 \pm \hbar\omega\right)$$

$$= \mp\frac{m}{\hbar^2 qk}, \tag{8.16}$$

so that

$$\frac{1}{\tau_{3D}^{\{e,a\}}(\mathbf{k})} = \frac{|C|^2 V}{4\pi^2\hbar}\left(\frac{2\pi m}{\hbar^2 k}\right)\left(n_q + \frac{1}{2} \pm \frac{1}{2}\right)\int_{q_{min}^{\{e,a\}}}^{q_{max}^{\{e,a\}}} dq\, \frac{q^3}{(q^2 + q_0^2)^2},$$

$$= \frac{|C|^2 V}{2\pi}\frac{m}{\hbar^3 k}\left(n_q + \frac{1}{2} \pm \frac{1}{2}\right)\int_{q_{min}^{\{e,a\}}}^{q_{max}^{\{e,a\}}} dq\, \frac{q^3}{(q^2 + q_0^2)^2}. \tag{8.17}$$

We consider the ranges of q associated with emission and absorption, $\{q_{min}^{\{e\}}, q_{max}^{\{e\}}\}$ and $\{q_{min}^{\{a\}}, q_{max}^{\{a\}}\}$ respectively. The range of q is determined by

$$q^2 - 2qk\cos\theta \pm \frac{2m\omega}{\hbar} = 0, \tag{8.18}$$

where the upper sign corresponds to emission and the lower sign to absorption. The roots of this equation are

$$q^{\{e\}} = k\cos\theta \pm \sqrt{k^2\cos^2\theta - \left(\frac{2m\omega}{\hbar}\right)}, \tag{8.19}$$

and

$$q^{\{a\}} = -k\cos\theta \pm \sqrt{k^2\cos^2\theta + \left(\frac{2m\omega}{\hbar}\right)}. \tag{8.20}$$

Hence, by taking $\cos\theta = \pm 1$, the minimum and maximum values are found:

$$\{q_{min}^{\{e\}}, q_{max}^{\{e\}}\} = \left\{k - \sqrt{k^2 - \left(\frac{2m\omega}{\hbar}\right)}, \; k + \sqrt{k^2 - \left(\frac{2m\omega}{\hbar}\right)}\right\}, \tag{8.21}$$

$$\{q_{min}^{\{a\}}, q_{max}^{\{a\}}\} = \left\{\sqrt{k^2 + \left(\frac{2m\omega}{\hbar}\right)} - k, \; \sqrt{k^2 + \left(\frac{2m\omega}{\hbar}\right)} + k\right\}. \tag{8.22}$$

Then the total scattering rate may be written as

$$\frac{1}{\tau_{3D}^{\{e,a\}}(\mathbf{k})} = \alpha\omega\frac{1}{k}\sqrt{\frac{2m\omega}{\hbar}}(n_q + 1)\int_{q_{min}^{\{e\}}}^{q_{max}^{\{e\}}} dq\, \frac{q^3}{(q^2 + q_0^2)^2}$$

$$+ \alpha\omega\frac{1}{k}\sqrt{\frac{2m\omega}{\hbar}}n_q\int_{q_{min}^{\{a\}}}^{q_{max}^{\{a\}}} dq\, \frac{q^3}{(q^2 + q_0^2)^2}, \tag{8.23}$$

where the relation

$$\frac{|C|^2 V}{2\pi} \frac{m}{\hbar^3 k} = \alpha\omega \frac{1}{k}\sqrt{\frac{2m\omega}{\hbar}} \tag{8.24}$$

with

$$\alpha = \frac{1}{2\hbar\omega} \frac{e^2}{\sqrt{\hbar/2m\omega}} \left[\frac{1}{\epsilon(\infty)} - \frac{1}{\epsilon(0)}\right] \tag{8.25}$$

has been used to replace $|C|^2$ by α. The scattering rate, $1/\tau_{3D}^{\{e,a\}}(\mathbf{k})$, may be evaluated analytically through use of the identity

$$\int_{q_{min}^{\{e,a\}}}^{q_{max}^{\{e,a\}}} dq \frac{q^3}{(q^2 + q_0^2)^2} = \frac{1}{2}\left[\ln\frac{q_{max}^{\{e,a\}2} + q_0^2}{q_{min}^{\{e,a\}2} + q_0^2} + \frac{q_0^2}{2}\frac{q_{min}^{\{e,a\}2} - q_{max}^{\{e,a\}2}}{\left(q_{max}^{\{e,a\}2} + q_0^2\right)\left(q_{min}^{\{e,a\}2} + q_0^2\right)}\right]. \tag{8.26}$$

Ignoring the terms of $\mathcal{O}(q_0^2)$ it follows that

$$\frac{1}{\tau_{3D}^{\{e,a\}}(\mathbf{k})} = \frac{\alpha\omega}{2}\frac{1}{k}\sqrt{\frac{2m\omega}{\hbar}}(n_q + 1)\ln\frac{\left(k + \sqrt{k^2 - 2m\omega/\hbar}\right)^2 + q_0^2}{\left(k - \sqrt{k^2 - 2m\omega/\hbar}\right)^2 + q_0^2}$$

$$+ \frac{\alpha\omega}{2}\frac{1}{k}\sqrt{\frac{2m\omega}{\hbar}}n_q\ln\frac{\left(\sqrt{k^2 + 2m\omega/\hbar} + k\right)^2 + q_0^2}{\left(\sqrt{k^2 - 2m\omega/\hbar} - k\right)^2 + q_0^2}. \tag{8.27}$$

Finally, in the limit as $q_0 \to 0$

$$\frac{1}{\tau_{3D}^{\{e,a\}}(\mathbf{k})} = \frac{\alpha\omega}{2}\frac{1}{k}\sqrt{\frac{2m\omega}{\hbar}}(n_q + 1)\ln\frac{k + \sqrt{k^2 - 2m\omega/\hbar}}{k - \sqrt{k^2 - 2m\omega/\hbar}}$$

$$+ \frac{\alpha\omega}{2}\frac{1}{k}\sqrt{\frac{2m\omega}{\hbar}}n_q\ln\frac{\sqrt{k^2 + 2m\omega/\hbar} + k}{\sqrt{k^2 - 2m\omega/\hbar} - k}. \tag{8.28}$$

8.1.2   Scattering rates in bulk würtzite semiconductors

The Fröhlich interaction Hamiltonian given in Section 7.1 for a polar uniaxial (UA) crystal may be written as

$$H_{\text{Fr}}^{\text{UA}} = \sum_q (-e)\phi(q)\, e^{i\mathbf{q}\cdot\mathbf{r}}(a_q + a_{-q}^\dagger) \tag{8.29}$$

$$= i\sum_q \sqrt{\frac{2\pi e^2\hbar}{V\omega_q}}\frac{1}{q}e^{i\mathbf{q}\cdot\mathbf{r}}(a_q + a_{-q}^\dagger)(\omega_{\perp,\text{TO}}^2 - \omega_q^2)(\omega_{\text{TO},\parallel}^2 - \omega_q^2)$$

$$\times \left\{ [\epsilon(0)_\perp - \epsilon(\infty)_\perp]\omega_{\perp,\text{TO}}^2(\omega_{\parallel,\text{TO}}^2 - \omega_q^2)^2\sin^2\theta \right.$$

$$\left. + [\epsilon(0)_\parallel - \epsilon(\infty)_\parallel]\omega_{\parallel,\text{TO}}^2(\omega_{\perp,\text{TO}}^2 - \omega_q^2)^2\cos^2\theta \right\}^{-1/2}$$

$$= i\sum_q \left\{ \frac{4\pi e^2\hbar V^{-1}}{(\partial/\partial\omega)[\epsilon(\omega)_\perp\sin^2\theta + \epsilon(\omega)_\parallel\cos^2\theta]} \right\}^{1/2}\frac{1}{q}e^{i\mathbf{q}\cdot\mathbf{r}}(a_q + a_{-q}^\dagger), \tag{8.30}$$

where

$$\epsilon_\perp(\omega) = \epsilon_\perp(\infty)\frac{\omega^2 - \omega_{\perp,\text{LO}}^2}{\omega^2 - \omega_{\perp,\text{TO}}^2}, \qquad \epsilon_\parallel(\omega) = \epsilon_\parallel(\infty)\frac{\omega^2 - \omega_{\parallel,\text{LO}}^2}{\omega^2 - \omega_{\parallel,\text{TO}}^2}, \tag{8.31}$$

and θ is the angle between the phonon wavevector $\mathbf{q}$ and the c-axis. Moreover, from Section 3.2, equation (3.21),

$$\epsilon_\perp(\omega)\sin^2\theta + \epsilon_\parallel(\omega)\cos^2\theta = 0. \tag{8.32}$$

As was discussed in Section 3.2, the high-frequency electronic response of a medium should not depend strongly on the crystalline structures; it is usually assumed (Loudon, 1964) that $\epsilon_\perp(\infty) \approx \epsilon_\parallel(\infty)$. Thus

$$\frac{\omega_{\text{LO},\perp}^2 - \omega^2}{\omega_{\text{TO},\perp}^2 - \omega^2}\sin^2\theta + \frac{\omega_{\text{LO},\parallel}^2 - \omega^2}{\omega_{\text{TO},\parallel}^2 - \omega^2}\cos^2\theta = 0 \tag{8.33}$$

or, equivalently,

$$\omega^4 - (\omega_1^2 + \omega_2^2)\omega^2 + \omega_{\text{TO},\perp}^2\omega_{\text{LO},\parallel}^2\cos^2\theta + \omega_{\text{LO},\perp}^2\omega_{\text{TO},\parallel}^2\sin^2\theta = 0 \tag{8.34}$$

where

$$\omega_1^2 = \omega_{\text{TO},\parallel}^2\sin^2\theta + \omega_{\text{TO},\perp}^2\cos^2\theta, \qquad \omega_2^2 = \omega_{\text{LO},\parallel}^2\cos^2\theta + \omega_{\text{LO},\perp}^2\sin^2\theta. \tag{8.35}$$

In Section 3.2 it was shown that when $|\omega_{\text{TO},\parallel} - \omega_{\text{TO},\perp}| \ll \omega_{\text{LO},\parallel} - \omega_{\text{TO},\parallel}$ and $\omega_{\text{LO},\perp} - \omega_{\text{TO},\perp}$ this equation has roots

$$\omega^2 = \tfrac{1}{2}\left\{ (\omega_1^2 + \omega_2^2) \pm [(\omega_1^2 - \omega_2^2) + 2\Delta\omega^2(\theta)] \right\}, \tag{8.36}$$

where

$$\Delta\omega^2(\theta) = 2\frac{(\omega^2_{LO,\parallel} - \omega^2_{LO,\perp})(\omega^2_{TO,\parallel} - \omega^2_{TO,\perp})}{\omega^2_2 - \omega^2_1}\sin^2\theta\cos^2\theta; \qquad (8.37)$$

thus, one root is

$$\Omega^2_{TO} = \omega^2 = \omega^2_{TO,\parallel}\sin^2\theta + \omega^2_{TO,\perp}\cos^2\theta$$
$$- \frac{(\omega^2_{LO,\parallel} - \omega^2_{LO,\perp})(\omega^2_{TO,\parallel} - \omega^2_{TO,\perp})}{\omega^2_2 - \omega^2_1}\sin^2\theta\cos^2\theta$$
$$\approx \omega^2_{TO,\parallel}\sin^2\theta + \omega^2_{TO,\perp}\cos^2\theta \qquad (8.38)$$

and the other root is

$$\Omega^2_{LO} = \omega^2 = \omega^2_{LO,\parallel}\cos^2\theta + \omega^2_{LO,\perp}\sin^2\theta$$
$$+ \frac{(\omega^2_{LO,\parallel} - \omega^2_{LO,\perp})(\omega^2_{TO,\parallel} - \omega^2_{TO,\perp})}{\omega^2_2 - \omega^2_1}\sin^2\theta\cos^2\theta$$
$$\approx \omega^2_{LO,\parallel}\cos^2\theta + \omega^2_{LO,\perp}\sin^2\theta. \qquad (8.39)$$

The inequalities assumed for the derivation of Ω^2_{TO} and Ω^2_{LO} are satisfied for the parameters of both GaN and AlN as may be verified from the numerical values of the parameters given in Section 3.3. For the case of GaN, the dependence for the phonon frequencies on θ is shown in Figure 8.1 for these infrared-active phonons. θ is the angle between the phonon wavevector $\mathbf{q}$ and the c-axis.

As in subsection 8.1.1, the transition rate for electron–optical-phonon scattering may be estimated by calculating the transition rate predicted by the Fermi golden rule for the perturbation Hamiltonian, H^{UA}_{Fr}. According to the Fermi golden rule, the transition rate, $S^{\{e,a\}}_{UA}(\mathbf{k}, \mathbf{k}')$, from an initial electron state $|\mathbf{k}\rangle$ and an initial phonon

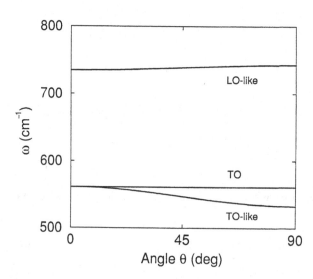

Figure 8.1. Angular dependence of the phonon frequencies ω for LO-like and TO-like infrared-active phonons in GaN. θ is the angle between the phonon wavevector $\mathbf{q}$ and the c-axis. From Lee *et al.* (1997), American Physical Society, with permission.

state $\left| N_q + \frac{1}{2} \pm \frac{1}{2} \right\rangle$ – denoted by $\left| \mathbf{k}, N_q + \frac{1}{2} \pm \frac{1}{2} \right\rangle$ – to a final electron state $\langle \mathbf{k}' |$ and a final phonon state $\left\langle N_q + \frac{1}{2} \pm \frac{1}{2} \right|$ – denoted by $\left\langle \mathbf{k}', N_q + \frac{1}{2} \pm \frac{1}{2} \right|$ – per unit time per unit volume is given by

$$S_{\text{UA}}^{\{e,a\}}(\mathbf{k}, \mathbf{k}') = \frac{2\pi}{\hbar} \sum_q \left| M_{\text{UA}}^{\{e,a\}}(q) \right|^2 \delta(E(k') - E(k) \pm \hbar\omega), \qquad (8.40)$$

where

$$M_{\text{UA}}^{\{e,a\}}(q) = \left\langle \mathbf{k}', N_q + \frac{1}{2} \pm \frac{1}{2} \left| H_{\text{Fr}}^{\text{UA}} \right| \mathbf{k}, N_q + \frac{1}{2} \pm \frac{1}{2} \right\rangle. \qquad (8.41)$$

As in subsection 8.1.1, the electron states are plane-wave states normalized in volume V,

$$|\mathbf{k}\rangle = e^{i\mathbf{k}\cdot\mathbf{r}}/\sqrt{V} \qquad \text{and} \qquad \langle\mathbf{k}'| = e^{-i\mathbf{k}'\cdot\mathbf{r}}/\sqrt{V}. \qquad (8.42)$$

For $\omega^2 = \Omega_{\text{TO}}^2$, the matrix element, $\left| M_{\text{UA}}^{\{e,a\},\text{TO}}(q) \right|^2$, is then given by (Lee *et al.*, 1997)

$$\left| M_{\text{UA}}^{\{e,a\},\text{TO}}(q) \right|^2$$

$$= \frac{2\pi e^2 \hbar}{V q^2 \Omega_{\text{TO}}} \frac{(\omega_{\perp,\text{TO}}^2 - \omega_{\parallel,\text{TO}}^2)^2 \sin^2\theta \cos^2\theta}{[\epsilon_\perp(0) - \epsilon_\perp(\infty)] \omega_{\perp,\text{TO}}^2 \cos^2\theta + [\epsilon_\parallel(0) - \epsilon_\parallel(\infty)] \omega_{\parallel,\text{TO}}^2 \sin^2\theta}$$

$$\times \left(n_q + \frac{1}{2} \pm \frac{1}{2} \right). \qquad (8.43)$$

This matrix element does not vanish, in general, since the TO-like mode is in reality not a pure TO mode in a uniaxial material. In the limit of an isotropic material, $\omega_{\perp,\text{TO}} = \omega_{\parallel,\text{TO}}$ and the transverse matrix element vanishes. Likewise, for $\omega^2 = \Omega_{\text{LO}}^2$, the matrix element $\left| M_{\text{UA}}^{\{e,a\},\text{LO}}(q) \right|^2$ is given by

$$\left| M_{\text{UA}}^{\{e,a\},\text{LO}}(q) \right|^2 = \frac{2\pi e^2 \hbar}{V q^2 \Omega_{\text{LO}}} \left[\frac{\sin^2\theta}{[1/\epsilon_\perp(\infty) - 1/\epsilon_\perp(0)] \omega_{\perp,\text{LO}}^2} \right.$$

$$\left. + \frac{\cos^2\theta}{[1/\epsilon_\parallel(\infty) - 1/\epsilon_\parallel(0)] \omega_{\parallel,\text{LO}}^2} \right]^{-1}$$

$$\times \left(n_q + \frac{1}{2} \pm \frac{1}{2} \right). \qquad (8.44)$$

In the isotropic limit, $\omega_{\perp,\text{LO}} = \omega_{\parallel,\text{LO}}$ and the longitudinal matrix element reduces to

$$\left| M_{\text{UA}}^{\{e,a\},\text{LO}}(q) \right|^2 = \frac{2\pi e^2 \hbar \omega_{\text{LO}}}{V q^2} [1/\epsilon_\perp(\infty) - 1/\epsilon_\perp(0)] \left(n_q + \frac{1}{2} \pm \frac{1}{2} \right). \quad (8.45)$$

Numerical evaluation of $|M_q| = \sqrt{\left| M_{\text{UA}}^{\{a\}}(q) \right|^2}$ indicates that the LO-like and TO-like contributions to the absorption matrix element for bulk GaN are as shown in Figure 8.2.

The scattering rates for LO-like and TO-like phonons have been evaluated numerically by Lee *et al.* (1997) for two cases: as a function of the incident angle of the electron with respect to the c-axis, θ_k, for an initial electron energy of 0.3 eV; and for $\theta_k = 0$ and $\theta_k = \pi/2$, for a range of electron energies. These results are depicted in Figures 8.3 and 8.4. In Figure 8.3 the scattering rates for emission and absorption of LO-like and TO-like phonons are presented as a function of the incident angle of the electron with respect to the c-axis, θ_k, for an initial electron energy of 0.3 eV. In Figure 8.4 the scattering rates for LO-like and TO-like absorption are presented for a range of incident electron energies for $\theta_k = 0$ and $\theta_k = \pi/2$.

8.2   Fröhlich potential in quantum wells

The Fröhlich interaction Hamiltonian of equation (5.34) in Section 5.2 describes the carrier–polar-optical-phonon interaction in the bulk. As discussed in subsection 7.3.2, the Fröhlich Hamiltonian for the two-dimensional slab takes the form (7.71). Moreover, the Fröhlich interaction Hamiltonian for the IF optical phonon modes in the dielectric slab is given by equation (7.72). The interaction Hamiltonians of subsection 7.3.1 describe the Fröhlich potential for a two-dimensional slab bounded by regions with $\epsilon = 1$, as discussed by Licari and Evrard (1977).

In this section, the Hamiltonians describing carrier–polar-optical-phonon interactions will be used to calculate carrier–LO-phonon scattering rates in the approximation of the Fermi golden rule.

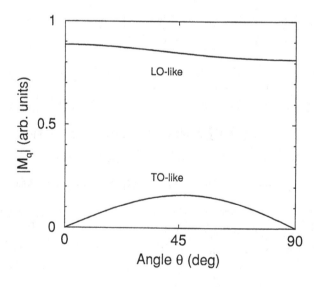

Figure 8.2. Matrix element $|M_q|$ for optical phonon absorption as a function of θ for GaN. θ is the angle between the phonon wavevector **q** and the c-axis. From Lee *et al.* (1997), American Physical Society, with permission.

8.2.1 Scattering rates in zincblende quantum-well structures

A particularly simple treatment of carrier–LO-phonon interactions in quantum-well structures is possible when the carrier confinement is considered in the limit of an infinitely deep confining potential for electrons – known as the so-called extreme quantum limit – and the phonons are taken to be bulk phonons. In this case the calculation of subsection 8.1.1 must be modified to account for the confinement of the charge carriers. As in subsection 8.1.1, the transition rate for electron–LO-phonon scattering may be estimated by calculating the transition rate predicted by the Fermi golden rule for the perturbation Hamiltonian H_{Fr} of Section 5.2. However, the carrier states are now those of an infinitely deep quantum well. According to the Fermi golden rule, the transition rate, $S_{2D}^{\{e,a\}}(\mathbf{k}, \mathbf{k}')$, from an initial electron state $|\mathbf{k}_{2D}\rangle$ and an initial phonon state $\left| N_q + \frac{1}{2} \pm \frac{1}{2} \right\rangle$ – denoted by $\left| \mathbf{k}_{2D}, N_q + \frac{1}{2} + \frac{1}{2} \right\rangle$ – to a final electron state $\langle \mathbf{k}'_{2D} |$ and a final phonon state $\left\langle N_q + \frac{1}{2} \pm \frac{1}{2} \right|$ – denoted by $\left\langle \mathbf{k}'_{2D}, N_q + \frac{1}{2} \pm \frac{1}{2} \right|$ – per unit time per unit heterostructure area A, is given by

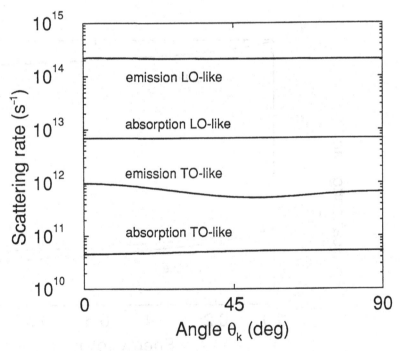

Figure 8.3. Scattering rates for emission and absorption of LO-like and TO-like phonons as a function of the incident angle of the electron with respect to the c-axis, θ_k, for an initial electron energy of 0.3 eV. From Lee *et al.* (1997), American Physical Society, with permission.

$$S_{2D}^{\{e,a\}}(\mathbf{k}, \mathbf{k}') = \frac{2\pi}{\hbar} \sum_q \left| M_{2D}^{\{e,a\}}(q) \right|^2 \delta(E(k') - E(k) \pm \hbar\omega), \qquad (8.46)$$

where

$$M_{2D}^{\{e,a\}}(q) = \left\langle \mathbf{k}'_{2D}, N_q + \tfrac{1}{2} \pm \tfrac{1}{2} \right| H_{\mathrm{Fr}} \left| \mathbf{k}_{2D}, N_q + \tfrac{1}{2} \pm \tfrac{1}{2} \right\rangle. \qquad (8.47)$$

In these expressions, the positive and negative signs are chosen to select the desired initial and final phonon states. In the case of phonon emission, the final phonon state contains one phonon more than the initial phonon state; accordingly, the initial and final states are selected to be $\left| \mathbf{k}_{2D}, N_q \right\rangle$ and $\left\langle \mathbf{k}'_{2D}, N_q + \tfrac{1}{2} \pm \tfrac{1}{2} \right|$ respectively. For phonon absorption processes, the initial and final states are chosen in a comparable manner. In the energy-conserving delta function in $S^{\{e,a\}}(\mathbf{k}, \mathbf{k}')$, the upper sign corresponds to phonon emission and the lower sign to phonon absorption. As in subsection 8.1.1, the electron and phonon states represent different spaces, so we have $|\mathbf{k}_{2D}\rangle |N_q\rangle \equiv |\mathbf{k}_{2D}, N_q\rangle$. The phonon states are as defined in Section 5.1 and the Fröhlich Hamiltonian is given by equation (5.34),

$$H_{\mathrm{Fr}} = \sum_q \frac{C}{q} \left(a_{\mathbf{q}} e^{i\mathbf{q}\cdot\mathbf{r}} - a_{\mathbf{q}}^{\dagger} e^{-i\mathbf{q}\cdot\mathbf{r}} \right), \qquad (8.48)$$

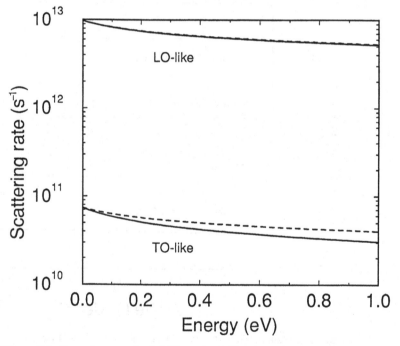

Figure 8.4. Scattering rates for LO-like and TO-like absorption are presented for a range of incident electron energies for $\theta_k = 0$ (solid line) and $\theta_k = \pi/2$ (broken line). From Lee *et al.* (1997), American Physical Society, with permission.

where

$$C = -i \sqrt{\frac{2\pi e^2 \hbar \omega_{LO}}{V} \left[\frac{1}{\epsilon(\infty)} - \frac{1}{\epsilon(0)} \right]}. \tag{8.49}$$

The electron states are taken as plane waves in the directions parallel to the heterointerfaces and as the ground state of an electron in an infinitely deep quantum well in the z-direction (Leburton, 1984),

$$|\mathbf{k}_{2D}\rangle = \frac{e^{i\mathbf{k}_\parallel \cdot \mathbf{r}_\parallel}}{\sqrt{A}} \sqrt{\frac{2}{L_z}} \sin \frac{\pi z}{L_z} \quad \text{and} \quad \langle \mathbf{k}'_{2D}| = \frac{e^{-i\mathbf{k}'_\parallel \cdot \mathbf{r}_\parallel}}{\sqrt{A}} \sqrt{\frac{2}{L_z}} \sin \frac{\pi z}{L_z}, \tag{8.50}$$

where A is the area of the heterointerface over which the electron wavefunction is normalized.

Following the procedures of subsection 8.1.1, the matrix element $M_{2D}^{\{e,a\}}(q)$ is then given by

$$M_{2D}^{\{e,a\}}(q) = \int d^3 r \frac{e^{-i\mathbf{k}'_\parallel \cdot \mathbf{r} + i\mathbf{k}_\parallel \cdot \mathbf{r}}}{\left(\sqrt{A}\right)^2} \frac{2}{L_z} \left(\sin^2 \frac{\pi z}{L_z} \right) \frac{C}{q}$$

$$\times \left\langle N_q + \tfrac{1}{2} \pm \tfrac{1}{2} \right| (a_q e^{i\mathbf{q}\cdot\mathbf{r}} - a_q^\dagger e^{-i\mathbf{q}\cdot\mathbf{r}}) \left| N_q + \tfrac{1}{2} \pm \tfrac{1}{2} \right\rangle$$

$$= \frac{C}{q} (\mp)(n_q + \tfrac{1}{2} \pm \tfrac{1}{2})^{1/2} \delta_{\mathbf{k}'_\parallel - \mathbf{k}_\parallel \pm \mathbf{q}_\parallel} F(q_z L_z), \tag{8.51}$$

where $F(q_z L_z)$ is defined as

$$F(q_z L_z) = \frac{2}{L_z} \int_0^{L_z} dz \, e^{-i q_z z} \sin^2 \frac{\pi z}{L_z}$$

$$= \frac{\pi^2}{\pi^2 - (q_z L_z/2)^2} \frac{\sin q_z L_z/2}{q_z L_z/2} e^{-i q_z L_z/2}. \tag{8.52}$$

Then, following the procedure of subsection 8.1.1, it follows that

$$S_{2D}^{\{e,a\}}(\mathbf{k}_\parallel, \mathbf{k}'_\parallel) = \frac{L_z |C|^2}{4\pi \hbar^2} (n_q + \tfrac{1}{2} \pm \tfrac{1}{2}) I_{2D}(\mathbf{q}_\parallel, L_z)$$

$$\times \delta_{\mathbf{k}'_\parallel - \mathbf{k}_\parallel \pm \mathbf{q}_\parallel} \delta(E(\mathbf{k}'_\parallel \mp \mathbf{q}_\parallel) - E(\mathbf{k}_\parallel) \pm \hbar \omega), \tag{8.53}$$

where

$$|C|^2 = \frac{2\pi}{V} e^2 \hbar \omega \left[\frac{1}{\epsilon(\infty)} - \frac{1}{\epsilon(0)} \right], \tag{8.54}$$

and

$$I_{2D}(\mathbf{q}_\|, L_z) = \int_{-\infty}^{+\infty} dq_z \frac{|F(q_z L_z)|^2}{q_\|^2 + q_z^2}$$

$$= \frac{\pi}{2} L_z \left\{ \frac{1}{S^2}[1 - G(S)] + \left[\frac{1}{2} + \left(2 - \frac{S^2}{S^2 + \pi^2} \right) G(S) \right] \frac{1}{S^2 + \pi^2} \right\}$$

$$(8.55)$$

with

$$G(S) = [1 - e^{-2S}]/2S \qquad \text{and} \qquad S = q_\| L_z/2. \qquad (8.56)$$

As in subsection 8.1.1, it has been assumed that the frequency of the phonon mode, ω_q, is independent of q and is equal to the zone-center phonon frequency, ω_q.

The total scattering rate, $1/\tau_{2D}^{\{e,a\}}(\mathbf{k}_\|)$, associated with $S_{2D}^{\{e,a\}}(\mathbf{k}_\|, \mathbf{k}_\|')$ results from integrating $S_{2D}^{\{e,a\}}(\mathbf{k}_\|, \mathbf{k}_\|')$ over $\mathbf{q}_\|$ and multiplying by the area of the heterostructure, V_{2D}:

$$\frac{1}{\tau_{2D}^{\{e,a\}}(\mathbf{k}_\|)} = \frac{|C|^2 L_z V_{2D}}{4\pi^2 \hbar} \left(n_q + \tfrac{1}{2} \pm \tfrac{1}{2}\right)$$

$$\times \int d\phi \int dq_\| \, I_{2D}(\mathbf{q}_\|, L_z) \delta(E(\mathbf{k}_\|' \mp \mathbf{q}_\|) - E(\mathbf{k}_\|) \pm \hbar\omega).$$

$$(8.57)$$

To evaluate this integral, it is necessary to consider the argument of the delta function. By assuming parabolic carrier bands and considering the case where a carrier of wavevector $\mathbf{k}_\|$ and mass m emits a phonon of wavevector $\mathbf{q}_\|$, so that the final carrier wavevector is $\mathbf{k}_\|' = \mathbf{k}_\| - \mathbf{q}_\|$, it follows that

$$E(\mathbf{k}_\|' \mp \mathbf{q}_\|) - E(\mathbf{k}_\|) \pm \hbar\omega$$

$$= \frac{\hbar^2(\mathbf{k}_\| - \mathbf{q}_\|)^2}{2m} + \frac{\hbar^2\pi^2}{2mL_z^2} - \frac{\hbar^2(\mathbf{k}_\|)^2}{2m} - \frac{\hbar^2\pi^2}{2mL_z^2} \pm \hbar\omega$$

$$= \frac{\hbar^2}{2m}[(\mathbf{k}_\|')^2 - (\mathbf{k}_\|)^2] \pm \hbar\omega$$

$$= \frac{\hbar^2}{2m}[(\mathbf{q}_\|)^2 - 2q_\| k_\| \cos\phi] \pm \hbar\omega, \qquad (8.58)$$

where ϕ is the angle between $\mathbf{k}_\|$ and $\mathbf{q}_\|$, the ground state energy for the carrier is $\hbar^2\pi^2/(2mL_z^2)$, and $q_\|$ and $k_\|$ represent the absolute magnitudes of $\mathbf{q}_\|$ and $\mathbf{k}_\|$ respectively. The argument of the delta function vanishes when

$$q_\|^2 - 2q_\| k_\| \cos\phi \pm \frac{2m\omega}{\hbar} = 0, \qquad (8.59)$$

where the upper sign corresponds to emission and the lower sign to absorption. The roots of this equation are

$$q_{\parallel\pm}^{\{e\}} = k_{\parallel} \cos\phi \pm \sqrt{k_{\parallel}^2 \cos^2\phi - \left(\frac{2m\omega}{\hbar}\right)}, \tag{8.60}$$

and

$$q_{\parallel\pm}^{\{a\}} = -k_{\parallel} \cos\phi \pm \sqrt{k_{\parallel}^2 \cos^2\phi + \left(\frac{2m\omega}{\hbar}\right)}. \tag{8.61}$$

Both roots of $q_{\parallel\pm}^{\{e\}}$ are allowed for ϕ when it is in the range $0 < \phi < \phi_{max}$, where $\phi_{max} = \arccos\sqrt{\hbar\omega/E}$ with $E = \hbar^2 k_{\parallel}^2/2m$ but with $q_{\parallel-}^{\{a\}}$ forbidden. For the case of emission,

$$E(\mathbf{k}_{\parallel}' \mp \mathbf{q}_{\parallel}) - E(\mathbf{k}_{\parallel}) \pm \hbar\omega = \frac{\hbar^2}{2m}\left(q_{\parallel} - q_{\parallel+}^{\{e,\}}\right)\left(q_{\parallel} - q_{\parallel-}^{\{e,\}}\right)$$
$$= \frac{\hbar^2}{2m}\left(q_{\parallel}^2 - 2q_{\parallel}k_{\parallel}\cos\phi + \frac{2m\omega}{\hbar}\right). \tag{8.62}$$

The integral over $q_{\parallel}$ is now performed by use of the relation

$$\int dy\, g(y)\delta(f(y) - a) = \frac{g(y)}{df/dy}\bigg|_{y=y_0}, \tag{8.63}$$

where y_0 is determined by $f(y_0) = a$. Then, taking

$$f(q_{\parallel}) = \frac{\hbar^2}{2m}\left(q_{\parallel}^2 - 2q_{\parallel}k_{\parallel}\cos\phi\right), \tag{8.64}$$

it follows that

$$\frac{df(q_{\parallel})}{dq_{\parallel}} = \frac{\hbar^2}{2m}\left(2q_{\parallel} - 2k_{\parallel}\cos\phi\right)$$
$$= \pm\frac{\hbar^2}{m}\sqrt{k_{\parallel}^2\cos^2\phi - \left(\frac{2m\omega}{\hbar}\right)}. \tag{8.65}$$

With these results, it follows that

$$\frac{1}{\tau_{2D}^{\{e\}}(E)}$$
$$= \frac{\alpha\omega}{\pi}(n_q + 1)\int_0^{\phi_{max}} d\phi\, \frac{I_{2D}(\mathbf{q}_{\parallel+}^{\{e\}}(E,\phi))q_{\parallel+}^{\{e\}}(E,\phi) + I_{2D}(\mathbf{q}_{\parallel-}^{\{e\}}(E,\phi))q_{\parallel-}^{\{e\}}(E,\phi)}{\sqrt{(E/\hbar\omega)\cos^2\phi - 1}}, \tag{8.66}$$

and by an analogous derivation (Leburton, 1984),

$$\frac{1}{\tau_{2D}^{\{a\}}(E)} = \frac{\alpha\omega}{\pi}n_q\int_0^{\pi} d\phi\, \frac{I_{2D}(\mathbf{q}_{\parallel+}^{\{a\}}(E,\phi))q_{\parallel+}^{\{a\}}(E,\phi)}{\sqrt{(E/\hbar\omega)\cos^2\phi + 1}}, \tag{8.67}$$

where

$$\alpha = \frac{1}{2\hbar\omega}\frac{e^2}{\sqrt{\hbar/2m\omega}}\left[\frac{1}{\epsilon(\infty)} - \frac{1}{\epsilon(0)}\right]. \tag{8.68}$$

8.2.2   Scattering rates in würtzite quantum wells

Komirenko *et al.* (2000a) have calculated energy-dependent electron scattering rates for electron–LO-phonon scattering in würtzite quantum wells. The Fröhlich potential used to calculate these scattering rates (Komirenko *et al.*, 1999; Lee *et al.*, 1998) is based on the Hamiltonians given in Appendix D and on the Loudon model of uniaxial crystals. In these calculations the Fermi golden rule is used to calculate the scattering rate for GaN free-standing quantum wells and for GaN quantum wells embedded in AlN. In all cases the c-axis is normal to the heterointerfaces and the GaN quantum wells are 50 ångstroms wide. The interacting electrons are restricted to energies close to the Γ point and only the lowest subband is considered. Figures 8.5 and 8.6 present the scattering rates calculated for bulk GaN, for a free-standing GaN quantum well, and for an AlN/GaN/AlN quantum well. Figure 8.5 depicts the total electron–optical-phonon scattering rates in bulk GaN and a free-standing GaN quantum well as functions of the electron energy, E_k, over a range of 0 to about 500 meV; this figure also presents the angular dependence of the scattering rate for bulk GaN crystals. Figure 8.6 depicts the total scattering rate as a function of the electron energy E_k over a range of 0 to 400 meV for a free-standing quantum well and a GaN quantum well embedded in AlN; this figure also indicates the contributions of the confined modes and the interface modes.

These results illustrate clearly the importance of including the effects of dimensional confinement in calculating the electron–optical-phonon scattering rates in dimensionally confined würtzite semiconductor structures.

8.3   Scattering of carriers by LO phonons in quantum wires

In this section, scattering rates for the scattering of charge carriers by LO phonons in quantum wires are calculated in the approximation of the Fermi golden rule.

8.3.1   Scattering rate for bulk LO phonon modes in quantum wires

A particularly simple treatment of carrier–LO-phonon interactions in quantum-wire structures is possible when the phonons are taken to be bulk phonons and the carrier confinement is considered in the limit of an infinitely deep confining potential for electrons, the so-called extreme quantum limit. In this case the calculation of subsection 8.1.1 must be modified to account for the confinement of the charge carriers. As in Sections 8.1.1 and 8.3.1, the transition rate for electron–LO-phonon scattering may be estimated by calculating the transition rate predicted by the Fermi golden rule for the perturbation Hamiltonian, H_{Fr}, of Section 5.2. However, the carrier states are now those of an infinitely deep quantum well. According to the Fermi golden rule, the transition rate, $S_{1D}^{\{e,a\}}(\mathbf{k}, \mathbf{k}')$, from an initial electron state $|k_{1D}\rangle$ and an initial phonon state $\left|N_q + \frac{1}{2} \pm \frac{1}{2}\right\rangle$ – denoted by $\left|k_{1D}, N_q + \frac{1}{2} \pm \frac{1}{2}\right\rangle$ – to a final electron state $\langle k'_{1D}|$ and a final phonon state $\left\langle N_q + \frac{1}{2} + \frac{1}{2}\right|$ – denoted by $\left\langle k'_{1D}, N_q + \frac{1}{2} + \frac{1}{2}\right|$ – per unit time per unit heterostructure area A is given by

$$S_{1D}^{\{e,a\}}(\mathbf{k}, \mathbf{k}') = \frac{2\pi}{\hbar} \sum_q \left| M_{1D}^{\{e,a\}}(q) \right|^2 \delta(E(k') - E(k) \pm \hbar\omega), \qquad (8.69)$$

where

$$M_{1D}^{\{e,a\}}(q) = \left\langle k'_{1D}, N_q + \frac{1}{2} + \frac{1}{2} \right| H_{Fr} \left| k_{1D}, N_q + \frac{1}{2} \pm \frac{1}{2} \right\rangle. \qquad (8.70)$$

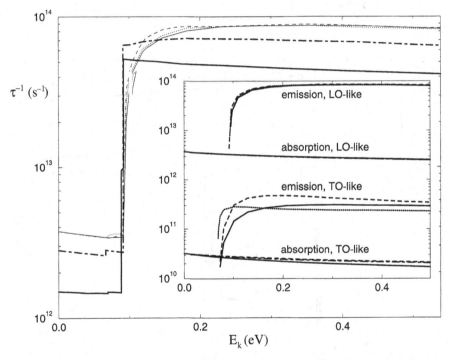

Figure 8.5. Main figure: total electron–optical-phonon scattering rates in bulk GaN (thin lines) and in a free-standing GaN quantum well (thick solid line) as functions of the electron energy E_k over a range of 0 to about 500 meV. Scattering rates at three particular angles for bulk GaN crystals are shown as follows: thin solid lines, $\theta_k = \pi/2$; thin dotted lines, $\theta_k = \pi/4$; thin broken lines, $\theta_k = 0$. The broken and dotted line indicates the scattering rate calculated for confined electrons and bulk phonons.

Inset: the LO-like and TO-like emission and absorption rates in bulk GaN for three different angles: solid lines, $\theta_k = \pi/2$; dotted lines, $\theta_k = \pi/4$; broken lines, $\theta_k = 0$. From Komirenko *et al.* (2000a), American Physical Society, with permission.

In these expressions, the positive and negative signs are chosen to select the desired initial and final phonon states. In the case of phonon emission, the final phonon state contains one phonon more than the initial phonon state; accordingly, the initial and final states are selected to be $\left|k_{1D}, N_q + \frac{1}{2} \pm \frac{1}{2}\right\rangle$ and $\left\langle k'_{1D}, N_q + \frac{1}{2} \pm \frac{1}{2}\right|$ respectively. For phonon absorption processes the initial and final states are chosen in similar manner. In the energy-conserving delta function in $S^{\{e,a\}}(\mathbf{k}, \mathbf{k}')$, the upper sign corresponds to phonon emission and the lower sign corresponds to phonon absorption. As in subsection 8.1.1, the electron and phonon states represent different spaces, so that we have $|k_{1D}\rangle |N_q\rangle \equiv |k_{1D}, N_q\rangle$. Again, the phonon states are as defined in Section 5.1 and the Fröhlich Hamiltonian is given by equation (5.34) of Section 5.2,

$$H_{Fr} = \sum_q \frac{C}{q} \left(a_\mathbf{q} e^{i\mathbf{q}\cdot\mathbf{r}} - a_\mathbf{q}^\dagger e^{-i\mathbf{q}\cdot\mathbf{r}} \right), \tag{8.71}$$

where

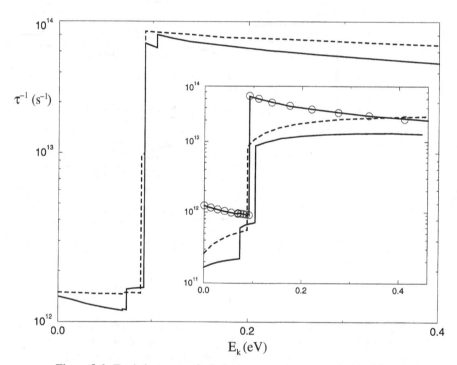

Figure 8.6. Total electron–optical-phonon scattering rates calculated for a free-standing GaN quantum well (broken line) and an AlN/GaN/AlN quantum well (solid line) as functions of the electron energy E_k over a range of 0 to about 400 meV.

Inset: the scattering rates for confined modes (line with circles) and for interface modes for the AlN/GaN/AlN structure (solid line) as well as for the free-standing GaN quantum well (broken line). From Komirenko *et al.* (2000a), American Physical Society, with permission.

$$C = -i\sqrt{\frac{2\pi e^2 \hbar \omega_{LO}}{V}\left[\frac{1}{\epsilon(\infty)} - \frac{1}{\epsilon(0)}\right]}. \tag{8.72}$$

The electron states are taken as plane-wave states along the quantum-wire axis, taken here as the x-axis, and as the ground states of an electron in an infinitely deep quantum wire in the y- and z-directions (Leburton, 1984),

$$|k_{1D}\rangle = \frac{e^{ik_xx}}{\sqrt{L_x}}\sqrt{\frac{2}{L_y}}\left(\sin\frac{\pi y}{L_z}\right)\sqrt{\frac{2}{L_z}}\left(\sin\frac{\pi z}{L_z}\right), \tag{8.73}$$

and

$$A_-(q_z)_\pm = \frac{-i}{\sqrt{2}}[a_\pm(q_x, q_y) - a_\pm(q_x, -q_y)], \tag{8.74}$$

where L_x is the length in the x-direction over which the free-electron wavefunction is normalized and L_y and L_z are the height and the width of the quantum wire; it is assumed that $L_x \gg L_y, L_z$.

Following the procedures of subsection 8.1.1, the matrix element, $M_{1D}^{\{e,a\}}(q)$, is then given by

$$M_{1D}^{\{e,a\}}(q) = \int d^3\mathbf{r}\, \frac{e^{-ik_x'x+ik_xx}}{(\sqrt{L_x})^2}\frac{2}{L_y}\left(\sin^2\frac{\pi y}{L_y}\right)\frac{2}{L_z}\left(\sin^2\frac{\pi z}{L_z}\right)$$

$$\times \frac{C}{q}\langle N_q + \tfrac{1}{2} + \tfrac{1}{2}\epsilon|(a_\mathbf{q}e^{i\mathbf{q}\cdot\mathbf{r}} - a_\mathbf{q}^\dagger e^{-i\mathbf{q}\cdot\mathbf{r}})|N_q\rangle$$

$$= \frac{C}{q}(\mp)(n_q + \tfrac{1}{2} \pm \tfrac{1}{2})^{1/2}\delta'_{k_x'-k_x\pm q_x}F(q_yL_y)F(q_zL_z), \tag{8.75}$$

where $F(q_yL_y)$ and $F(q_zL_z)$ are defined by

$$F(q_yL_y) = \frac{2}{L_y}\int_0^{L_y} dy\, e^{-iq_yy}\sin^2\frac{\pi y}{L_y}$$

$$= \frac{\pi^2}{\pi^2 - (q_yL_y/2)^2}\frac{\sin q_yL_y/2}{(q_yL_y/2)}e^{-iq_yL_y/2},$$

$$F(q_zL_z) = \frac{2}{L_z}\int_0^{L_z} dz\, e^{-iq_zz}\sin^2\frac{\pi z}{L_z} \tag{8.76}$$

$$= \frac{\pi^2}{\pi^2 - (q_zL_z/2)^2}\frac{\sin q_zL_z/2}{(q_zL_z/2)}e^{-iq_zL_z/2}.$$

Then, following the procedure of subsection 8.1.1, it follows that

$$S_{1D}^{\{e,a\}}(k_x, k_x') = \frac{L_z|C|^2}{4\pi\hbar^2}(n_q + \tfrac{1}{2} \pm \tfrac{1}{2})I_{1D}(q_x, L_y, L_z)$$

$$\times \delta_{k_x'-k_x\pm q_x}\delta(E(k_x \mp q_x) - E(k_x) \pm \hbar\omega), \tag{8.77}$$

where

$$|C|^2 = \frac{2\pi}{V} e^2 \hbar\omega \left[\frac{1}{\epsilon(\infty)} - \frac{1}{\epsilon(0)}\right] \tag{8.78}$$

and

$$I_{1D}(q_x, L_y, L_z) = \int_{-\infty}^{+\infty} dq_y \int_{-\infty}^{+\infty} dq_z \frac{|F(q_y L_y)|^2 |F(q_z L_z)|^2}{q_x^2 + q_y^2 + q_z^2}, \tag{8.79}$$

which has been evaluated numerically (Leburton, 1984).

As in subsection 8.1.1, it has been assumed that the frequency of the phonon mode is independent of q and equal to the zone-center phonon frequency, ω_q.

8.3.2 Scattering rate for confined LO phonon modes in quantum wires

The bulk Fröhlich interaction Hamiltonian, H_{Fr}^{3D}, of Section 5.2 may be used to derive straightforwardly the Fröhlich interaction Hamiltonian for a one-dimensional quantum wire, H_{Fr}^{1D}, by an analysis similar to that of subsection 7.3.2, where the Fröhlich interaction Hamiltonian for a quantum well, H_{Fr}^{2D}, was obtained from H_{Fr}^{3D} by imposing the boundary condition that $H_{Fr}^{3D}(x, y, z)\big|_{z=L_z/2} = 0$. In particular, the analysis of subsection 7.3.2 may be extended by applying not just the boundary condition $H_{Fr}^{3D}(x, y, z)\big|_{z=L_z/2} = 0$ but also the boundary condition $H_{Fr}^{3D}(x, y, z)\big|_{y=L_y/2} = 0$. Starting with the canonical form of the bulk Fröhlich interaction Hamiltonian,

$$H_{Fr}^{3D} = -\sqrt{\frac{2\pi e^2 \hbar\omega_{LO}}{V} \left[\frac{1}{\epsilon(\infty)} - \frac{1}{\epsilon(0)}\right]} \sum_q \frac{1}{q} (a_q + a_{-q}^\dagger) e^{-iq\cdot r}, \tag{8.80}$$

we write the sum over q as a sum over positive values of q_z in equation (8.84):

$$H_{Fr} = \sum_{q_\parallel, q_z > 0} V_q e^{-iq_\parallel \cdot r} \big[e^{-iq_z z} \big(a_{q_\parallel, q_z} + a_{-q_\parallel, -q_z}^\dagger \big)$$
$$+ e^{iq_z z} \big(a_{q_\parallel, -q_z} + a_{-q_\parallel, q_z}^\dagger \big) \big], \tag{8.81}$$

where

$$V_q = -\sqrt{\frac{2\pi e^2 \hbar\omega_{LO}}{V} \left[\frac{1}{\epsilon(\infty)} - \frac{1}{\epsilon(0)}\right]}. \tag{8.82}$$

Then, defining

$$a_+(q_\parallel) = \frac{1}{\sqrt{2}} (a_{q_\parallel, q_z} + a_{q_\parallel, -q_z}) \qquad a_-(q_\parallel) = \frac{-i}{\sqrt{2}} (a_{q_\parallel, q_z} - a_{-q_\parallel, q_z}), \tag{8.83}$$

with

$$A_+(q_z)_\pm = \frac{1}{\sqrt{2}}[a_\pm(q_x, q_y) + a_\pm(q_x, -q_y)],$$

$$A_-(q_z)_\pm = \frac{-i}{\sqrt{2}}[a_\pm(q_x, q_y) - a_\pm(q_x, -q_y)],$$

(8.84)

expanding $e^{\pm iq_y y}$ and $e^{\pm iq_z z}$, and taking $q_y = \pm m\pi/L_y$ and $q_z = \pm m\pi/L_z$ to enforce the electromagnetic boundary conditions that the confined phonon modes vanish at $y = \pm L_y/2$ and $z = \pm L_z/2$, leads to the result (Stroscio, 1989)

$$
\begin{aligned}
H_{Fr}^{1D} \\
= 2\alpha' \sum_{q_x} e^{-iq_x x} & \left\{ \sum_{m=1,3,5,\dots} \sum_{n=1,3,5,\dots} \lambda_{mn} \cos\frac{m\pi y}{L_y} \cos\frac{n\pi z}{L_z} [A_+(q_x)_+ + A_+^\dagger(-q_x)_+] \right. \\
& + \sum_{m=1,3,5,\dots} \sum_{n=2,4,6,\dots} \lambda_{mn} \cos\frac{m\pi y}{L_y} \sin\frac{n\pi z}{L_z} [A_+(q_x)_- + A_+^\dagger(-q_x)_-] \\
& + \sum_{m=2,4,6,\dots} \sum_{n=1,3,5,\dots} \lambda_{mn} \sin\frac{m\pi y}{L_y} \cos\frac{n\pi z}{L_z} [A_-(q_x)_+ + A_-^\dagger(-q_x)_+] \\
& + \left. \sum_{m=2,4,6,\dots} \sum_{n=2,4,6,\dots} \lambda_{mn} \sin\frac{m\pi y}{L_y} \sin\frac{n\pi z}{L_z} [A_-(q_x)_- + A_-^\dagger(-q_x)_-] \right\},
\end{aligned}
$$

(8.85)

where

$$\lambda_{mn} = \left[q_x^2 + \left(\frac{m\pi}{L_y}\right)^2 + \left(\frac{n\pi}{L_z}\right)^2 \right]^{-1/2}$$

(8.86)

and

$$\alpha' = \left\{ \frac{2\pi e^2}{V} \hbar\omega \left[\frac{1}{\epsilon(\infty)} - \frac{1}{\epsilon(0)} \right] \right\}^{1/2}.$$

(8.87)

Equation (8.85) may be written more compactly as (Campos et al., 1992),

$$H_{Fr}^{1D} = 2\alpha' \sum_{q_x} e^{-iq_x x} \sum_{m=1}^\infty \sum_{n=1}^\infty \xi_m(y)\xi_n(y)\lambda_{mn}\beta_{mn},$$

(8.88)

where

$$\beta_{mn} = A_{mn}(q_x) + A_{mn}^\dagger(-q_x),$$

(8.89)

$$
\begin{aligned}
A_{mn}(q_x) = \tfrac{1}{2}[(-1)^{m+1} a(q_x, q_y, q_z) + a(q_x, q_y, -q_z) \\
+ (-1)^{n+m} a(q_x, -q_y, q_z) + (-1)^{n+1} a(q_x, -q_y, -q_z)],
\end{aligned}
$$

(8.90)

$$\xi_n(t) = \left(\sin \frac{n\pi t}{L_t} + \frac{\pi}{2}\delta_n \right) \tag{8.91}$$

with $\delta_n = 1$ for odd n and $\delta_n = 0$ for even n.

The Hamiltonian (8.88) includes contributions only from the confined modes of the dielectric continuum model with electrostatic boundary conditions – also known as the slab modes. The interface and half-space modes are not included and must be considered separately. For calculations based on the guided modes, the contributions to the Hamiltonian in the interior of the quantum wire may be written in terms of the Hamiltonian for the slab modes by taking

$$\xi_n(t) = \left(\cos \frac{n\pi t}{L_t} + \frac{\pi}{2}\delta_n \right). \tag{8.92}$$

Of course, it is necessary to augment this expression with terms containing the evanescent contributions of the guided modes in the regions surrounding the quantum wire. In the present account, scattering rates will be calculated within the dielectric continuum model with electrostatic boundary conditions. As discussed in Appendix C, intrasubband and intersubband transitions rates are the same in either the slab or guided models as long as complete sets of orthogonal phonon modes are used in performing the calculation.

As in the previous subsection, the transition rate, $S_{1D}^{\{e,a\}}(\mathbf{k}, \mathbf{k}')$, from an initial electron state $|k_{1D}\rangle$ and an initial phonon state $\left| N_q + \frac{1}{2} \pm \frac{1}{2} \right\rangle$ – denoted by $\left| k_{1D}, N_q + \frac{1}{2} \pm \frac{1}{2} \right\rangle$ – to a final electron state $\langle k'_{1D}|$ and a final phonon state $\left\langle N_q + \frac{1}{2} + \frac{1}{2} \right|$ – denoted by $\left\langle k'_{1D}, N_q + \frac{1}{2} + \frac{1}{2} \right|$ – per unit time per unit length, L, is given by

$$S_{1D}^{\{e,a\}}(\mathbf{k}, \mathbf{k}') = \frac{2\pi}{\hbar} \sum_q \left| M_{1D}^{\{e,a\}}(q) \right|^2 \delta(E(k') - E(k) \pm \hbar\omega), \tag{8.93}$$

where

$$M_{1D}^{\{e,a\}}(q) = \left\langle k'_{1D}, N_q + \frac{1}{2} + \frac{1}{2} \left| H_{Fr}^{1D} \right| k_{1D}, N_q + \frac{1}{2} \pm \frac{1}{2} \right\rangle. \tag{8.94}$$

The electron wavefunction is taken to be

$$|k_{1D}\rangle = \frac{e^{ik_x x}}{\sqrt{L_x}} \sqrt{\frac{2}{L_y}} \left(\cos \frac{\pi y}{L_z} \right) \sqrt{\frac{2}{L_z}} \left(\cos \frac{\pi z}{L_z} \right), \tag{8.95}$$

within the finitely long quantum wire with cross-sectional area defined by $-L_y < y < L_y/2$, $-L_z < z < L_z/2$ and is taken to vanish outside this region. In this so-called extreme quantum limit, the electron energy for the ground state is

$$E_{1D} = \frac{\hbar^2 k_x^2}{2m^*} + \frac{\hbar^2 \pi^2}{2m^*} \left(\frac{1}{L_y^2} + \frac{1}{L_z^2} \right). \tag{8.96}$$

Then the procedures of the last subsection lead straightforwardly to

$$M_{1D,m,n}^{\{e,a\}}(q) = \mp 2\alpha' \delta_{k_x - k_x' \mp q_x} \lambda_{mn} 2 P_{mn} \left(n_q + \tfrac{1}{2} \pm \tfrac{1}{2} \right)^{1/2}, \tag{8.97}$$

where λ_{mn} is given in (8.96) and where (Stroscio, 1989)

$$P_{mn} = \int_{-L_y/2}^{L_y/2} \frac{dy}{L_y/2} \int_{-L_z/2}^{L_z/2} \frac{dz}{L_z/2} \cos^2 \frac{\pi y}{L_y} \cos^2 \frac{\pi z}{L_z}$$

$$\times \begin{cases} \cos \dfrac{m\pi y}{L_y} \cos \dfrac{n\pi z}{L_z} \\[2mm] \cos \dfrac{m\pi y}{L_y} \sin \dfrac{n\pi z}{L_z} \\[2mm] \sin \dfrac{m\pi y}{L_y} \cos \dfrac{n\pi z}{L_z} \\[2mm] \sin \dfrac{m\pi y}{L_y} \sin \dfrac{n\pi z}{L_z} \end{cases} \tag{8.98}$$

The four expressions for P_{mn} correspond to the four terms of (8.85). The numerical values for the dominant values of P_{mn} are: $P_{11} = (8/3\pi)^2$, $P_{13} = P_{31} = (1/5)P_{11}$, $P_{15} = P_{51} = -(1/35)(8/3\pi)^2$, $P_{33} = (1/25)(8/3\pi)^2$, $P_{35} = P_{53} = -(1/175)(8/3\pi)^2$, and $P_{55} = (1/1225)(8/3\pi)^2$. The transition rate is then given by

$$\frac{1}{\tau_{1D}^{\{e,a\}}(k_x)} = \frac{|\alpha'|^2 V}{4\pi^2 \hbar} \left(n_q + \tfrac{1}{2} \pm \tfrac{1}{2} \right) \int_{-\infty}^{+\infty} dq_x \, I_{1D}(q_x, L_y, L_z) \delta(E(k_x \pm q_x) - E(k_x) \pm \hbar\omega), \tag{8.99}$$

where

$$I_{1D}(q_x, L_y, L_z) = \frac{(2\pi)^2}{L_y L_z} \sum_{m=1,3,5,\dots} \sum_{n=1,3,5,\dots} 16 |P_{mn}|^2 \lambda_{mn}. \tag{8.100}$$

Figure 8.7 provides numerical values for $I_{1D}(q_x, L_y, L_z)$ for wires with selected cross-sectional dimensions.

Rewriting the argument of the delta function $\delta(E(k_x \pm q_x) - E(k_x) \pm \hbar\omega)$ through the use of

$$E(k_x \pm q_x) - E(k_x) \pm \hbar\omega = \frac{\hbar^2}{2m^*} (q_x - q_+^{\{e,a\}})(q_x - q_-^{\{e,a\}}), \tag{8.101}$$

where

$$q_\pm^{\{e,a\}} = \epsilon k_x \pm \left[k_x^2 - \epsilon \left(\frac{2m^*\omega}{\hbar} \right) \right]^{1/2} \tag{8.102}$$

with $\epsilon = 1$ for emission and $\epsilon = -1$ for absorption, it follows that

$$\frac{1}{\tau_{1D}^{\{e,a\}}(k_x)} = \frac{\alpha\omega}{2\pi}\left(n_q + \frac{1+\epsilon}{2}\right)\frac{I_{1D}\left(q_+^{\{e,a\}}\right) + I_{1D}\left(q_-^{\{e,a\}}\right)}{\sqrt{E/\hbar\omega - \epsilon}},\tag{8.103}$$

with

$$\alpha = \frac{1}{2\hbar\omega}\frac{e^2}{\sqrt{\hbar/2m\omega}}\left[\frac{1}{\epsilon(\infty)} - \frac{1}{\epsilon(0)}\right].\tag{8.104}$$

8.3.3   Scattering rate for interface-LO phonon modes

The Fröhlich interaction Hamiltonian, H_{IF}^{1D}, associated with the interface phonons in rectangular polar-semiconductor quantum wires has been derived analytically using an approximation method (Stroscio *et al.*, 1990; Kim *et al.*, 1991) involving the separation of variables (Marcatili, 1969) and has been treated exactly by the use of numerical techniques (Knipp and Reinecke, 1992). As a result, it is known that the approximate separation-of-variables approach fails to predict so-called 'corner' modes considered by Knipp and Reinecke – the well-known 'corner' problem. Unlike the confined optical phonon modes of the last subsection, the interface modes do not vanish on the interfaces of the quantum wire. Therefore, it is clear that difficulties will be encountered in the vicinity of the corners when attempting to have consistent boundary conditions on adjacent sides of the quantum wire. Nevertheless,

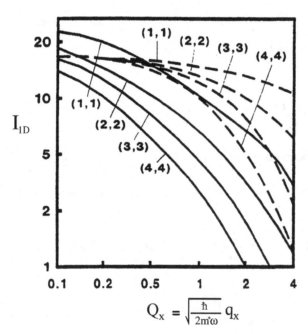

Figure 8.7. Numerical values for $I_{1D}(q_x, L_y, L_z)$ for wires with selected cross-sectional dimensions defined by the four pairs $(L_y/L_0, L_z/L_0) =$ (1, 1), (2, 2), (3, 3), (4, 4), where $L_0 = [\hbar/(2m^*\omega)]^{1/2}$. The solid and broken lines represent results obtained for bulk phonons and for the lowest-order confined modes respectively.

the approximate solution based on separation of variables will be presented since it provides physical insight into the nature of the interface modes. Subsequent to deriving the IF modes in this approximation, the exact treatment of Knipp and Reinecke (1992) will be presented and the corner modes discovered by Knipp and Reinecke will be described.

Taking the quantum wire to have infinite extent in the x-direction, the electrostatic potential, $\Phi(\mathbf{r})$, may be written as a one-dimensional Fourier transform:

$$\Phi(\mathbf{r}) = \sum_{q_x} \Phi(q_x, y, z) e^{-iq_x x}. \tag{8.105}$$

Then

$$\mathbf{E}(\mathbf{r}) = -\nabla\Phi(\mathbf{r}) = \sum_{q_x} \mathbf{E}(q_x, y, z) e^{-iq_x x},$$

$$\mathbf{P}(\mathbf{r}) = \varkappa(\omega)\mathbf{E}(\mathbf{r}) = \sum_{q_x} \mathbf{P}(q_x, y, z) e^{-iq_x x}. \tag{8.106}$$

The normalization condition is given in terms of $u_i(q_x, y, z)$ by

$$\int_{-L_y/2}^{+L_y/2} dy \int_{-L_z/2}^{+L_z/2} dz \left|\sqrt{n_i \mu_i L} u_i(q_x, y, z)\right|^2 = \frac{\hbar}{2\omega}, \tag{8.107}$$

where the subscript i refers to a general material region i. From the results of subsection 7.3.4 for a quantum well, it is clear that

$$\frac{|\mu_i n_i u_i(q_x, y, z)|^2}{\left\{ q_x^2 |\Phi_i(q_x, y, z)|^2 + \left|\dfrac{\partial \Phi_i(q_x, y, z)}{\partial z}\right|^2 \right\}} = \frac{1}{4\pi} \epsilon_i(\infty) \frac{\omega_{LO,i}^2 - \omega_{TO,i}^2}{(\omega^2 - \omega_{TO,i}^2)^2}$$

$$= \frac{1}{4\pi} \frac{1}{2\omega} \frac{\partial \epsilon_i(\omega)}{\partial \omega}, \tag{8.108}$$

holds for the case of a quantum wire and it follows that

$$\sum_i \frac{1}{4\pi} \frac{1}{2\omega} \frac{\partial \epsilon_i(\omega)}{\partial \omega} \int_{R_i} dz \left\{ q_x^2 |\Phi_i(q_x, y, z)|^2 + \left|\dfrac{\partial \Phi_i(q_x, y, z)}{\partial z}\right|^2 \right\} = \frac{\hbar}{2\omega L}, \tag{8.109}$$

where the sum over i is over the two material regions in the problem at hand: $i = 1$ inside and $i = 2$ outside the quantum wire. The Fröhlich interaction Hamiltonian, H_{IF}^{1D}, is then given by

$$H_{IF}^{1D} = -e \sum_{q_x} \Phi(q_x, y, z) e^{-iq_x x} [A_{IF}(q_x) + A_{IF}^\dagger(-q_x)], \tag{8.110}$$

where $A_{IF}(q_x)$ and $A_{IF}^\dagger(-q_x)$ are the annihilation and creation operators for the appropriate IF-phonon modes of the quantum wire and where in the separation-of-variables approximation $\Phi(q_x, y, z)$ has the form

$$\Phi_s(q_x, y, z) = \begin{cases} C\dfrac{\cosh\alpha y}{\cosh\alpha L_y/2}\dfrac{\cosh\beta y}{\cosh\beta L_z/2} & |y| \leq L_y/2, \quad |z| \leq L_z/2, \\[2mm] C\dfrac{\cosh\alpha y}{\cosh\alpha L_y/2}e^{\beta L_z/2}e^{-\beta|z|} & |y| \leq L_y/2, \quad |z| \geq L_z/2, \\[2mm] Ce^{\alpha L_y/2}e^{-\alpha|y|}\dfrac{\cosh\beta y}{\cosh\beta L_z/2} & |y| \geq L_y/2, \quad |z| \leq L_z/2, \\[2mm] Ce^{\alpha L_y/2}e^{-\alpha|y|}e^{\beta L_z/2}e^{-\beta|z|} & |y| \geq L_y/2, \quad |z| \geq L_z/2, \end{cases}$$
(8.111)

for the symmetric IF mode and

$$\Phi_s(q_x, y, z) = \begin{cases} +C\dfrac{\sinh\alpha y}{\sinh\alpha L_y/2}\dfrac{\sinh\beta y}{\sinh\beta L_z/2} & |y| \leq L_y/2, \quad |z| \leq L_z/2, \\[2mm] \mp C\dfrac{\sinh(\alpha y)}{\sinh(\alpha L_y/2)}e^{\beta L_z/2}e^{-\beta|z|} & |y| \leq L_y/2, \quad |z| \geq L_z/2, \\[2mm] \mp Ce^{\alpha L_y/2}e^{-\alpha|y|}\dfrac{\sinh(\beta y)}{\sinh(\beta L_z/2)} & |y| \geq L_y/2, \quad |z| \leq L_z/2, \\[2mm] \pm Ce^{\alpha L_y/2}e^{-\alpha|y|}e^{\beta L_z/2}e^{-\beta|z|} & |y| \geq L_y/2, \quad |z| \geq L_z/2, \end{cases}$$
(8.112)

for the antisymmetric IF modes; in these potentials, α and β satisfy the relation

$$\alpha^2 + \beta^2 - q_x^2 = 0, \tag{8.113}$$

and in (8.112) the plus sign is taken when $yz > 0$ and the minus sign when $yz < 0$. As usual, the dispersion relations are obtained by imposing the boundary conditions. For the symmetric solution this dispersion relation is

$$\epsilon_1(\omega)\tanh\alpha L_y/2 + \epsilon_2(\omega) = 0, \tag{8.114}$$

and for the antisymmetric mode the dispersion relation is

$$\epsilon_1(\omega)\coth\alpha L_y/2 + \epsilon_2(\omega) = 0, \tag{8.115}$$

where

$$\alpha L_y = \beta L_z \tag{8.116}$$

must also be satisfied for both modes. These dispersion relations are of the same forms as those for a quantum well, as is expected since the separation-of-variables approximation has been assumed thus far in this treatment of the optical phonons in quantum wires. The numerical solutions for these dispersion relations are depicted in Figure 8.8 for the case where $L_y = L_z = d$. In this figure the dispersion relations are given for a GaAs quantum wire in two cases: a free-standing quantum wire having frequencies $\omega_{S,f}$ and $\omega_{A,f}$ for the symmetric and antisymmetric interface modes respectively, and a GaAs quantum wire embedded in AlAs having frequencies for the symmetric and antisymmetric interface modes of $\omega_{S,\pm}$ and $\omega_{A,\pm}$ respectively. The subscripts $\pm$ designate the high-frequency (+) and low-frequency (−) modes for both the symmetric and antisymmetric cases.

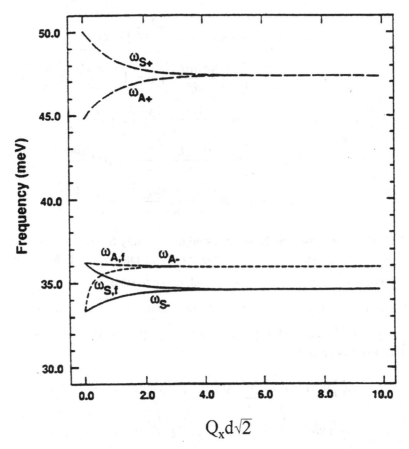

Figure 8.8. Frequency versus wavevector for the IF optical phonons of a free-standing GaAs quantum wire having frequencies $\omega_{S,f}$ and $\omega_{A,f}$ for the symmetric and antisymmetric interface modes respectively, and a GaAs quantum wire embedded in AlAs having frequencies $\omega_{S,\pm}$ and $\omega_{A,\pm}$ for the symmetric and antisymmetric interface modes respectively. The subscripts $\pm$ designate the high-frequency (+) and low-frequency (−) modes for both the symmetric and antisymmetric cases. From Kim *et al.* (1991), American Institute of Physics, with permission.

The normalization condition leads to the following equation in terms of the normalization constant, C, for the symmetric mode:

$$
\begin{aligned}
4\pi \frac{\hbar}{2\omega} L^{-1} C^{-2} = {}& \epsilon_1(\infty) \frac{\omega_{LO,1}^2 - \omega_{TO,1}^2}{\omega^2 - \omega_{TO,1}^2} \left(\cosh \frac{\alpha L_y}{2} \cosh \frac{\beta L_z}{2} \right)^{-2} \\
& \times \left[\frac{\alpha^2}{2} \left(\frac{L_z}{\alpha} \sinh \alpha L_y + \frac{\sinh \alpha L_y \sinh \beta L_z}{\alpha \beta} \right) \right. \\
& \left. + \frac{\beta^2}{2} \left(\frac{L_y}{\beta} \sinh \beta L_z + \frac{\sinh \alpha L_y \sinh \beta L_z}{\alpha \beta} \right) \right] \\
& + \epsilon_2(\infty) \frac{\omega_{LO,2}^2 - \omega_{TO,2}^2}{\omega^2 - \omega_{TO,2}^2} \left(\cosh \frac{\alpha L_y}{2} \right)^{-2} \\
& \times \left[\alpha^2 \left(\frac{\sinh \alpha L_y}{\alpha \beta} \right) + \beta^2 \left(\frac{\sinh \alpha L_y}{\alpha \beta} + \frac{L_y}{\beta} \right) \right] \\
& + \epsilon_2(\infty) \frac{\omega_{LO,2}^2 - \omega_{TO,2}^2}{\omega^2 - \omega_{TO,2}^2} \left(\cosh \frac{\beta L_z}{2} \right)^{-2} \\
& \times \left[\alpha^2 \left(\frac{\sinh \beta L_z}{\alpha \beta} + \frac{L_z}{\alpha} \right) + \beta^2 \left(\frac{\sinh \alpha L_y}{\alpha \beta} \right) \right] \\
& + \epsilon_2(\infty) \frac{\omega_{LO,2}^2 - \omega_{TO,2}^2}{\omega^2 - \omega_{TO,2}^2} \left[\frac{2(\alpha^2 + \beta^2)}{\alpha \beta} \right].
\end{aligned} \tag{8.117}
$$

A similar expression may be derived straightforwardly for the antisymmetric mode.

As in the previous subsection, the transition rate, $S_{1D,IF}^{\{e,a\}}(\mathbf{k}, \mathbf{k}')$, from an initial electron state $|k_{1D}\rangle$ and an initial phonon state $\left|N_q + \frac{1}{2} \pm \frac{1}{2}\right\rangle$ – denoted by $\left|k_{1D}, N_q + \frac{1}{2} \pm \frac{1}{2}\right\rangle$ – to a final electron state $\langle k'_{1D}|$ and a final phonon state $\left\langle N_q + \frac{1}{2} + \frac{1}{2}\right|$ – denoted by $\left\langle k'_{1D}, N_q + \frac{1}{2} + \frac{1}{2}\right|$ – per unit time per unit heterostructure area A is given by

$$
S_{1D,IF}^{\{e,a\}}(\mathbf{k}, \mathbf{k}') = \frac{2\pi}{\hbar} \sum_q \left| M_{1D,IF}^{\{e,a\}}(q_x) \right|^2 \delta(E(k') - E(k) \pm \hbar\omega), \tag{8.118}
$$

$$
M_{1D,IF}^{\{e,a\}}(q_x) = \left\langle k'_{1D}, N_q + \frac{1}{2} + \frac{1}{2} \right| H_{IF}^{1D} \left| k_{1D}, N_q + \frac{1}{2} \pm \frac{1}{2} \right\rangle. \tag{8.119}
$$

The electron wavefunction and the electron energy for the ground state are as given in the last subsection. Then the procedures of the last subsection lead straightforwardly to the equation

$$
\frac{1}{\tau_{1D,IF}^{\{e,a\}}(k_x)} = \frac{\alpha'' \omega}{2\pi} \left(n_q + \frac{1+\epsilon}{2} \right) \frac{I_{1D,IF}(q_+^{\{e,a\}}) + I_{1D,IF}(q_-^{\{e,a\}})}{\sqrt{E/\hbar\omega - \epsilon}}, \tag{8.120}
$$

with $\epsilon = 1$ for emission and $\epsilon = -1$ for absorption,

$$q_{\pm}^{\{e,a\}} = \epsilon k_x \pm \left[k_x^2 - \epsilon \left(\frac{2m^*\omega}{\hbar} \right) \right]^{1/2}$$

(8.121)

and

$$\alpha'' = \frac{1}{2\hbar\omega} \frac{e^2}{\sqrt{\hbar/2m\omega}}.$$

(8.122)

Furthermore,

$$I_{1D,IF} = \frac{(2\pi)^2}{\omega^2} \left(\frac{L}{4\pi} \right) C^2 P_s^2,$$

(8.123)

where

$$P_s = \frac{1}{\cosh \alpha L_y/2 \cosh \beta L_z/2}$$

$$\times \int_{-L_y/2}^{L_y/2} \frac{dy}{L_y/2} \int_{-L_z/2}^{L_z/2} \frac{dz}{L_z/2} \cos^2 \frac{\pi y}{L_y} \cos^2 \frac{\pi z}{L_z}$$

$$\times \cosh \alpha y \cosh \beta z$$

$$= \frac{1}{\cosh \alpha L_y/2 \cosh \beta L_z/2}$$

$$\times \frac{4 \sinh(\alpha L_y/2) \sinh (\beta L_z/2)}{\alpha\beta L_y L_z \left[(\alpha L_y/2\pi)^2 + 1 \right] \left[(\beta L_z/2\pi)^2 + 1 \right]}.$$

(8.124)

Only the symmetric mode contributes to the scattering rate for the ground state electronic wavefunction in the extreme quantum limit.

The scattering rate $1/\tau_{1D,IF}^{\{e,a\}}(k_x)$ has the same form as $1/\tau_{1D}^{\{e,a\}}(k_x)$, (8.103), of the last subsection. Indeed, α and α'' differ only by the factor $\left[\epsilon^{-1}(\infty) - \epsilon^{-1}(0) \right]$, and I_{1D} is replaced by $I_{1D,IF}$. Otherwise the expressions are identical. Therefore, it follows that the scattering rate due to both confined optical phonons and interface optical phonons is given by

$$\frac{1}{\tau_{wire}^{\{e,a\}}(k_x)} = \frac{\alpha''\omega}{2\pi} \left(n_q + \frac{1+\epsilon}{2} \right) \frac{I_{wire}\left(q_+^{\{e,a\}} \right) + I_{wire}\left(q_-^{\{e,a\}} \right)}{\sqrt{E/\hbar\omega - \epsilon}},$$

(8.125)

where

$$I_{wire} = \left[\frac{1}{\epsilon(\infty)} - \frac{1}{\epsilon(0)} \right] I_{1D} + I_{1D,IF}.$$

(8.126)

The electron–optical-phonon scattering rates for emission and absorption for selected phonon modes are displayed in Figure 8.9 for a 40 Å × 40 Å GaAs quantum
wire embedded in AlAs. The interface phonons are labeled as SO (surface-optical)
modes and the confined phonons are labeled simply as LO modes.

The one-dimensional density-of-states peaks are prominent features of the curves
representing the phonon emission rates. The electrons are in the ground states in the
y- and z-directions. The maximum scattering rates for SO-phonon emission occur
at the energies of the symmetric high-frequency interface mode (about 50 meV) and
the symmetric low-frequency interface mode (about 33 meV).

Knipp and Reinecke (1992) examined the accuracy of using the separation-
of-variables approximation to the dielectric continuum model for calculating the
Fröhlich potential of the interface optical phonons in quantum wires. They consid-

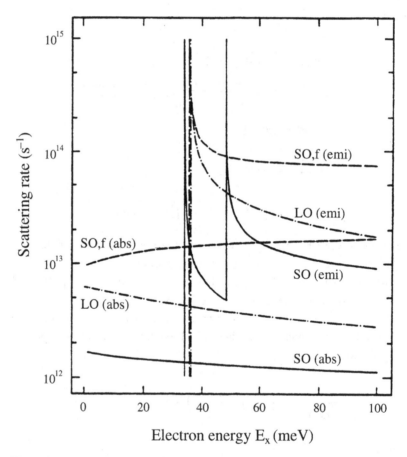

Figure 8.9. The electron–optical-phonon scattering rates for emission and
absorption for selected phonon modes for a 40 Å × 40 Å GaAs quantum wire
embedded in AlAs. The interface phonons are labeled as SO (surface-optical) modes
and the confined phonons are labeled simply as LO modes. From Kim *et al.* (1991),
American Institute of Physics, with permission.

ered a variety of wire geometries, as illustrated by the cross-sectional areas of Figure 8.10.

As indicated in Figure 8.10, the material inside the quantum wire has dielectric constant $\epsilon_1(\omega)$ and the surrounding material has dielectric constant $\epsilon_2(\omega)$. For the cases of quantum wires with circular and elliptical cross sections these workers obtained elegant analytical results. For quantum wires with nearly rectangular cross sections having rounded corners they used numerical techniques to determine the Fröhlich potentials.

An interesting feature of the calculations of Knipp and Reinecke is the localization of interface modes near the corners of the nearly rectangular quantum wires in the vicinity of the rounded corners, as illustrated in Figure 8.11. In Figure 8.11 the Fröhlich potential in the plane of the quantum wire – denoted by ϕ – is depicted for the mode of highest azimuthal symmetry $(m = 0)$ for a quantum wire with $R/r = 2$ and for a corner curvature $a = r/10$. The Fröhlich potential shown in Figure 8.11 represents the case where $qr = 2$.

The Fröhlich potential in the plane of an elliptical quantum wire with $R/r = 2$ is shown in Figure 8.12. The potential shown in Figure 8.12 is again the Fröhlich potential for the mode of highest azimuthal symmetry $(m = 0)$, with $qr = 2$. The potential is enhanced in the regions near the vertices of the rectangle into which the ellipse fits.

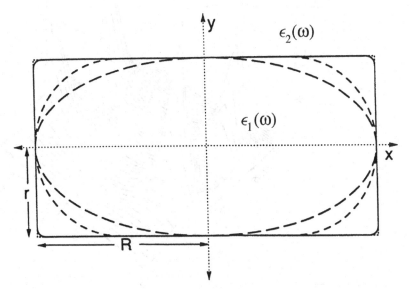

Figure 8.10. Cross sections considered by Knipp and Reinecke (1992). The long-broken line represents an elliptical cross section. The shorter broken line designates a cross section that is basically elliptical but has flattened sides. The solid and dotted lines (visible at the corners) represent 'rectangles' with rounded corners. The geometrical parameters R and r are as indicated in this figure. From Knipp and Reinecke (1992), American Physical Society, with permission.

8.3.4   Collective effects and non-equilibrium phonons in polar quantum wires

Campos *et al.* (1992) and Das Sarma *et al.* (1992) considered the influence of collective carrier screening on carrier–optical-phonon scattering in quantum wires. Since the Fröhlich Hamiltonian represents the interaction between a carrier and the fields produced by polar optical phonons, it is clear that the screening of such a Coulomb-like interaction by carriers is possible.

To estimate the magnitude of such a collective screening of the Fröhlich interaction, Campos *et al.* and Das Sarma *et al.* calculated the power loss per carrier caused by carrier–optical-phonon interactions in a GaAs quantum wire for the case where dynamical screening, phonon confinement, and the presence of the so-called 'hot-phonon' effect are all taken into account. The energy loss rate of a carrier in such a polar-semiconductor quantum wire is determined by both the rate at which the carrier's energy is lost by phonon emission and the rate at which the carrier gains

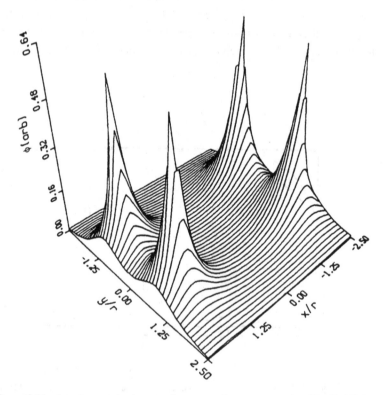

Figure 8.11. Interface modes in a nearly rectangular quantum wire. The Fröhlich potential ϕ in the plane of the quantum wire is depicted for the mode of highest azimuthal symmetry ($m = 0$) for a quantum wire with $R/r = 2$ and for a corner curvature $a = r/10$; $qr = 2$. From Knipp and Reinecke (1992), American Physical Society, with permission.

energy from phonon absorption. This latter rate can be significant in quantum wires and quantum wells, since the phonons emitted by energetic carriers can accumulate in these structures. Indeed, the phonon densities in many dimensionally confined semiconductor devices are typically well above those of the equilibrium phonon population and there is an appreciable probability that these non-equilibrium – or 'hot' – phonons will be reabsorbed. The net loss of energy by a carrier in such a situation depends on the rates for both phonon absorption and emission.

The lifetimes of the optical phonons are also important in determining the total energy loss rate for such carriers. As discussed in Chapter 6, the longitudinal optical (LO) phonons in GaAs and in many other polar materials decay into acoustic phonons through the Klemens' channel. Over a wide range of temperatures and phonon wavevectors, the lifetimes of these longitudinal optical phonons in GaAs vary from a few picoseconds to about 10 ps (Bhatt *et al.*, 1994). Typical lifetimes for other polar semiconductors are also of this magnitude. The LO phonons undergoing decay into acoustic phonons are not available for absorption by the carriers and,

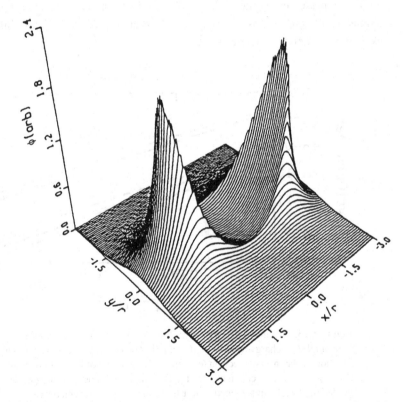

Figure 8.12. The Fröhlich potential in the plane of an elliptical quantum wire with $R/r = 2$. This potential was derived analytically by Knipp and Reinecke (1992) for the mode of highest azimuthal symmetry ($m = 0$), with $qr = 2$. From Knipp and Reinecke (1992), American Physical Society, with permission.

as a result of the Klemens' channel, the carrier energy loss is more rapid than it would be otherwise; this phenomenon is referred to as the 'hot-phonon-bottleneck' effect. In fact, the numerical calculations of Campos *et al.* and Das Sarma *et al.* for a 50 Å × 50 Å GaAs quantum wire indicate, over a temperature range from 50 K to 300 K, for wire dimensions from 50 Å × 50 Å to 500 Å × 500 Å, and for a linear density range of 10^4 to 10^6 cm^{-1}, that the hot-phonon-bottleneck effect changes the net power loss by an order of magnitude or more while the effect of carrier screening results in a change of only, roughly, a factor of two.

Figure 8.13 illustrates these findings for the case of a linear charge density of 10^6 cm^{-1} (Campos *et al.*, 1992). The power loss P per carrier is given as a function of the inverse electron temperature for both the slab modes and the guided modes. The curve of highest power loss (long-broken line) is calculated for the slab modes with no screening effects. The thinner solid line represents the power loss with screening effects included for a phonon decay time τ_{phonon} of 0 ps; the thicker solid line represents the power loss with screening effects included for a phonon decay time τ_{phonon} of 7 ps. For the case of the guided modes, the unscreened power loss function is represented by the broken-and-dotted line, and the screened power loss functions for $\tau_{phonon} = 0$ ps and $\tau_{phonon} = 7$ ps are represented by the thinner and thicker short-broken lines respectively.

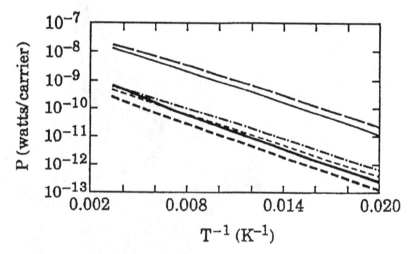

Figure 8.13. Results of Campos *et al.* for a 50 Å × 50 Å GaAs quantum wire for the case of a linear charge density of 10^6 cm^{-1}. The power loss per carrier is given as a function of the inverse electron temperature. The curve of highest power loss (long-broken line) is calculated for the slab modes with no screening effects. For the guided modes, the unscreened power loss function is represented by the broken-and-dotted line and the screened power loss functions for $\tau_{phonon} = 0$ ps and $\tau_{phonon} = 7$ ps are represented by the thinner and thicker short-broken lines respectively. From Campos *et al.* (1992), American Physical Society, with permission.

Thus while dynamical screening effects can be noticeable the effects of the hot-photon bottleneck can lead to a change of an order of magnitude or more in the carrier energy loss rates associated with carrier–LO-photon scattering in quantum wires. The screening calculations of Campos *et al.* (1992) and Das Sarma *et al.* (1992) involve non-trivial numerical calculations. For additional information on dynamical screening and hot-phonon effects the reader is referred to these papers.

8.3.5 Reduction of interface–phonon scattering rates in metal–semiconductor structures

The first calculations of carrier–LO-phonon scattering rates in quantum wires made of polar semiconductors (Stroscio *et al.*, 1990; Kim *et al.*, 1991) revealed that the scattering of carriers with interface optical phonon modes dominates over the scattering associated with confined and half-space phonons. As a means of reducing the unwanted inelastic scattering caused by the interface LO phonons in semiconductor quantum wires, it was suggested (Stroscio *et al.*, 1992) that the large carrier–LO-phonon scattering caused by interface phonons could be eliminated by encapsulation of the quantum wire in a metal. This proposal is of little use for most metal–semiconductor systems, but such a metal–semiconductor interface would lead to a depletion of the semiconductor when the cross-sectional dimensions – L_y and L_z – of the quantum wire are small enough (i.e., less than about $q^{-1} = 1/0.02$ Å) for the scattering rate due to interface phonons to be appreciable.

However, in some metal–semiconductor systems, such as the InAs–Al system, the Fermi level is pinned so that there is no depletion of carriers in the semiconductor but an accumulation of carriers near the metal–semiconductor interface. In cases where the semiconductor is not depleted, quantum wires with $L_y, L_z \lesssim q^{-1} = 1/0.02$ Å will exhibit large carrier–LO-interface phonon scattering rates detrimental to high-mobility transport.

To reduce these effects, the interface optical phonons may indeed be modified or practically eliminated through the use of metal boundaries at the semiconductor interfaces. The highly mobile carriers in a metal generally screen electric fields and these fields penetrate only very short distances, of order δ, into the metal; δ is known as the Thomas–Fermi screening length and is generally of the order of an ångstrom or so. The significance of this strong screening of electric fields near the metal–semiconductor interface is that the phonon potential ϕ must vanish near the interface. In such a situation, it is clear that both the symmetric and antisymmetric interface optical phonons will have very small amplitudes in quantum-wire and in quantum-well structures.

To estimate the magnitude of this reduction, Bhatt *et al.* (1993a) considered a metal–semiconductor–semiconductor (MSS) double-heterointerface structure with the metal–semiconductor interface at $z = -d$ and the semiconductor–

semiconductor interface at $z = d$. For this heterostructure, the interface phonon potential is of the form

$$\phi(r) = \phi_0 e^{i\mathbf{q}\cdot\boldsymbol{\rho}} \times \begin{cases} A e^{z/\delta} & z < -d, \\ B e^{qz} + C e^{-qz} & |z| < d, \\ e^{-q(z-d)} & z > d, \end{cases} \qquad (8.127)$$

where the constants A, B, and C as well as the dispersion relation for the mode are determined by the conditions that ϕ_i and $\epsilon_i \partial\phi_i/\partial z$ are continuous at the two heterointerfaces. It is straightforward to show by imposing these boundary conditions that

$$A = \left[e^{qd} \left(\frac{\epsilon_1}{\epsilon_1 - \epsilon_2} \right) \left(1 + \frac{\epsilon_m}{q\delta\epsilon_1} \right) \right]^{-1},$$

$$B = \frac{1}{2} e^{-qd} \left(1 - \frac{\epsilon_2}{\epsilon_1} \right), \qquad (8.128)$$

$$C = \frac{1}{2} e^{qd} \left(1 + \frac{\epsilon_2}{\epsilon_1} \right)$$

and that the dispersion relation is

$$\frac{\epsilon_1 - \epsilon_2}{\epsilon_1 + \epsilon_2} e^{-2qd} \left(\frac{q\delta\epsilon_1 - \epsilon_m}{q\delta\epsilon_1 + \epsilon_m} \right) - 1 = 0. \qquad (8.129)$$

In these results, the dielectric function of the metal is given by

$$\epsilon_m = 1 - \frac{\omega_p^2 \tau^2}{1 - \omega^2 \tau^2}, \qquad (8.130)$$

where ω_p is the plasma frequency for the carriers in the metal and τ is the dielectric relaxation time for the carriers in the metal.

Since the ratio $|A/B|$ is a measure of the amplitude of the IF optical phonon mode in the metal and since

$$\left| \frac{A}{B} \right| = \frac{2}{1 + \epsilon_m/q\delta\epsilon_1} \ll 1, \qquad (8.131)$$

for typical metal–semiconductor structures, it follows that the fields associated with the IF optical phonon mode are damped dramatically by electron screening as it enters the metal. Thus, to a very good approximation, $\phi(r) = 0$ at a metal–semiconductor interface. Indeed, numerical calculations (Bhatt et al., 1993a) show that $|A/B|$ is in the range 4×10^{-3}–10^{-8} when δ is in the range 1–10 Å and $q = 0.01$ Å^{-1}. Since the IF optical phonon modes, in general, have non-zero amplitudes at heterointerfaces, it follows that these modes do not satisfy the boundary condition $\phi(r) = 0$. Accordingly, these results imply that the IF modes suffer dramatic reductions in amplitude when semiconductor–semiconductor heterostructures are replaced with metal–semiconductor structures. Indeed, for the metal–semiconductor–metal (MSM) structure of Figure 8.14, it follows trivially that the optical phonon potentials should vanish at both $z = 0$ and $z = d$.

Clearly, neither the symmetric nor the antisymmetric optical phonon modes can satisfy these conditions. The MSM structure of Figure 8.14 does, however, support confined LO phonons. It is straightforward to extend these conclusions to the case of a semiconductor quantum wire embedded in a metal. In closing we note that Constantinou (1993) also considered the reduction of IF optical phonon mode amplitudes in metal–semiconductor systems. The concept of using a metal–semiconductor interface to reduce electron–interface–phonon interactions in quantum wires was considered further by Dutta *et al.* (1993).

8.4   Scattering of carriers and LO phonons in quantum dots

As discussed in Sections 5.2 and 7.1, the Fröhlich Hamiltonian for a bulk polar semiconductor may be written as

$$H_{\text{Fr}}^{(3D)} = \sum_q \frac{C}{|q|} \left(a_{\mathbf{q}} - a_{-\mathbf{q}}^\dagger \right) e^{i\mathbf{q}\cdot\mathbf{r}}, \tag{8.132}$$

where

$$C = -i \sqrt{\frac{2\pi e^2 \hbar \omega_{\text{LO}}}{V} \left[\frac{1}{\epsilon(\infty)} - \frac{1}{\epsilon(0)} \right]}. \tag{8.133}$$

Consider a quantum dot with the geometry of a free-standing cube each of whose sides is of length L (de la Cruz *et al.*, 1993). Then by writing the sum over q as a

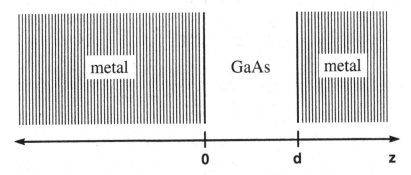

Figure 8.14. A metal–semiconductor–metal heterostructure in which the metal–semiconductor interfaces are at $z = 0$ and $z = d$. Interface LO phonon modes are not present in such a structure since the LO phonon potential vanishes at each metal–semiconductor interface. From Bhatt *et al.* (1993a), American Institute of Physics, with permission.

sum over positive values of q_x, q_y, and q_z, expanding the exponentials in a manner similar to that for a free-standing quantum wire (Stroscio, 1989), and requiring that $q_x = \pm m\pi/L$, $q_y = \pm n\pi/L$, and $q_z = \pm p\pi/L$, so that the Fröhlich potential vanishes on the faces of the box, it follows that the Fröhlich Hamiltonian for the quantum box must have the form

$$
\begin{aligned}
H_{\text{Fr}}^{(0D)} = 2^{3/2} C \Bigg\{ & \sum_{m=1,3\ldots} \sum_{n=1,3\ldots} \sum_{p=1,3\ldots} \beta(m,n,p) \\
& \times \cos\frac{m\pi x}{L} \cos\frac{n\pi y}{L} \cos\frac{p\pi z}{L} \left[A_{++}(q)_+ + A_{++}^\dagger(-q)_+ \right] \\
& + \sum_{m=2,4\ldots} \sum_{n=1,3\ldots} \sum_{p=1,3\ldots} \beta(m,n,p) \sin\frac{m\pi x}{L} \cos\frac{n\pi y}{L} \cos\frac{p\pi z}{L} \\
& \times \left[A_{+-}(q)_+ + A_{+-}^\dagger(-q)_+ \right] \\
& + \sum_{m=1,3\ldots} \sum_{n=2,4\ldots} \sum_{p=1,3\ldots} \beta(m,n,p) \cos\frac{m\pi x}{L} \sin\frac{n\pi y}{L} \cos\frac{p\pi z}{L} \\
& \times \left[A_{-+}(q)_+ + A_{-+}^\dagger(-q)_+ \right] \\
& + \sum_{m=2,4\ldots} \sum_{n=2,4\ldots} \sum_{p=1,3\ldots} \beta(m,n,p) \sin\frac{m\pi x}{L} \sin\frac{n\pi y}{L} \cos\frac{p\pi z}{L} \\
& \times \left[A_{--}(q)_+ + A_{--}^\dagger(-q)_+ \right] \\
& + \sum_{m=1,3\ldots} \sum_{n=1,3\ldots} \sum_{p=2,4\ldots} \beta(m,n,p) \cos\frac{m\pi x}{L} \cos\frac{n\pi y}{L} \sin\frac{p\pi z}{L} \\
& \times \left[A_{++}(q)_- + A_{++}^\dagger(-q)_- \right] \\
& + \sum_{m=2,4\ldots} \sum_{n=1,3\ldots} \sum_{p=2,4\ldots} \beta(m,n,p) \sin\frac{m\pi x}{L} \cos\frac{n\pi y}{L} \sin\frac{p\pi z}{L} \\
& \times \left[A_{+-}(q)_- + A_{+-}^\dagger(-q)_- \right] \\
& + \sum_{m=1,3\ldots} \sum_{n=2,4\ldots} \sum_{p=2,4\ldots} \beta(m,n,p) \cos\frac{m\pi x}{L} \sin\frac{n\pi y}{L} \sin\frac{p\pi z}{L} \\
& \times \left[A_{-+}(q)_- + A_{-+}^\dagger(-q)_- \right] \\
& + \sum_{m=2,4\ldots} \sum_{n=2,4\ldots} \sum_{p=2,4\ldots} \beta(m,n,p) \sin\frac{m\pi x}{L} \sin\frac{n\pi y}{L} \sin\frac{p\pi z}{L} \\
& \times \left[A_{--}(q)_- + A_{--}^\dagger(-q)_- \right] \Bigg\},
\end{aligned}
$$

$$(8.134)$$

where

$$
\beta(m,n,p) = \left[\left(\frac{m\pi}{L}\right)^2 + \left(\frac{n\pi}{L}\right)^2 + \left(\frac{p\pi}{L}\right)^2 \right]^{1/2},
$$

$$(8.135)$$

and where the phonon annihilation operators for the zero-dimensional confined modes are

$$A_{++}(q)_{\pm} = \frac{1}{\sqrt{2}} \left[A_{+}(q_x)_{\pm} + A_{+}(-q_x)_{\pm} \right],$$

$$A_{+-}(q)_{\pm} = \frac{-i}{\sqrt{2}} \left[A_{+}(q_x)_{\pm} - A_{+}(-q_x)_{\pm} \right],$$

$$A_{-+}(q)_{\pm} = \frac{1}{\sqrt{2}} \left[A_{-}(q_x)_{\pm} + A_{-}(-q_x)_{\pm} \right],$$

$$A_{--}(q)_{--} = \frac{-i}{\sqrt{2}} \left[A_{-}(q_x)_{\pm} - A_{-}(-q_x)_{\pm} \right].$$

(8.136)

The operators $A_{\pm}(q_x)_{\pm}$ can be written in terms of the annihilation operator for a three-dimensional semiconductor, in a manner analogous to that employed for the quantum wire (Stroscio, 1989), through use of the relations

$$A_{+}(q_x)_{\pm} = \frac{1}{\sqrt{2}} \left[a_{\pm}(q_x, q_y) + a_{\pm}(q_x, -q_y) \right],$$

$$A_{-}(q_x)_{\pm} = \frac{-i}{\sqrt{2}} \left[a_{\pm}(q_x, q_y) - a_{\pm}(q_x, -q_y) \right],$$

(8.137)

where $a_{\pm}(q_x, q_y)$ are defined by

$$a_{+}(q_x, q_y) = \frac{1}{\sqrt{2}} \left[a(q_x, q_y, q_z) + a(q_x, q_y, -q_z) \right],$$

$$a_{-}(q_x, q_y) = \frac{-i}{\sqrt{2}} \left[a(q_x, q_y, q_z) - a(q_x, q_y, -q_z) \right].$$

(8.138)

For the case of a GaAs quantum dot in vacuum, the electron energy levels are generally separated by more than the typical longitudinal optical phonon energy, $\hbar\omega_{LO}$, unless the dimensions of the dot are such that L is greater than several hundred ångstroms. Phonon confinement effects are generally not significant for such large confinement lengths. Specifically, for such a GaAs quantum 'box' the confined phonons and the interface phonons have energies within a few meV of the bulk phonon energy, 36 meV. In the case where the quantum dot is embedded, in, say, AlAs, there would be interface phonons with energies close to 50 meV but even these energies are small compared to typical separations between electron energy states unless the length L is several hundred ångstroms or more. For holes, the de Broglie wavelength is considerably shorter and energy level separations are smaller than for the case of electrons. Accordingly, for transitions involving holes the quantum 'box' may be small enough for phonon confinement effects to be significant. It will be instructive to summarize key features of the calculation of de la Cruz (1993) because it illustrates the general procedure within the effective-mass approximation and the Fermi golden rule approximation (Appendix E).

For a 'box' in vacuum with infinitely high potential barriers at its boundaries, defined by $-L/2 \leq x, y, z \leq L/2$, the carrier wavefunctions are represented by

$$|i, j, l\rangle = \sqrt[3]{\frac{2}{L}} \cos \frac{i\pi x}{L} \cos \frac{j\pi y}{L} \cos \frac{l\pi z}{L}, \qquad \text{for odd } i, j, l,$$

$$|i, j, l\rangle = \sqrt[3]{\frac{2}{L}} \sin \frac{i\pi x}{L} \sin \frac{j\pi y}{L} \sin \frac{l\pi z}{L}, \qquad \text{for even } i, j, l. \tag{8.139}$$

The carrier eigenenergies for this system in the effective-mass approximation, with effective mass m^*, are given by

$$E_{i,j,l} = \frac{\hbar^2}{2m^*} \left[\left(\frac{i\pi}{L}\right)^2 + \left(\frac{j\pi}{L}\right)^2 + \left(\frac{l\pi}{L}\right)^2 \right]. \tag{8.140}$$

The transition probability given by the Fermi golden rule is then

$$W^{\{e,a\}} = \frac{2\pi}{\hbar} \left| M^{\{e,a\}} \right|^2 \delta(E_{i',j',l'} - E_{i,j,l} \pm \hbar\omega_{LO}), \tag{8.141}$$

with

$$M^{\{e,a\}} = \langle i', j', l'; N_q + \tfrac{1}{2} \pm \tfrac{1}{2} | H_{Fr}^{(0D)} | i, j, l; N_q + \tfrac{1}{2} \pm \tfrac{1}{2} \rangle \tag{8.142}$$

where, as discussed several times previously, the signs are chosen to describe the desired phonon process. The energy-conservation constraint required by the δ-function may be expressed as

$$\left(i'^2 - i^2\right) + \left(j'^2 - j^2\right) + \left(l'^2 - l^2\right) \pm \frac{2m^* L^2}{\hbar \pi^2} \omega_{LO} = 0. \tag{8.143}$$

For the system under consideration, non-zero matrix elements correspond to transitions between states of the same parity. Accordingly, the transitions corresponding to the smallest quantum boxes are those between the $|1, 1, 1\rangle$ states and states where the carrier quantum numbers, i', j', l' and i, j, l, are selected from $1, 3$, and 5. For these latter states, all the degenerate states for a given set of carrier quantum numbers are denoted by $\{|i, j, l\rangle\}$. The transition probabilities for a selection of such transitions were calculated by de la Cruz (1993) for GaAs at room temperature. These probabilities are summarized in Table 8.1, where, in the case of emission, the carrier transition occurs between the indicated set of degenerate states $\{|i, j, l\rangle\}$ and the $|1, 1, 1\rangle$ state. Likewise, for the case of absorption, the initial state is the $|1, 1, 1\rangle$ state and the final state is indicated by the specific set $\{|i, j, l\rangle\}$ corresponding to the transition.

Table 8.1. Transition probabilities W for LO phonon emission and absorption resulting from the Fröhlich interaction in a GaAs quantum box of side L at room temperature. The $|1, 1, 1\rangle$ state is the final state in emission and the initial state in absorption.

L (nm)	$W_{emission}$ $(\times 10^{13} \text{ s}^{-1})$	Initial states	$W_{absorption}$ $(\times 10^{12} \text{ s}^{-1})$	Final states
352	2.51	$\{|3, 1, 1\rangle\}$	5.77	$\{|3, 1, 1\rangle\}$
498	1.71	$\{|3, 3, 1\rangle\}$	3.93	$\{|3, 3, 1\rangle\}$
610	1.02	$\{|3, 3, 3\rangle\}$	2.33	$\{|3, 3, 3\rangle\}$
610	1.69	$\{|5, 1, 1\rangle\}$	3.88	$\{|5, 1, 1\rangle\}$
704	1.20	$\{|1, 3, 5\rangle\}$	2.76	$\{|1, 3, 5\rangle\}$
787	0.81	$\{|5, 3, 3\rangle\}$	1.86	$\{|5, 3, 3\rangle\}$
863	0.91	$\{|5, 5, 1\rangle\}$	2.09	$\{|5, 5, 1\rangle\}$
1056	0.86	$\{|5, 5, 5\rangle\}$	1.56	$\{|5, 5, 5\rangle\}$

Chapter 9

Carrier–acoustic-phonon scattering

O praise ye the Lord, all things that give sound, each jubilant
chord re-echo around; Loud organs, his glory tell in deep tone
and, sweet harp, the story of what he hath done.
Henry Williams Baker, 1875

9.1 Carrier–acoustic-phonon scattering in bulk zincblende structures

In this section the strain tensor introduced in Section 7.2 will be related to both the
deformation-potential interaction and the piezoelectric interaction.

9.1.1 Deformation-potential scattering in bulk zincblende structures

The deformation-potential interaction of Section 5.3 applies to the case of an
isotropic medium. In the general case of a carrier band with energy $E_\alpha(\mathbf{k})$, the
deformation potential depends on the strain tensor and the deformation-potential
constant, $E^\alpha_{\mu\nu1}$, must also be represented as a second-rank tensor; accordingly,

$$H_{\alpha,\text{def}} = E^{\text{strained}}_\alpha(\mathbf{k}) - E^{\text{unpert}}_\alpha(\mathbf{k}) = E^\alpha_{\mu\nu1}S_{\mu\nu}, \qquad (9.1)$$

where $S_{\mu\nu}$ is the strain tensor defined in Section 7.2, (7.27), and the deformation
potential is the difference in the energy of band α when strain is present, $E^{\text{strained}}_\alpha(\mathbf{k})$,
and the energy of band α when there is no strain, $E^{\text{unpert}}_\alpha(\mathbf{k})$. Since the phonon field is
always present, the case where there is no strain is unphysical; it corresponds to the
situation where there is no deformation potential. For cubic crystals and for carriers
experiencing a spherically symmetric energy surface, symmetry considerations
make it evident that the crystalline potential should be proportional to $\nabla V/V$, as
in Section 5.3. This conclusion holds for carriers at a non-degenerate Γ point but

not for the cases of other symmetry points such as the X or L points. Accordingly, for the case of a cubic crystal and for carriers at the Γ point,

$$H_{\alpha,\text{def}} = E_1^\alpha \nabla V / V = E_1^\alpha \nabla \cdot \mathbf{u}, \tag{9.2}$$

or, in the notation of Section 7.2,

$$H_{\alpha,\text{def}} = E_1^\alpha \triangle, \tag{9.3}$$

since here $\triangle = \nabla \cdot \mathbf{u}$. Consider an acoustic mode of wavevector $\mathbf{q}$ and a displacement field which varies as $e^{-i\mathbf{q}\cdot\mathbf{r}}$. Then $\triangle(\mathbf{r}) = \nabla \cdot \mathbf{u} = -i\mathbf{q}\cdot\mathbf{u}$ and it follows that longitudinal acoustic (LA) but not transverse acoustic (TA) modes lead to non-zero values of $H_{\alpha,\text{def}}$. However, for cubic crystals with carriers at energy minima at other symmetry points – such as X or L points – $E_{\mu\nu 1}^\alpha$ does not reduce to a scalar but instead takes the form

$$E_{\mu\nu 1}^\alpha = E_{1l}^\alpha \delta_{\mu\nu} + E_{1t}^\alpha i_\mu i_\nu, \tag{9.4}$$

where the minimum is on an axis along the unit vector $\hat{\mathbf{i}}$. For this more general case, the E_{1t}^α term contributes to the deformation potential and, in general, the TA and optical phonons produce deformation potentials leading to intravalley processes.

9.1.2 Piezoelectric scattering in bulk semiconductor structures

As described in Section 5.4, piezoelectric scattering results from the interaction of carriers with the macroscopic piezoelectric polarization produced by strain fields, in polar crystals lacking an inversion symmetry. These strain fields may be generated either externally as, for example, by a mismatch between the lattice constants in pseudomorphic systems, or internally, as by an acoustic phonon. In the notation of Section 7.2, the piezoelectric polarization of the cubic crystal, considered in Section 5.4 takes the general form

$$P_\lambda^{\text{piezo}} = e_{\lambda,\mu\nu} S_{\mu\nu}, \tag{9.5}$$

where $S_{\mu\nu}$ is the strain tensor and $e_{\lambda,\mu\nu}$ is the generalized tensor describing the piezoelectric coupling of the polarization, P_λ^{piezo}, to the strain field $S_{\mu\nu}$. In practice, the components of $e_{\lambda,\mu\nu}$ may be determined through knowledge of the strain and electromagnetic fields and the static dielectric constant, $\epsilon_{\lambda\mu}^0$, via the relation

$$e_{\lambda,\mu\nu} S_{\mu\nu} = P_\lambda - \frac{1}{4\pi}(\epsilon_{\lambda\mu}^0 - \delta_{\lambda\mu}) E_\mu. \tag{9.6}$$

As described previously (Vogl, 1980), $e_{\lambda,\mu\nu}$ is manifestly a bulk quantity. In general, P_λ and E_λ depend on the geometry of the crystal structure but the components of $e_{\lambda,\mu v}$ are bulk-like quantities determined by the local electronic structure of the crystal.

The piezoelectric interaction may be modeled in terms of the interaction of carriers with the macroscopic electric potential produced by the piezoelectric field. To calculate this potential for a bulk semiconductor, the phonon displacement is taken to vary as $e^{-i\mathbf{q}\cdot\mathbf{r}}$, and it follows that $\nabla \cdot \mathbf{u} = i\mathbf{q} \cdot \mathbf{u}$. In a region without free charges $\nabla \cdot \mathbf{D} = 0$. Hence, applying q_λ to

$$D_\lambda = 4\pi e_{\lambda,\mu v} S_{\mu v} + \epsilon^0_{\lambda\mu} E_\mu \tag{9.7}$$

it follows that

$$q_\lambda \epsilon^0_{\lambda\mu} E_\mu = -4\pi q_\lambda e_{\lambda,\mu v} S_{\mu v}. \tag{9.8}$$

For the phonon displacement field under consideration, $S_{\mu v} = \frac{1}{2} i (q_\mu u_v + u_v q_\mu)$ and, by defining the potential associated with the piezoelectric potential through $-i\mathbf{q} V^{\text{piezo}}(\mathbf{q}) = \mathbf{E}$, it follows that

$$V^{\text{piezo}}(\mathbf{q}) = 4\pi \frac{q_\lambda e_{\lambda,\mu v} q_\mu}{q_\lambda \epsilon^0_{\lambda\mu} q_\mu} u_v. \tag{9.9}$$

For isotropic, cubic, trigonal, tetragonal, and hexagonal structures $q_\lambda \epsilon^0_{\lambda\mu} q_\mu$ reduces to relatively simple expressions since in these cases $\epsilon^0_{\lambda\mu}$ may be represented as a 3×3 matrix with all non-diagonal elements equal to zero. For cubic structures, $q_\lambda \epsilon^0_{\lambda\mu} q_\mu$ reduces to $\epsilon^0 \mathbf{q} \cdot \mathbf{q}$ since the three diagonal elements of $\epsilon^0_{\lambda\mu}$ are all equal to ϵ^0.

For non-cubic crystals it is convenient to use the tensor notation of Section 7.2 to determine P^{piezo}_λ from $P^{\text{piezo}}_\lambda = e_{\lambda,\mu v} S_{\mu v}$. As in Section 7.2, $S_{\mu v}$ may be represented as a six-element vector, P^{piezo}_λ is a three-element vector, and $e_{\lambda,\mu v}$ is a 3×6 matrix; for zincblende structures (Auld, 1973), the 3×6 matrix is

$$\begin{pmatrix} 0 & 0 & 0 & e_{x4} & 0 & 0 \\ 0 & 0 & 0 & 0 & e_{x4} & 0 \\ 0 & 0 & 0 & 0 & 0 & e_{x4} \end{pmatrix} \tag{9.10}$$

and for würtzite crystals this matrix takes the form

$$\begin{pmatrix} 0 & 0 & 0 & 0 & e_{x5} & 0 \\ 0 & 0 & 0 & e_{x5} & 0 & 0 \\ e_{x1} & e_{x2} & e_{x3} & 0 & 0 & 0 \end{pmatrix}. \tag{9.11}$$

9.2 Carrier–acoustic-phonon scattering in two-dimensional structures

For a single band and for a non-degenerate Γ point the results of subsection 9.1.1 reduce to

$$H_{\mathrm{def}} = E_1 \nabla V / V = E_1 \nabla \cdot \mathbf{u} = E_1 \Delta, \qquad (9.12)$$

and with the expression for the quantized displacement derived in (7.145), Section 7.6 (Bannov, 1995), it follows that

$$
\begin{aligned}
H_{\mathrm{def}} &= E_1 \sum_{\mathbf{q}_\parallel, n} \sqrt{\frac{\hbar}{2L^2 \rho \omega_n(\mathbf{q}_\parallel)}} (a_{n,q} + a^\dagger_{n,-q}) \nabla \cdot \left[\mathbf{w}_n(\mathbf{q}_\parallel, z) \, e^{i\mathbf{q}_\parallel \cdot \mathbf{r}_\parallel} \right] \\
&= \sum_{\mathbf{q}_\parallel, n} e^{i\mathbf{q}_\parallel \cdot \mathbf{r}_\parallel} \Gamma(\mathbf{q}_\parallel, n, z)(a_{n,q} + a^\dagger_{n,-q})
\end{aligned}
\qquad (9.13)
$$

where

$$\Gamma(\mathbf{q}_\parallel, n, z) = \sqrt{\frac{E_1^2 \hbar}{2L^2 \rho \omega_n(\mathbf{q}_\parallel)}} \left[i\mathbf{q}_\parallel \cdot \mathbf{w}_n(\mathbf{q}_\parallel, z) + \frac{\partial w_{nz}(\mathbf{q}_\parallel, z)}{\partial z} \right]. \qquad (9.14)$$

9.3 Carrier–acoustic-phonon scattering in quantum wires

In this section electron–acoustic-phonon scattering rates will be considered for two different quantum-wire geometries, cylindrical and rectangular. In both cases, the scattering rates will be calculated by applying the Fermi golden rule and by treating the acoustic phonon modes as modes of an elastic continuum with the normalization condition of Chapter 7.

9.3.1 Cylindrical wires

For the case of a cylindrical quantum wire, the quantization of the acoustic phonons may be performed by taking

$$\mathbf{u}(\mathbf{r}) = \frac{1}{\sqrt{N}} \sum_{\mathbf{q}} [\mathbf{u}(\mathbf{q}, \mathbf{r}) a_{\mathbf{q}} + \text{c.c.}] \qquad (9.15)$$

and by normalizing the acoustic phonon Fourier amplitude, $\mathbf{u}(\mathbf{q}, r, \varphi, z)$, according to

$$\frac{1}{\pi a^2} \int_0^{2\pi} d\varphi \int_0^a dr \, r \mathbf{u}(\mathbf{q}, r, \varphi, z) \cdot \mathbf{u}^*(\mathbf{q}, r, \varphi, z) = \frac{\hbar}{2M\omega_q}. \qquad (9.16)$$

Here a is the radius of the quantum wire, N is the number of unit cells in the normalization volume V, $a_{\mathbf{q}}$ is the phonon annihilation operator, $\mathbf{q}$ is the phonon wavevector, ω_q is the angular frequency of the phonon mode, M is the mass of the ions in the unit cell, and r, φ, z are the usual cylindrical coordinates. As discussed in Chapter 7, the elastic continuum model provides an adequate description of such a quantum wire as long as the diameter of the quantum wire is several times larger than the linear dimension of the unit cell. Within the elastic continuum model, the equation describing the acoustic phonon modes is (Yu *et al.*, 1996)

$$\rho \frac{\partial^2 \mathbf{u}}{\partial t^2} = c_{44} \nabla^2 \mathbf{u} + (c_{12} + c_{44}) \nabla (\nabla \cdot \mathbf{u})$$

$$+ c^* \left(\hat{\mathbf{i}}_1 \frac{\partial^2 u_1}{\partial x_1^2} + \hat{\mathbf{i}}_2 \frac{\partial^2 u_2}{\partial x_2^2} + \hat{\mathbf{i}}_3 \frac{\partial^2 u_3}{\partial x_3^2} \right) \tag{9.17}$$

where $c^* = c_{11} - c_{12} - 2c_{44}$. For the isotropic case of Section 7.6, $c^* = c_{11} - c_{12} - 2c_{44} = 0$, and

$$\rho \frac{\partial^2 \mathbf{u}}{\partial t^2} = c_{44} \nabla^2 \mathbf{u} + (c_{44} - c_{11}) \nabla (\nabla \cdot \mathbf{u})$$

$$= \rho c_t^2 \nabla^2 \mathbf{u} + \rho (c_l^2 - c_t^2) \nabla (\nabla \cdot \mathbf{u}), \tag{9.18}$$

or, equivalently,

$$\frac{\partial^2 \mathbf{u}}{\partial t^2} = c_l^2 \nabla (\nabla \cdot \mathbf{u}) - c_t^2 \nabla \times \nabla \times \mathbf{u}, \tag{9.19}$$

where $c_l = \sqrt{c_{11}/\rho}$ is the sound speed for longitudinal acoustic waves and $c_t = \sqrt{c_{44}/\rho}$ is the sound speed for transverse acoustic waves. In considering the acoustic phonon modes in a cylinder it is convenient to introduce two equivalent quantities, the Young's modulus velocity v_0 and the Poisson ratio σ, through

$$c_0 = c_l \sqrt{\frac{(3c_l^2 - c_t^2)}{c_l^2 - c_t^2}} \quad \text{and} \quad \sigma = \frac{\frac{1}{2}(c_l/c_t)^2 - 1}{(c_l/c_t)^2 - 1}. \tag{9.20}$$

In this subsection, the electron–acoustic-phonon scattering rate is evaluated for an isotropic, cylindrical quantum wire. From subsection 9.1.1, the deformation-potential interaction for non-degenerate carriers at a Γ point may be written as $H_{\text{def}} = E_1 \nabla \cdot \mathbf{u}$. In this case, only longitudinal acoustic modes contribute to the scattering rate. As discussed by Auld (1973), the classical longitudinal modes for an isotropic cylindrical structure have no azimuthal components and are of the form

$$u_r(r, z) = \left\{ \frac{d}{dr} [B J_0(q_l r) + A J_0(q_t r)] \right\} e^{i(qz - \omega t)},$$

$$u_z(r, z) = i \left\{ q B J_0(q_l r) - \frac{q_t^2}{q} A J_0(q_t r) \right\} e^{i(qz - \omega t)}, \tag{9.21}$$

where q is the z-component of the wavevector, ω is the angular frequency, and J_0 is the lowest-order Bessel function of the first kind, and where q_l and q_t are given by

$$q_{l,t}^2 = \frac{\omega^2}{v_{l,t}^2} - q^2. \tag{9.22}$$

We define dimensionless velocities, wavevectors, and frequencies by dividing by c_0, $q_0 = \pi/a$, and $\omega_0 = q_0 c_0$ respectively. That is, we take $\tilde{c}_{l,t} = c_{l,t}/c_0$, $\tilde{q} = q/q_0 = q(a/\pi)$, and $\tilde{\omega} = \omega/\omega_0 = \omega(a/\pi c_0)$. Upon evaluating

$$\frac{1}{\pi a^2} \int_0^{2\pi} d\varphi \int_0^a dr\, r[u_r(r, z)u_r^*(r, z) + u_z(r, z)u_z^*(r, z)] = \frac{\hbar}{2M\omega_q},$$

(9.23)

expressing $\mathbf{u}(\mathbf{r})$ in terms of $\mathbf{u}(\mathbf{q}, \mathbf{r})$, and assuming $H_{\text{def}} = E_1 \nabla \cdot \mathbf{u}$ it follows that

$$H_{\text{def}} = iE_1 \sum_n \sum_q \sqrt{\frac{\hbar}{2NM\omega}}$$
$$\times \frac{a}{\sqrt{2s_n}} \beta(q^2 + q_l^2) J_0(q_l r)(a_q e^{iqz} - a_q^\dagger e^{-iqz}),$$

(9.24)

where

$$s_n = \left\{ \frac{\tilde{q}_t^4}{\tilde{q}^2} \frac{\pi^2}{2} [J_1^2(\tilde{q}_t\pi) + J_0^2(\tilde{q}_t\pi)] \right.$$
$$- 2\tilde{q}_t^2 \beta \frac{\pi}{\tilde{q}_l^2 - \tilde{q}_t^2} [\tilde{q}_l J_1(\tilde{q}_l\pi) J_0(\tilde{q}_t\pi) - \tilde{q}_t J_1(\tilde{q}_t\pi) J_0(\tilde{q}_l\pi)]$$
$$+ \tilde{q}^2 \beta^2 \frac{\pi^2}{2} [J_1^2(\tilde{q}_l\pi) + J_0^2(\tilde{q}_l\pi)] + \tilde{q}_t^2 \frac{\pi^2}{2} [J_1^2(\tilde{q}_t\pi) - J_0(\tilde{q}_t\pi) J_2(\tilde{q}_t\pi)]$$
$$+ 2\tilde{q}_t\tilde{q}_l \beta \frac{\pi}{\tilde{q}_l^2 - \tilde{q}_t^2} [\tilde{q}_t J_0(\tilde{q}_t\pi) J_1(\tilde{q}_l\pi) - \tilde{q}_l J_0(\tilde{q}_l\pi) J_1(\tilde{q}_t\pi)]$$
$$\left. + \tilde{q}_l^2 \beta^2 \frac{\pi^2}{2} [J_1^2(\tilde{q}_l\pi) - J_0(\tilde{q}_l\pi) J_2(\tilde{q}_l\pi)] \right\};$$

(9.25)

here $\beta = B/A$.

In classical acoustics (Auld, 1973), there are two particularly simple types of boundary condition. They are referred to frequently as the cardinal boundary conditions and are the free-surface boundary condition (FSBC) and the clamped-surface boundary condition (CSBC). In the free-surface boundary condition, at the boundary between the cylindrical wire and the vacuum the normal components of the stress tensor are zero and the displacement is unrestricted. In the clamped-surface boundary condition, at the boundary between the cylindrical wire and the vacuum the normal components of the stress tensor are unrestricted and the displacement is zero. The dispersion relation for the free-surface boundary condition is then given by

$$(\tilde{q}^2 - \tilde{q}_t^2)T_1(\tilde{q}_l\pi) - 2\tilde{q}_t^2(\tilde{q}^2 + \tilde{q}_t^2) + 4\tilde{q}^2\tilde{q}_t^2 T_1(\tilde{q}_t\pi) = 0,$$

(9.26)

and the dispersion relation for the clamped-surface boundary condition is given by

$$\tilde{q}_l^2 T_1(\tilde{q}_t\pi) + \tilde{q}^2 T_1(\tilde{q}_l\pi) = 0,$$

(9.27)

where

$$T_1(x) = x J_0(x)/J_1(x). \tag{9.28}$$

For the free-surface boundary condition β is given by

$$\beta = -\frac{\tilde{q}_t(\tilde{q}^2 - \tilde{q}_t^2)}{2\tilde{q}^2\tilde{q}_l} \frac{J_1(\tilde{q}_t\pi)}{J_1(\tilde{q}_l\pi)}, \tag{9.29}$$

and for the clamped-surface boundary condition β is given by

$$\beta = -\frac{\tilde{q}_t}{\tilde{q}_l} \frac{J_1(\tilde{q}_t\pi)}{J_1(\tilde{q}_l\pi)}. \tag{9.30}$$

The electron–acoustic-phonon scattering rates may be evaluated straightforwardly, in the limit where electrons are assumed to be confined in the cylinder by an infinite potential at its surface. In this case (Wang and Lei, 1994)

$$|k\rangle = \frac{e^{ikz}}{\sqrt{L_z}}\psi(r, \theta), \tag{9.31}$$

where

$$\psi(r, \theta) = \frac{1}{\sqrt{\pi}Y_1^0 a} J_0\left(\frac{X_1^0}{a}r\right), \tag{9.32}$$

X_1^0 being the position of the first zero of $J_0(x)$ and Y_1^0 being equal to $J_1(X_1^0)$; L_z, the length of the cylinder, is assumed to be much larger than a. The energy of the ground state electron energy is $[(\hbar k)^2 + (X_1^0/a)^2]/2m^*$, where m^* is the electron's effective mass. Then it follows that the matrix element of the deformation potential, H_{def}, takes the form

$$M_n^{\{e,a\}} = \left\langle k', N_n + \tfrac{1}{2} + \tfrac{1}{2}\epsilon \right| H_{\text{def}} |k, N_n\rangle$$

$$= E_1\sqrt{\frac{\hbar}{2NM\omega_n}} \frac{a}{\sqrt{2s_n}}\beta_n(q^2 + q_l^2)$$

$$\times F(q_l a)\sqrt{N_n + \tfrac{1}{2} + \tfrac{1}{2}\epsilon}\,\delta_{-k'+k-\epsilon q}, \tag{9.33}$$

where N_n is the phonon occupation number of Section 5.1, $\epsilon = 1, -1$ in the cases of phonon emission and absorption respectively and F is given by

$$F_n(t) = \int_0^1 dx\, x \left| J_0(X_1^0 x) \right|^2 J_0(tx). \tag{9.34}$$

Then, assuming the Fermi golden rule,

$$\frac{1}{\tau^{\{e,a\}}(k)} = \frac{2\pi}{\hbar} \left| M_n^{\{e,a\}} \right|^2 \delta(E(k') - E(k) + \epsilon\hbar\omega)$$

$$= \sum_n \frac{\pi}{|Y_1^0|^4 \, \hbar^2} \frac{m^* |E_1|^2}{\rho}$$

$$\times \frac{1}{c_0 \tilde{c}_l^4} \frac{1}{a^2} \left[I_{1c_n}(\tilde{q}_+^{\{e,a\}}) + I_{1c_n}(\tilde{q}_-^{\{e,a\}}) \right], \qquad (9.35)$$

with

$$I_{1c_n}(\tilde{q}) = \frac{|\beta_n(\tilde{q})|^2}{|s_n(\tilde{q})|} |\tilde{\omega}_n|^3 |F_n(\tilde{q}_l\pi)|^2 \left(N_n + \tfrac{1}{2} + \tfrac{1}{2}\epsilon \right)$$

$$\times \frac{1}{(\tilde{q} + \tilde{k}) - (\epsilon m^* c_0 a / \pi\hbar)(d\tilde{\omega}/d\tilde{q})}. \qquad (9.36)$$

From energy- and momentum-conservation conditions it follows that $\tilde{q}_\pm^{\{e,a\}}$ is the solution of $(\hbar^2/2m^*)(\tilde{q}^2 \pm 2\tilde{q}\tilde{k}) - \epsilon\hbar\omega = 0$; accordingly, the plus sign is taken for forward scattering and the negative sign for backward scattering. As usual, n labels the quantized acoustic phonon modes.

For the piezoelectric scattering of carriers in a cylindrical quantum wire, the normalization procedure is of course the same as that for scattering by a deformation potential but the interaction Hamiltonian is based on the interaction potential of subsection 9.1.2. For cubic materials this piezoelectric interaction potential takes the form

$$V^{\text{piezo}}(\mathbf{q}) = 4\pi \frac{q_\lambda e_{\lambda,\mu\nu} q_\mu}{\epsilon^0 \mathbf{q} \cdot \mathbf{q}} u_\nu. \qquad (9.37)$$

In cylindrical coordinates the piezoelectric tensor of subsection 9.1.2 takes the form (Auld, 1973)

$$\mathbf{e} = \begin{pmatrix} 0 & 0 & 0 & e_{x4}A(\varphi) & 2e_{x4}B(\varphi) & 0 \\ 0 & 0 & 0 & -2e_{x4}B(\varphi) & e_{x4}A(\varphi) & 0 \\ 2e_{x4}B(\varphi) & -2e_{x4}B(\varphi) & 0 & 0 & 0 & e_{x4}A(\varphi) \end{pmatrix}$$

$$(9.38)$$

where $A(\varphi) = \cos^2\varphi - \sin^2\varphi$ and $B(\varphi) = \cos\varphi\sin\varphi$. Moreover, the polarization vector $\mathbf{P}$ and strain components S_ν are given by

$$\mathbf{P} = \begin{pmatrix} P_1 \\ P_2 \\ P_3 \end{pmatrix} \qquad (9.39)$$

and

$$S_1 = S_{rr} = \frac{\partial u_r}{\partial r}, \qquad S_2 = S_{\varphi\varphi} = \frac{u_r}{r} + \frac{1}{r}\frac{\partial u_\varphi}{\partial \varphi},$$

$$S_3 = S_{zz} = \frac{\partial u_z}{\partial z}, \qquad S_4 = 2S_{z\varphi} = 2S_{\varphi z} = \frac{\partial u_\varphi}{\partial z} + \frac{1}{r}\frac{\partial u_z}{\partial \varphi},$$

$$S_5 = 2S_{rz} = 2S_{zr} = \frac{\partial u_r}{\partial z} + \frac{\partial u_z}{\partial r}, \tag{9.40}$$

$$S_6 = 2S_{r\varphi} = 2S_{\varphi r} = \frac{1}{r}\frac{\partial u_r}{\partial \varphi} + \frac{\partial u_\varphi}{\partial r} - \frac{u_\varphi}{r}.$$

Accordingly, the relation $\mathbf{P} = \mathbf{e}\mathbf{S}$ implies that (Stroscio and Kim, 1993)

$$P_1 = 2e_{x4}(\cos^2\varphi - \sin^2\varphi)S_{z\varphi} + 4e_{x4}\cos\varphi\sin\varphi\, S_{rz},$$

$$P_2 = -4e_{x4}\cos\varphi\sin\varphi\, S_{z\varphi} + 2e_{x4}(\cos^2\varphi - \sin^2\varphi)S_{rz}, \tag{9.41}$$

$$P_3 = 2e_{x4}\cos\varphi\sin\varphi\,(S_{rr} - S_{\varphi\varphi}) + 2e_{x4}(\cos^2\varphi - \sin^2\varphi)S_{r\varphi}.$$

From classical acoustic theory, the lowest-order azimuthally symmetric torsional mode of a cylindrical structure of radius a has the form (Auld, 1973)

$$u_\varphi^{\text{wire}} = \gamma r e^{-i\omega t/v_t}, \tag{9.42}$$

where $v_t = (c_{44}/\rho)^{1/2}$ and γ is a normalization factor to be determined. The solutions for a free-standing cylindrical structure satisfy the boundary condition of vanishing stress components on the surface of the cylinder. That is, the vibrational modes are such that the cylinder undergoes distortions until there is no remaining force on its surface to cause additional distortion. For such a case the displacement field on the surface of the cylinder is unrestricted. For the case of a free-standing structure, the displacement field of lowest-order azimuthally symmetric torsional mode is characterized by vanishing u_r and u_z and $q = q_z = \omega/v_t$. For a wire of radius a and length L the classical solution for this mode is found by requiring the $z\varphi$-component of the stress to be zero at the ends of the cylinder located at $z = 0$ and $z = L$, so that the classical mode is given by

$$u_\varphi^{\text{finite wire}} = \gamma' r \cos\frac{v\pi}{L}z, \tag{9.43}$$

where $v\pi/L = \omega/v_t$ and γ' is the normalization constant. The normalization condition,

$$\frac{1}{\pi a^2 L}\int_0^L dz \int_0^{2\pi} d\varphi \int_0^a dr\, r\mathbf{u}(\mathbf{q}, r, \varphi, z)\cdot\mathbf{u}^*(\mathbf{q}, r, \varphi, z) = \frac{\hbar}{2M\omega_q}, \tag{9.44}$$

yields

$$u_\varphi^{\text{finite wire}} = \sqrt{\frac{\hbar}{2M\omega_q}\frac{2}{a}}\, r \cos\frac{v\pi}{L}z. \tag{9.45}$$

Strictly speaking, this case of a quantum wire of finite length also corresponds to a quantum dot since dimensional confinement is considered in all three dimensions. For a quantum wire of infinite length the normalization introduced at the beginning of this section requires the condition

$$\gamma = \frac{1}{a}\left(\frac{\hbar}{M\omega_q}\right)^{1/2},$$

(9.46)

so that

$$u_\varphi^{\text{finite wire}} = \sqrt{\frac{\hbar}{M\omega_q}}\frac{1}{a}re^{-i\omega t/v_t}.$$

(9.47)

These normalized displacement fields determine the components of $\mathbf{S}$ and, therefore, $\mathbf{P}$; and the piezoelectric interaction potential $V^{\text{piezo}}(\mathbf{q})$ is thus determined for the lowest-order azimuthally symmetric torsional modes of quantum wires and cylindrical quantum wires of finite length. As illustrated for the deformation potential, the carrier–acoustic-phonon scattering rate may be estimated by perturbation theory. In particular, the Fermi golden rule of Appendix E provides a convenient formalism for such calculations.

9.3.2 Rectangular wires

The calculation of the carrier–acoustic-phonon scattering rate in a rectangular quantum wire proceeds along the lines of the analogous calculation of subsection 9.3.1 for a cylindrical quantum wire. The deformation-potential interaction for acoustic phonons interacting with carriers in a rectangular quantum wire was derived in subsection 7.6.3, (7.196), where it was shown that

$$H_{\text{def}}^\alpha = \frac{E_1^\alpha}{\sqrt{L}}\sum_{n,m,\gamma} A_\gamma\frac{\omega_\gamma}{c_l^2 q_1}\cos q_1 x\cos hy\,[a_{n,m}(\gamma)+a_{n,m}(-\gamma)]\,e^{i\gamma z},$$

(9.48)

with

$$A_\gamma^2 = \frac{2\hbar}{M\omega_\gamma B_\gamma},$$

(9.49)

$$hd = (n+\tfrac{1}{2})\pi, \qquad n = 0, 1, 2, \ldots.$$

In the extreme quantum limit where the carriers are assumed to be confined within the rectangular boundaries of the quantum wire by an infinitely high potential, the carrier wavefunction is

$$\psi_k(x, y, z) = |k\rangle = \frac{1}{\sqrt{ad}}\cos\frac{\pi x}{2a}\cos\frac{\pi y}{2d}e^{ikz},$$

(9.50)

and the ground-state carrier eigenenergy is

$$E = \frac{\hbar^2}{2m}\left[\left(\frac{\pi^2}{(2a)^2} + \frac{\pi^2}{(2d)^2}\right) + k^2\right]. \tag{9.51}$$

To estimate the carrier–acoustic-phonon scattering rate, it is necessary to evaluate the matrix element of the deformation-potential interaction. This matrix element is given by

$$\begin{aligned}
M_n^{\{e,a\}} &= \left\langle k', N_n + \tfrac{1}{2} + \tfrac{1}{2}\epsilon \,\middle|\, H_{\text{def}} \,\middle|\, k, N_n \right\rangle \\
&= \frac{E_1^\alpha}{\sqrt{L}} \sum_{n,m,\gamma} A \frac{\omega_\gamma^2}{c_l^2 q_1} \langle k' | \cos q_1 x \cos hy\, e^{i\gamma z} | k \rangle \\
&\quad \times \left\langle N + \tfrac{1}{2} + \tfrac{1}{2}\epsilon \,\middle|\, [a_{n,m}(\gamma) + a_{n,m}(-\gamma)] \,\middle|\, N \right\rangle \\
&= \frac{E_1^\alpha}{\sqrt{L}} \sum_{n,m,\gamma} A \frac{\omega_\gamma^2}{c_l^2 q_1} \frac{1}{ad} \delta_{k-k'+\gamma} \\
&\quad \times \int_{-a}^{a} dx \cos^2 \frac{\pi x}{2a} \cos q_1 x \int_{-d}^{d} dy \cos^2 \frac{\pi y}{2d} \cos(hy) \\
&\quad \times \left\langle N + \tfrac{1}{2} + \tfrac{1}{2}\epsilon \,\middle|\, [a_{n,m}(\gamma) + a_{n,m}(-\gamma)] \,\middle|\, N \right\rangle \\
&= \frac{E_1^\alpha}{\sqrt{L}} \sum_{n,m,\gamma} A \frac{\omega_\gamma^2}{c_l^2 q_1} \frac{\pi^2 \sin q_1 a}{q_1 a(\pi^2 - q_1^2 a^2)} \frac{1}{\pi\left(n+\tfrac{1}{2}\right)\left[1 - \left(n+\tfrac{1}{2}\right)^2\right]} \\
&\quad \times \left\langle N + \tfrac{1}{2} + \tfrac{1}{2}\epsilon \,\middle|\, [a_{n,m}(\gamma) + a_{n,m}(-\gamma)] \,\middle|\, N \right\rangle \\
&= \frac{E_1^\alpha}{\sqrt{L}} \sum_{n,m,\gamma} A \frac{\omega_\gamma^2}{c_l^2 q_1} \frac{\pi^2 \sin(q_1 a)}{q_1 a(\pi^2 - q_1^2 a^2)} \\
&\quad \times \frac{1}{\pi\left(n+\tfrac{1}{2}\right)\left[1 - \left(n+\tfrac{1}{2}\right)^2\right]} \left(n + \tfrac{1}{2} \pm \tfrac{1}{2}\right)^{1/2} \delta_{k-k'\pm\gamma}. \tag{9.52}
\end{aligned}$$

We have used

$$\frac{1}{a} \int_{-a}^{a} dx \cos^2 \frac{\pi x}{2a} \cos q_1 x = \frac{\pi^2 \sin q_1 a}{q_1 a(\pi^2 - q_1^2 a^2)} \tag{9.53}$$

and

$$\frac{1}{d} \int_{-d}^{d} dy \cos^2 \frac{\pi y}{2d} \cos hy = \frac{\pi^2 \sin hd}{hd(\pi^2 - h^2 d^2)} = \frac{1}{\pi\left(n+\tfrac{1}{2}\right)\left[1 - \left(n+\tfrac{1}{2}\right)^2\right]}. \tag{9.54}$$

As usual the plus signs in the phonon-occupancy and momentum-conservation terms correspond to phonon absorption and the negative signs to phonon emission. Then the scattering rate predicted by the Fermi golden rule is given by

$$\frac{1}{\tau} = \frac{L}{2\pi} \int_{-\infty}^{+\infty} d\gamma \, \frac{2\pi}{\hbar} \left| M_n^{\{e,a\}} \right|^2 \delta(E(k') - E(k) \pm \hbar\omega_\gamma)$$

$$= \sum_{n,m} \frac{L}{2\pi} \int_{-\infty}^{+\infty} d\gamma \, \frac{2\pi}{\hbar} \left(\frac{E_1^\alpha}{\sqrt{L}} A_\gamma \frac{\omega_\gamma^2}{c_l^2 q_1} \right)^2 \left[\frac{\pi^2 \sin q_1 a}{q_1 a(\pi^2 - q_1^2 a^2)} \right]^2$$

$$\times \left\{ \frac{1}{\pi \left(n + \tfrac{1}{2}\right) \left[1 - \left(n + \tfrac{1}{2}\right)^2\right]} \right\}^2 \left(n + \tfrac{1}{2} \pm \tfrac{1}{2}\right)$$

$$\times \delta \left(\frac{\hbar^2}{2m} (\gamma^2 \mp 2k\gamma) \pm \hbar\omega_\gamma \right), \tag{9.55}$$

where the sum over γ has been converted to an integral, n is the Bose–Einstein occupation number for the acoustic phonons, and L is the normalization length. Yu *et al.* (1994) evaluated this deformation-potential scattering rate numerically for GaAs quantum wires at 77 K for three different cross sections, 28.3 Å × 56.6 Å, 100 Å × 200 Å, and 50 Å × 200 Å. The results of these numerical calculations

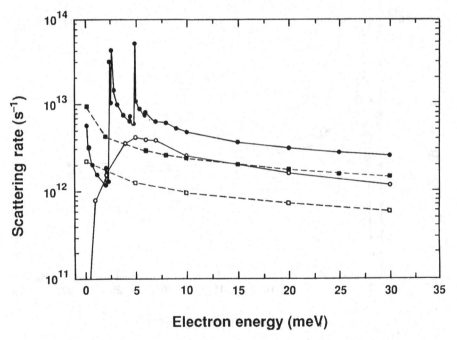

Figure 9.1. Deformation-potential emission and absorption scattering rates for bulk and confined acoustic phonons in a 28.3 Å × 56.6 Å GaAs quantum wire at 77 K. Confined modes: –●––●––●–, emission; –■ – –■ – –■–, absorption. Bulk modes: –○––○––○–, emission; – □ – –□ – –□ –, absorption. Energy thresholds for the width modes are at 0.03, 2.36, 2.55, 4.90, 7.30, and 7.40 meV. Energy thresholds for thickness modes are at 2.06, 4.44, 5.90, 9.87, 14.5, and 15.1 meV. One-dimensional density-of-states peaks are evident in the emission rate for the quantum wire. From Yu *et al.* (1994), American Physical Society, with permission.

are shown in Figures 9.1, 9.2, and 9.3 respectively. In each of these cases the acoustic phonon emission and absorption rates are shown for the GaAs quantum wire and for bulk GaAs. The distinct one-dimensional density-of-states peaks are evident for the case of phonon emission in each of the quantum-wire structures. These peaks occur at the threshold electron energy for the emission of one of the various width or thickness modes. The peak magnitudes of the scattering rates in these quantum wires suggest that carrier–acoustic-phonon scattering via the deformation potential is significant in quantum-wire elements of nanoscale devices. The calculations of Yu *et al.* (1994) are based on the assumption that the wires have perfectly uniform cross sections. In envisioned practical situations it is unlikely that the dimensional tolerances associated with the cross-sectional areas will be small enough to realize subatomic dimensional control. It is thus unlikely that the measured one-dimensional density-of-states peaks will be as pronounced as they are in Figures 9.1, 9.2, and 9.3.

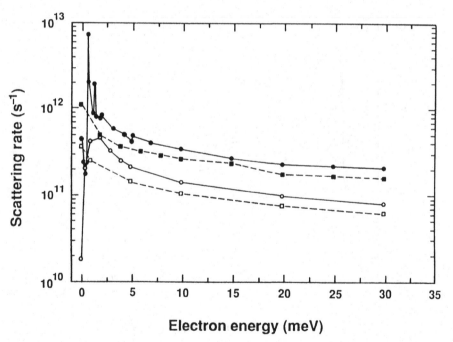

Figure 9.2. Deformation-potential emission and absorption scattering rates for bulk and confined acoustic phonons in a 100 Å × 200 Å GaAs quantum wire at 77 K. Confined modes: –●–●–●–, emission; – ■ – – ■ – – ■ –, absorption. Bulk modes: –○–○–○–, emission; – □ – – □ – – □ –, absorption. Energy thresholds for the confined modes are at 0.03, 0.65, 0.75, 1.39, 2.06, and 2.12 meV. Energy thresholds for the bulk modes are at 0.59, 1.26, 1.68, 2.80, 4.11, and 4.28 meV. One-dimensional density-of-states peaks are evident in the emission rate for the quantum wire. From Yu *et al.* (1994), American Physical Society, with permission.

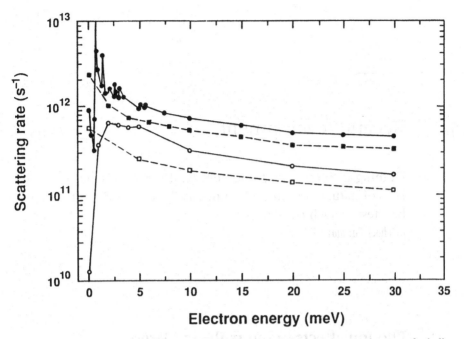

Figure 9.3. Deformation-potential emission and absorption scattering rates for bulk and confined acoustic phonons in a 50 Å × 200 Å GaAs quantum wire at 77 K. Confined modes: -●—●—●-, emission; – ■ – – ■ – – ■ –, absorption. Bulk modes: –○—○—○-, emission; – □ – – □ – – □ –, absorption. Energy thresholds for the confined modes are at 0.03, 0.65, 0.75, 1.39, 2.06, and 2.12 meV. Energy thresholds for the bulk modes are at 0.61, 2.58, 3.05, 5.56, 8.24, and 8.43 meV. One-dimensional density-of-states peaks are evident in the emission rate for the quantum wire. From Yu *et al.* (1994), American Physical Society, with permission.

Chapter 10

Recent developments

Nothing is too wonderful to be true, if it is consistent with the laws of nature, and in such things as these, experiment is the best test of such consistency.

Michael Faraday, 1849

10.1 Phonon effects in intersubband lasers

The effects of dimensional confinement on optical phonons, phonon-assisted electron intersubband transition rates (Teng *et al.*, 1998), and gain (Kisin *et al.*, 1997) have been evaluated in a series of studies on semiconductor lasers. Many of the novel semiconductor lasers – such as the tunneling injection laser and the quantum cascade laser – contain quantum wells with confinement dimensions of about 50 Å or less. An example of such a semiconductor laser structure is given in Figure 10.1. In this laser, the conduction band is engineered so that the upper and lower energy levels are E_3 and E_2 respectively, with a third level, E_1, such that phonon-assisted tunneling from level E_2 to level E_1 is promoted. The IF optical phonons are of special importance in such heterostructures. Two properties of the IF optical phonons account for their special significance (Stroscio, 1996) in narrow-well semiconductor lasers: (a) in narrow wells the interface phonons have appreciable interaction potentials throughout the quantum well since, as was demonstrated in Chapter 7, the interface optical phonons have potentials near the heterointerfaces of the form $ce^{-q|z|}e^{-i\mathbf{q}\cdot\boldsymbol{\rho}}$, where, for example, q has values of very roughly 0.02 Å^{-1} for a typical intrasubband transition in GaAs; and (b) the energies associated with phonon-assisted processes in these heterostructures can be substantially different from those in semiconductor lasers without significant dimensional confinement, since the interface phonons may have frequencies ω_q which are significantly different from those of the other phonons in the quantum well.

As an example, optical phonon (LO) transitions play an especially important role in novel intersubband lasers operating at infrared wavelengths, since selected intersubband transition rates are critical to establishing and maintaining a population inversion. In many specific cases, such lasers (Faist *et al.*, 1996a, b; Zhang *et al.*, 1996) can be designed only if quantum-well regions have thicknesses as small as 25–50 Å. In such polar-semiconductor structures, the dominant electron energy relaxation processes are due to the electron–LO-phonon interaction. In this subsection these processes will be considered and their role in the operation of novel semiconductor lasers will be examined.

Teng *et al.* (1998) determined the optical phonon (LO) modes and the electron–LO-phonon intersubband transition rates due to the electron–LO-phonon interactions in two generic heterostructures similar to the regions where phonon-induced transitions occur in a variety of dimensionally confined semiconductor lasers. These structures are depicted in Figure 10.2, where materials 1, 2, and 3 are GaAs, $Al_{0.25}Ga_{0.75}As$ and $Al_{0.4}Ga_{0.6}As$ respectively.

The heterostructure of Figure 10.2(*a*) has three interfaces and an asymmetrical potential profile. Figure 10.2(*b*) is a four-interface structure with a symmetrical potential. The dielectric constants of regions 1, 2, and 3 are taken as $\epsilon_1(\omega)$, $\epsilon_2(\omega)$, and $\epsilon_3(\omega)$ respectively. The dielectric continuum model of optical phonons, Sections 7.1 and 7.3 and Appendix C, leads immediately to the following LO phonon

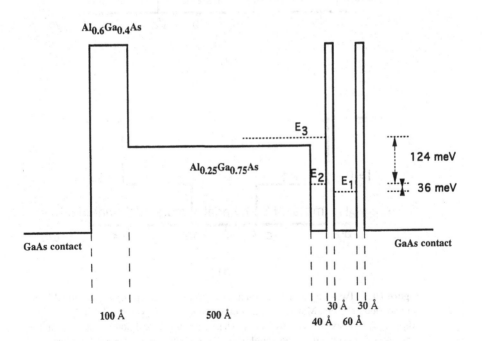

Figure 10.1. Step quantum-well laser with narrow quantum wells.

potentials: in polar material 1, where $i = 2$, the confined LO modes are given by

$$\phi(z) \propto \begin{cases} \cos \dfrac{m\pi}{a} z, & m = 1, 3, 5, \ldots \\[2mm] \sin \dfrac{m\pi}{a} z, & m = 2, 4, 6, \ldots \quad |z| < \dfrac{a}{2}. \end{cases} \tag{10.1}$$

The confined LO modes in polar material 2, where $i = 1$, are given by

$$\phi(z) \propto \begin{cases} \cos \dfrac{2m\pi}{b-a}\left(z + \dfrac{b+a}{4}\right), & m = 1, 3, 5, \ldots \\[2mm] \sin \dfrac{2m\pi}{b-a}\left(z + \dfrac{b+a}{4}\right), & m = 2, 4, 6, \ldots \end{cases}$$

$$-\frac{b}{2} < z < -\frac{a}{2}. \tag{10.2}$$

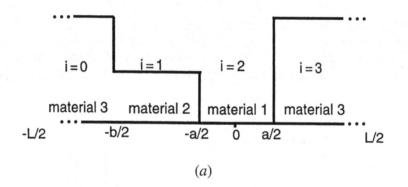

(a)

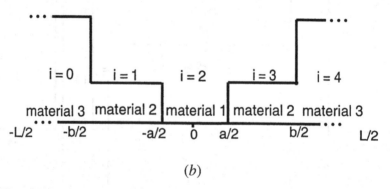

(b)

Figure 10.2. Two generic heterostructures common to dimensionally confined semiconductor lasers. Materials 1, 2, and 3 are GaAs, $Al_{0.25}Ga_{0.75}As$ and $Al_{0.4}Ga_{0.6}As$ respectively. Structure (a) has an asymmetrical potential profile and structure (b) a symmetrical potential profile. From Teng et al. (1998), American Institute of Physics, with permission.

For the half-space LO modes in polar material 3, where $i = 0, 3$,

$$\phi(z) \propto \begin{cases} \sin \dfrac{2m\pi}{L}\left(z + \dfrac{b}{2}\right), & m = 1, 2, 3, \ldots & z \leq -\dfrac{b}{2} \\[4mm] \sin \dfrac{2m\pi}{L}\left(z - \dfrac{a}{2}\right), & m = 1, 2, 3, \ldots & z \geq \dfrac{a}{2}. \end{cases} \tag{10.3}$$

Moreover, the interface LO phonon modes are as given in subsection 7.3.4 and have the form

$$\phi_i(\mathbf{q}, z) \propto c'_{i-}e^{-qz} + c'_{i+}e^{+qz}, \tag{10.4}$$

where the values of $c'_{i\pm}$ are determined by the boundary conditions:

$$\begin{aligned} \phi_1(\mathbf{q}, z = -L/2) &= 0 & \Longrightarrow \quad c'_{0-} &= 0, \\ \phi_3(\mathbf{q}, z = L/2) &= 0 & \Longrightarrow \quad c'_{3+} &= 0, \\ \phi_i(\mathbf{q}, z_i) &= \phi_{i-1}(\mathbf{q}, z_i), \\ \epsilon_i \frac{\partial \phi_i(\mathbf{q}, z_i)}{\partial z} &= \epsilon_{i-1}\frac{\partial \phi_{i-1}(\mathbf{q}, z_i)}{\partial z}, \end{aligned} \tag{10.5}$$

where L is taken to be large compared with other dimension scales in the system, and z_i designates the z-coordinate at the location of the heterointerface between regions i and $i - 1$. The conditions on the continuity of the potential and the continuity of the normal component of the electric displacement may be used to relate recursively the non-zero coefficients $c'_{i\pm}$ to the dielectric constants of the various materials composing the heterostructure. The boundary condition at $z = L/2$ requires that $c'_{3+} = 0$ and the dispersion relation for the interface modes for the heterostructure of Figure 10.2(a) is

$$\begin{aligned} 0 = \frac{1}{\epsilon_1 \epsilon_2 \epsilon_3} \Big[& (\epsilon_3 - \epsilon_1)(\epsilon_1 + \epsilon_2)(\epsilon_2 - \epsilon_3)\, e^{-q(a+b)} \\ & + (\epsilon_3 - \epsilon_1)(\epsilon_1 - \epsilon_2)(\epsilon_2 + \epsilon_3)\, e^{-2qa} \\ & + (\epsilon_3 + \epsilon_1)(\epsilon_1 - \epsilon_2)(\epsilon_2 - \epsilon_3)\, e^{-q(b-a)} \\ & + (\epsilon_3 + \epsilon_1)(\epsilon_1 + \epsilon_2)(\epsilon_2 + \epsilon_3) \Big], \end{aligned} \tag{10.6}$$

where the generalized Lyddane–Sachs–Teller relationship is used to write $\epsilon_n(\omega)$ in terms of $\epsilon_n(\infty)$, equations (7.5), (7.6); for binary materials,

$$\epsilon_n(\omega) = \epsilon_n(\infty)\frac{\omega^2 - \omega_{LO,n}^2}{\omega^2 - \omega_{TO,n}^2}, \tag{10.7}$$

and for ternary materials of the form $A_y B_{1-y} C$

$$\epsilon_n(\omega) = \epsilon_n(\infty)\left(\frac{\omega^2 - \omega_{LO,n,a}^2}{\omega^2 - \omega_{TO,n,a}^2}\right)\left(\frac{\omega^2 - \omega_{LO,n,b}^2}{\omega^2 - \omega_{TO,n,b}^2}\right), \tag{10.8}$$

where, as discussed in Section 7.1, the subscript a denotes frequencies associated with the dipole pairs AC and the subscript b denotes frequencies associated with the dipole pairs BC. The condition that $c'_{3+}(\omega) = 0$ is met for 10 frequencies and, accordingly, there are 10 interface modes: six of these correspond to GaAs-like modes with energies in the range of 32 to 37 meV and four are AlAs-like modes with energies close to 46 meV. The dispersion relations for these 10 modes are shown in Figure 10.3.

As in Teng *et al.* (1998), the dielectric constant is taken as

$$\epsilon(\infty) = 10.89 - 2.73x, \tag{10.9}$$

where x represents the content of the Al in $Al_xGa_{1-x}As$. The phonon energies used in determining these dispersion relations are given in Table 10.1.

For the structure of Figure 10.2(b), the boundary condition determines that $c'_{4+}/c'_{0+} = 0$ and the dispersion relations for the interface modes – as determined by the previously described iteration procedure – are given by

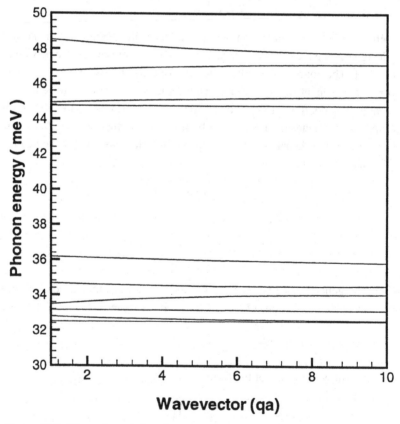

Figure 10.3. Dispersion relations for the interface LO phonon modes of the structure depicted in Figure 10.2(a). From Teng *et al.* (1998), American Institute of Physics, with permission.

$$0 = \frac{1}{\epsilon_1 \epsilon_2^2 \epsilon_3} \left[-(\epsilon_3 - \epsilon_2)^2 (\epsilon_1 + \epsilon_2)^2 e^{2qb} - 2(\epsilon_3^2 - \epsilon_2^2)(\epsilon_2^2 - \epsilon_1^2) e^{q(a+b)} \right.$$
$$+ (\epsilon_3^2 - \epsilon_2^2)^2 (\epsilon_2 - \epsilon_1)^2 e^{2q(b-a)} + 2(\epsilon_3^2 - \epsilon_2^2)(\epsilon_2^2 - \epsilon_1^2) e^{q(b-a)}$$
$$\left. - (\epsilon_3 + \epsilon_2)^2 (\epsilon_2 - \epsilon_1)^2 e^{2qa} + (\epsilon_3 + \epsilon_2)^2 (\epsilon_1 + \epsilon_2)^2 \right].$$

$$(10.10)$$

The 14 solutions of this equation correspond to six GaAs-like interface modes and eight AlAs-like interface modes. The potentials for the six GaAs-like modes are plotted in Figure 10.4 for $qa = 0.5$.

These LO phonon modes are either symmetric or antisymmetric, as is to be expected from the symmetry of the heterostructure. They must still be normalized. The normalization condition of subsection 7.3.4, equation (7.116),

$$\sum_i \frac{1}{4\pi} \frac{1}{2\omega} \frac{\partial \epsilon_i(\omega)}{\partial \omega} \int_{R_i} dz \left\{ q^2 |\Phi_i(\mathbf{q}, z)|^2 + \left| \frac{\partial \Phi_i(\mathbf{q}, z)}{\partial z} \right|^2 \right\} = \frac{\hbar}{2\omega L^2},$$

$$(10.11)$$

provides the necessary condition to determine the normalization constant for each mode. Applying this condition leads to the following normalized LO phonon potentials: in polar material 1, where $i = 2$, the confined LO modes are given by

$$\phi(z) = \left[\frac{4\pi \hbar}{L^2} \frac{1}{\partial \epsilon_i(\omega)/\partial \omega} \right]^{1/2} \left[\frac{1}{q^2 + (m\pi/a)^2} \right]^{1/2}$$
$$\times \left(\frac{2}{a} \right)^{1/2} \begin{cases} \cos \frac{m\pi}{a} z, & m = 1, 3, 5, \dots \\ \sin \frac{m\pi}{a} z, & m = 2, 4, 6, \dots \end{cases} \quad |z| < \frac{a}{2}. \quad (10.12)$$

In polar material 2 where $i = 1$, the confined LO modes are given by

Table 10.1. Values of phonon energies adopted for these calculations.

Phonon energy (meV)	GaAs	AlAs	$Al_x Ga_{1-x} As$
$\hbar\omega_{LO,a}$	36.25		$36.25 - 6.55x + 1.79x^2$
$\hbar\omega_{LO,b}$		50.09	$44.63 + 8.78x - 3.32x^2$
$\hbar\omega_{TO,a}$	33.29		$33.29 - 0.64x - 1.16x^2$
$\hbar\omega_{TO,b}$		44.88	$44.63 + 0.55x - 0.30x^2$

$$\phi(z) = \left[\frac{4\pi\hbar}{L^2} \frac{1}{\partial\epsilon_i(\omega)/\partial\omega} \right]^{1/2} \left\{ \frac{1}{q^2 + [2m\pi/(b-a)]^2} \right\}^{1/2}$$

$$\times \left(\frac{4}{b-a} \right)^{1/2} \begin{cases} \cos \dfrac{2m\pi}{b-a} \left(z + \dfrac{b+a}{4} \right) & m = 1, 3, 5, \ldots \\[2mm] \sin \dfrac{2m\pi}{b-a} \left(z + \dfrac{b+a}{4} \right) & m = 2, 4, 6, \ldots \end{cases}$$

$$-\frac{b}{2} < z < -\frac{a}{2}. \tag{10.13}$$

For the half-space LO modes in polar material 3, where $i = 0, 3$,

$$\phi(z) = \left[\frac{4\pi\hbar}{L^2} \frac{1}{\partial\epsilon_i(\omega)/\partial\omega} \right]^{1/2} \left[\frac{1}{q^2 + (2m\pi/L)^2} \right]^{1/2}$$

$$\times \left(\frac{4}{L} \right)^{1/2} \begin{cases} \sin \dfrac{2m\pi}{L} \left(z + \dfrac{b}{2} \right) & m = 1, 2, 3, \ldots \quad z \leq -\dfrac{b}{2} \\[2mm] \sin \dfrac{2m\pi}{L} \left(z - \dfrac{a}{2} \right) & m = 1, 2, 3, \ldots \quad z \geq \dfrac{a}{2}. \end{cases}$$

$$\tag{10.14}$$

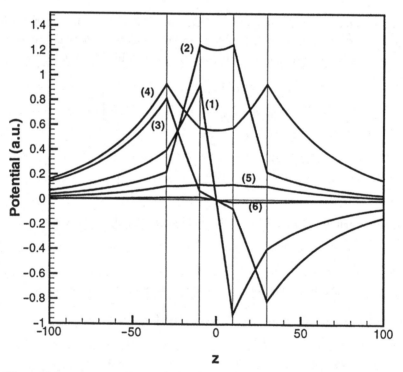

Figure 10.4. Potential profiles for the six GaAs-like interface LO phonon modes for the structure of Figure 10.2(b). The mode frequencies are (1) 35.86 meV, (2) 34.91 meV, (3) 34.26 meV, (4) 34.01 meV, (5) 33.17 meV, and (6) 33.07 meV. From Teng *et al.* (1998), American Institute of Physics, with permission.

As in subsection 7.3.4, the interface LO phonon mode potentials are of the form, equations (7.117), (7.118),

$$\Phi_i(\mathbf{q}, z) = A(c'_{i-}e^{-qz} + c'_{i+}e^{+qz}) = A\Psi_i(\mathbf{q}, z), \qquad (10.15)$$

where the normalization constant A is given by

$$A = \left(\frac{\hbar}{2\omega L^2}\right)^{1/2}\left(\sum_i \frac{1}{4\pi}\frac{1}{2\omega}\frac{\partial\epsilon_i(\omega)}{\partial\omega}\int_{R_i} dz\right.$$

$$\left.\times\left\{q^2\left|\Psi_i(\mathbf{q}, z)\right|^2 + \left|\frac{\partial\Psi_i(\mathbf{q}, z)}{\partial z}\right|^2\right\}\right)^{-1/2}. \qquad (10.16)$$

Finally, from subsection 7.3.4, equation (7.119),

$$H_{IF} = e\Phi_i(\mathbf{r}) = e\sum_{\mathbf{q}} e^{-i\mathbf{q}\cdot\boldsymbol{\rho}}\Phi_i(\mathbf{q}, z)(a^\dagger_{-q} + a_q)$$

$$= e\sum_{\mathbf{q}} e^{-i\mathbf{q}\cdot\boldsymbol{\rho}}A\Psi_i(\mathbf{q}, z)(a^\dagger_{-q} + a_q). \qquad (10.17)$$

For a particular phonon mode j the Fröhlich interaction Hamiltonian is given by

$$H_j = e\phi_j(\mathbf{r}) = e\sum_{\mathbf{q}} e^{-i\mathbf{q}\cdot\boldsymbol{\rho}}\phi_j(\mathbf{r})(a^\dagger_{-q} + a_q) \qquad (10.18)$$

and, from Section 8.1,

$$S^{\{e,a\}}_{n,n'}(\mathbf{k}, \mathbf{k}') = \frac{2\pi}{\hbar}\sum_q \left|M^{\{e,a\}}_{n,n'}(q)\right|^2 \delta(E(k') + E_{n'} - E(k) - E_n \pm \hbar\omega),$$

$$(10.19)$$

with

$$M^{\{e,a\}}_{n,n'}(q) = \left\langle n', \mathbf{k}', N_q + \tfrac{1}{2} \pm \tfrac{1}{2}\middle|H_j\middle|n, \mathbf{k}, N_q + \tfrac{1}{2} \pm \tfrac{1}{2}\right\rangle, \qquad (10.20)$$

where the initial and final energies each have been written as a sum of the in-plane and subband energies, $E(k) + E_n$ and $E(k') + E_{n'}$ respectively; $E(k) = \hbar^2k^2/2m$. Since the in-plane energy associated with the two-dimensional wavevector $\mathbf{k}$ has been separated from the subband energy, the subscript 2D of Section 8.1 has been omitted. Phonon-assisted transitions between the initial and final carrier states are depicted in Figure 10.5.

Teng *et al.* (1998) solved the Schrödinger equation numerically to determine the electron wavefunctions, for selected phonon modes, needed to evaluate the matrix elements for phonon-assisted transitions. These have focused on the emission process to gain insights into phonon-assisted processes of importance in narrow-well semiconductor lasers. As expected (Stroscio, 1996), their results show that the half-space modes and confined LO modes yield phonon-assisted rates that are small relative to those associated with the interface LO phonon modes. Moreover, it is found that certain interface modes make contributions about an order of magnitude larger than most of the remaining interface LO phonon modes. Figure 10.6 shows the transition rate for the dominant phonon-assisted transition in the symmetric structure of Figure 10.2(b). This maximum transition is associated with the emission of the antisymmetric interface longitudinal optical phonon mode having an energy close to

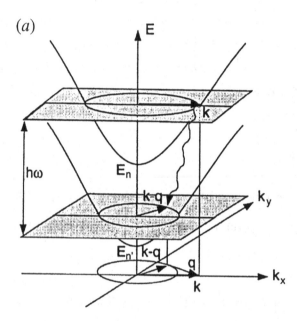

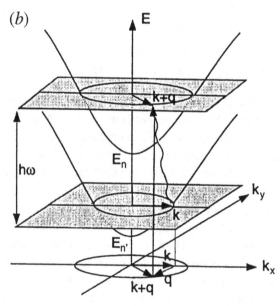

Figure 10.5. Phonon-assisted transitions between states $|n, \mathbf{k}\rangle$ and $|n', \mathbf{k}'\rangle$. (a) An emission event, between the state $|n, \mathbf{k}\rangle$ (upper plane) and the state $|n, \mathbf{k}' = \mathbf{k} - \mathbf{q}\rangle$ (lower plane). (b) An absorption event between $|n, \mathbf{k}\rangle$ (lower plane) and $|n, \mathbf{k}' = \mathbf{k} + \mathbf{q}\rangle$ (upper plane). The potential function is parabolic in the wavevector. From Teng *et al.* (1998), American Institute of Physics, with permission.

50 meV. As is apparent from Figure 10.6, the rate is substantially larger than the rate associated with the emission of a bulk phonon. Clearly, in the design of narrow-well semiconductor lasers it is essential that the interface modes be considered. The usual approximation of treating the phonons as bulk phonons is inadequate.

10.2 Effect of confined phonons on gain of intersubband lasers

In the previous section, it was demonstrated that it is essential to take into account the spectrum of dimensionally confined phonon modes in nanostructure-based intersubband semiconductor lasers. In the present section, it will be shown that the gain of an intersubband narrow-well semiconductor laser can be modeled accurately only if the realistic heterostructure phonon spectrum is taken into account (Kisin *et al.*, 1997). The system considered in calculating the gain of an intersubband quantum-well laser is an $Al_xGa_{1-x}As/GaAs/Al_xGa_{1-x}As$ double-heterostructure of width a. The electron wavefunctions and dispersion relations for this system

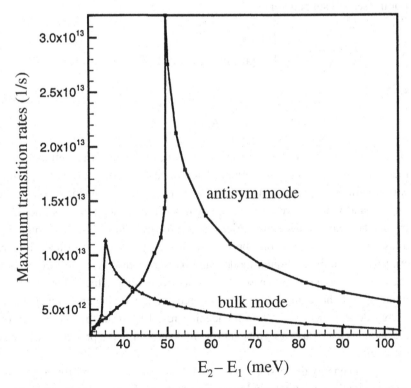

Figure 10.6. For the symmetric structure of Figure 10.2(*b*), the maximum transition rates for the antisymmetric interface mode and the bulk mode are depicted as functions of the difference $E_2 - E_1$ between the final and initial energies.

are modeled (Kisin *et al.*, 1997) with a four-band Kane model. The electronic wavefunctions (Gorfinkel *et al.*, 1996) for the states in subband n are taken to be

$$\Psi_{\mathbf{k}}^{(n)}(\boldsymbol{\rho}, z) = \frac{1}{\sqrt{Sa}} e^{i\mathbf{k}\cdot\boldsymbol{\rho}} \psi_{\mathbf{k}}^{(n)}(z), \tag{10.21}$$

where, as defined previously, $\boldsymbol{\rho} = (x, y)$ and S is the area of the heterostructure. For intersubband lasers – such as the quantum-cascade laser – involving only transitions in the conduction band, the four-component Kane wavefunction envelopes, $\psi_{\mathbf{k}}^{(n)}$, may be described simply in terms of scalar wavefunctions. For the lowest subband, $n = 1$,

$$\psi_1(z) = \sqrt{2}C_1 \begin{cases} \cos k_{z1}z & |z| < a/2, \\ \cos(k_{z1}a/2)\,e^{-\lambda_1(|z|-a/2)} & |z| > a/2, \end{cases} \tag{10.22}$$

with

$$k_{z1}\tan k_{z1}a/2 = \lambda_1 E_{w1}/E_{b1}. \tag{10.23}$$

For the second subband, $n = 2$,

$$\psi_2(z) = \sqrt{2}C_2 \begin{cases} \sin k_{z2}z & |z| < a/2, \\ \operatorname{sign} z \sin k_{z2}a/2\,e^{-\lambda_2(|z|-a/2)} & |z| > a/2, \end{cases} \tag{10.24}$$

with

$$k_{z2}\tan k_{z2}a/2 = -\lambda_2 E_{w2}/E_{b2}. \tag{10.25}$$

Here E_{wn} and E_{bn} are the energies for the nth subband of the well and the barrier measured with respect to the valence band in each material. For these approximate wavefunctions, it is clear that phonon-assisted transitions from the second to the lowest subband will involve only the antisymmetric interface phonons, the odd-parity confined phonon modes, and barrier modes. However, intrasubband phonon-assisted transitions will include contributions from the symmetric interface phonons, the even-parity confined phonon modes, and barrier modes. As will become apparent, such phonon-assisted processes have line-broadening effects that exert a major influence on the gain of the laser. Moreover, it will become apparent that the energy dependence of the line broadening resulting from the energy spectrum of the phonons also plays a significant role in determining the properties of the gain of the laser.

Subband energy dispersion curves for 60-ångstrom-wide and 100-ångstrom-wide quantum wells are depicted in Figure 10.7 along with examples of intersubband and intrasubband phonon-assisted processes; in Figure 10.7, $x = 0.3$ and the conduction band and valence band off-sets are $\Delta_c = 300$ meV and $\Delta_v = 150$ meV respectively.

The phonon spectrum used in the calculation of the intersubband optical gain in the quantum-well system under consideration includes the symmetric interface modes, the antisymmetric interface modes, the two lowest-order confined modes, and the half-space modes, which are referred to as the barrier modes in this section. The Fröhlich interaction Hamiltonians used here are those of the dielectric continuum model with electrostatic boundary conditions, as discussed in Chapter 7 and in Appendix C.

These modes may be obtained straightforwardly from the modes for the double-heterojunction, uniaxial structure of Appendix D by taking the limit where both $\epsilon(\omega)_\perp$ and $\epsilon(\omega)_\parallel$ are equal to $\epsilon(\omega)$. Figures 10.8(a), (b) and 10.9(a), (b) depict the contributing scattering rates (Kisin et al., 1997) for the scalar wavefunctions presented previously and for the phonon modes of the dielectric continuum model with electrostatic boundary conditions.

In calculating these results, the parameters have been selected as follows: $x = 0.4$, $\Delta_c = 300$ meV, $\Delta_v = 200$ meV, and $a = 60$ Å. The energy spectrum of Figure 10.8(a) illustrates clearly that there are several distinct thresholds.

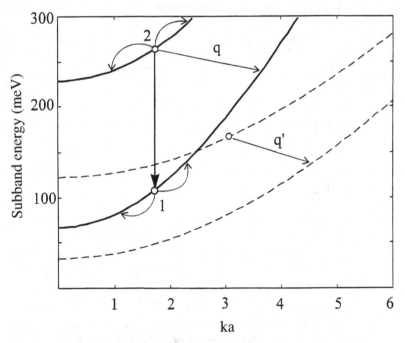

Figure 10.7. Energy dispersion curves for 60-ångstrom-wide (solid lines) and 100-ångstrom-wide (broken lines) Al$_{0.3}$Ga$_{0.7}$As/GaAs/Al$_{0.3}$Ga$_{0.7}$As quantum wells. Typical intersubband and intrasubband transitions are shown for both quantum wells. The energy gap for the GaAs well, E_g(GaAs), is taken as 1.4 eV and the ratio of the effective mass to the electron mass is taken as 0.067 for GaAs. From Kisin et al. (1997), American Institute of Physics, with permission.

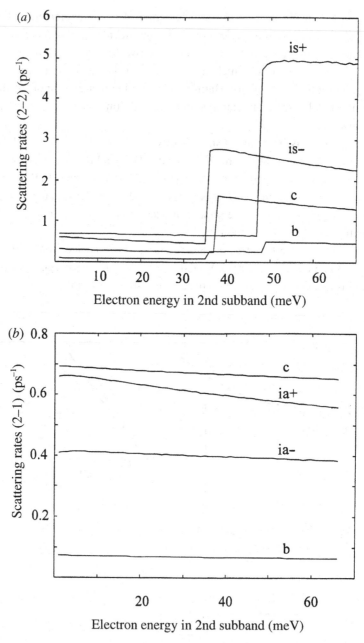

Figure 10.8. (*a*) Intrasubband scattering rates in the second subband as a function of the electron energy in the second subband. The high- and low-frequency symmetric interface LO phonon modes, the confined modes, and the barrier modes are labeled by is +, is −, c and b respectively. (*b*) Intersubband scattering rates from the second to the lowest subband as a function of the electron energy in the second subband. The high- and low-frequency antisymmetric interface LO phonon modes, the confined modes, and the barrier modes are labeled by ia+, ia−, c and b respectively. From Kisin *et al.* (1997), American Institute of Physics, with permission.

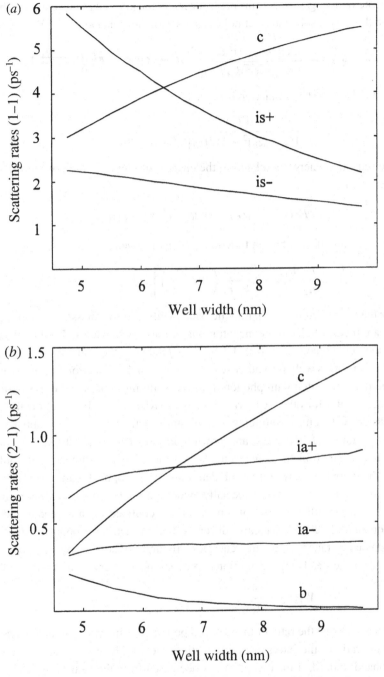

Figure 10.9. (*a*) Intrasubband scattering rates in the lowest subband as a function of well thickness. The initial electron energy in the first subband is 60 meV. (*b*) Intersubband scattering rates for transitions from the second to the first subband as a function of well thickness. The initial electron energy in the second subband is 10 meV. From Kisin *et al.* (1997), American Institute of Physics, with permission.

The optical gain spectrum $g(\Omega)$ for intersubband transitions between the second and the lowest subband (Gelmont *et al.*, 1996; Gorfinkel *et al.*, 1996) is given by

$$g(\Omega) = \frac{4\pi n_2 e^2 |z_{12}|^2 \Omega}{\hbar a c \sqrt{\epsilon(\infty)} k_B T_e} \int_0^\infty d\varepsilon\, \tau_\varepsilon(\Omega)\, e^{-\varepsilon/k_B T_e} \left(1 - \frac{f_1}{f_2}\right), \qquad (10.26)$$

where the line-shape function is defined by

$$\tau_\varepsilon(\Omega) = \frac{W(\varepsilon)}{[\Omega - \Omega_\varepsilon]^2 + [W(\varepsilon)]^2}, \qquad (10.27)$$

the transition energy is related to the energy difference $\hbar\Omega_0$ between the subbands at $k = 0$ by

$$\hbar\Omega_\varepsilon = E_2(k) - E_1(k) = \hbar\Omega_0 + \varepsilon_2(k) - \varepsilon_1(k), \qquad (10.28)$$

and the ratio of distribution functions, f_1/f_2, is given by

$$\frac{f_1}{f_2} = \frac{n_1}{n_2} \frac{m_1}{m_2} \exp\left[\frac{\varepsilon}{k_B T_e}\left(1 - \frac{m_1}{m_2}\right)\right], \qquad (10.29)$$

for the simplified case of parabolic bands with effective masses m_1 and m_2; n_1 and n_2 are the areal electron concentrations of the two subbands. The broadening term in the denominator of $\tau_\varepsilon(\Omega)$, $[W(\varepsilon)]^2$, is related to the phonon-assisted scattering rate, $1/\tau(\varepsilon)$, as well as to other scattering rates such as electron–plasmon scattering. The rate associated with phonon-assisted scattering processes generally dominates over the rates for other scattering processes (Kisin *et al.*, 1997) and is the only rate considered herein. Room-temperature gain spectra for a second subband electron concentration of 10^{11} cm^{-2} are shown in Figures 10.10 and 10.11.

It is clear from these gain curves that the effects of dimensional confinement of the phonons play a role in determining the optical gain of a narrow-well intersubband laser. This is especially evident in Figure 10.10(*b*), where gain curves based on the full spectrum of dimensionally confined phonon modes and on bulk phonons only are dramatically different and, in fact, have opposite signs over an appreciable range of photon energies. To understand this result, it is useful to consider the steady state condition expressing the conservation of particle flow:

$$\frac{n_2}{\tau_{12}} = W_{12} n_2 = \frac{n_1}{\tau_{\text{out}}}. \qquad (10.30)$$

This shows that the ratio of the subband populations in equilibrium depends directly on the ratio of the intersubband transition rate and the escape rate from the lower subband. Indeed, if the escape rate is decreased from the value consistent with this equilibrium condition then the population inversion is not expected to be maintained and the gain will become negative. For the parameters used in calculating the results of Figure 10.10(*b*), $\tau_{12} = 0.56$ ps and it is now clear why the gain curves for $\tau_{\text{out}} = 0.55$ ps are so sensitive to the energy spectrum of the phonon modes.

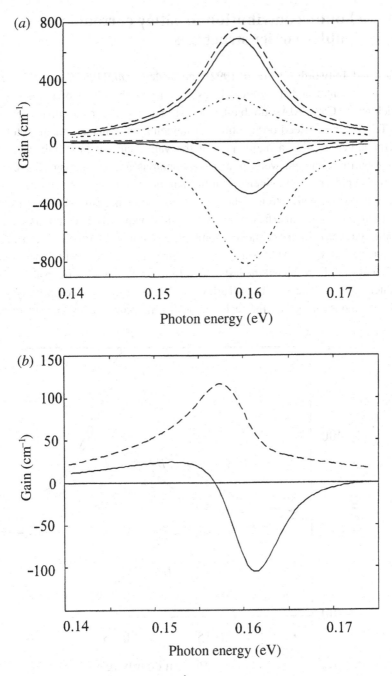

Figure 10.10. Optical gain in cm^{-1} for different rates of electron escape, τ_{out}, from the lowest subband: (a) τ_{out} is 0.4 ps and 0.6 ps; (b) τ_{out} is 0.55 ps. The broken lines are calculated assuming that only bulk GaAs LO phonons contribute; the broken-and-dotted lines assuming that only bulk GaAlAs phonons contribute. The solid lines are calculated from the full spectrum of dimensionally confined phonon modes. From Kisin *et al.* (1997), American Institute of Physics, with permission.

10.3   Phonon contribution to valley current in double-barrier structures

Turley and Teitworth (1991a, b, 1992) and Turley *et al.* (1993) performed consistent experimental and theoretical investigations of the effect of phonon-assisted tunneling in GaAs/AlAs double-barrier structures. The experiments of Turley and Teitsworth provided observations of tunneling currents and magnetotunneling spectra indicating that a major part of the valley current in certain GaAs/AlAs double-barrier quantum wells is due to phonon-assisted tunneling. These experiments also present clear evidence that the symmetric interface phonon mode makes a dominant contribution to the phonon-assisted tunneling current in these devices. The dominance of the interface mode is consistent with early calculations suggesting that the symmetric interface modes would play an important role in carrier transport in short-period GaAs/AlAs superlattices (Stroscio *et al.*, 1991a) and it is also consistent with the results of the last two sections on narrow-well intersubband semiconductor lasers. The results of Turley and Teitsworth provided early systematic studies demonstrating that the technologically important valley current in certain

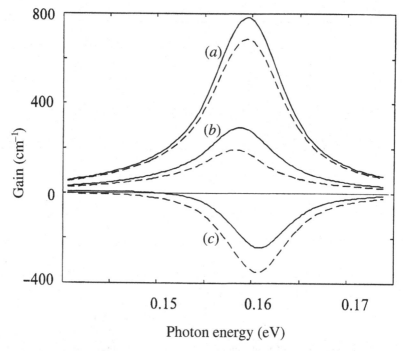

Figure 10.11. Optical gain in cm^{-1} for two different sets of phonon modes: broken lines, all phonon modes are included; solid lines, all modes but the barrier mode are included. Results are given for three different values of the electron escape rate, τ_{out}: (*a*) 0.4 ps; (*b*) 0.5 ps; (*c*) 0.6 ps. From Kisin *et al.* (1997), American Institute of Physics, with permission.

double-barrier resonant tunneling devices (DBRTDs) is, in fact, due partially to phonon-assisted tunneling. This is an important conclusion since DBRTDs have been the subject of intense study for almost two decades because of their perceived potential for playing a major role in new generations of high-performance electronic devices. Moreover, the careful and systematic results of Turley and Teitsworth demonstrate convincingly the utility of the dielectric continuum model of confined phonons.

Current–voltage measurements at 4.2 K by Turley *et al.* (1993) are shown in Figure 10.12. The double-barrier structure in this case was an 80-ångstrom-wide GaAs quantum well with a 33-ångstrom-wide AlAs emitter barrier and a 45-ångstrom-wide AlAs collector barrier. All these layers are nominally undoped. In Figure 10.12 the principal resonant tunneling peak is at about 0.54 volts and a distinct phonon-assisted tunneling peak is seen at about 0.81 volts.

The magnetotransport experiments of Turley *et al.* provided even more revealing data. Figures 10.13(*a*), (*b*) depict measured magnetic field versus applied voltage diagrams. The two 'Landau fans' converge at 0.67 volts and at 0.81 volts. Moreover, Figure 10.14 features two straight vertical lines to which other lines converge at −0.45 volts and −0.52 volts. By calibrating their experimental voltage scale, Turley *et al.* (1993) related the electronic energy scale ΔE_z and the applied voltage V through the relation $V = \alpha \Delta E_z$, where $\alpha = 5.49 \pm 0.15$ V/eV, and concluded that the Landau fans converging at 0.67 volts in Figures 10.13(*a*), (*b*) correspond to confined LO phonon modes with energies in the range 36.0 ± 0.9 meV. Likewise, the Landau fans converging at 0.81 volts in Figures 10.13(*a*), (*b*) as well as that

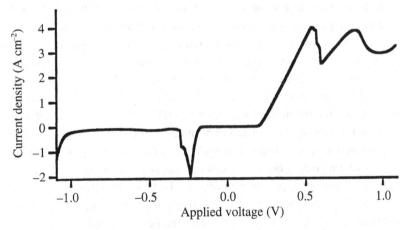

Figure 10.12. Measured current–voltage curve at 4.2 K for a double-barrier structure with an 80-ångstrom-wide GaAs quantum well having a 33-ångstrom-wide AlAs emitter barrier and a 45-ångstrom-wide AlAs collector barrier. The principal tunneling peak is at about 0.54 volts and a strong phonon-assisted tunneling peak occurs at 0.81 volts. From Turley *et al.* (1993), American Physical Society, with permission.

converging at −0.45 volts in Figure 10.14 correspond to symmetric interface modes with energies in the range 50.3 ± 1.3 meV. In both cases the expected phonon energies fall well within the error bars of Turley *et al.* (1993). The remaining Landau fan, converging at −0.52 in Figure 10.14, corresponds to emission of phonons with a changed Landau-level index.

Phonon emission processes have been observed in the measured currents in other double-barrier-based heterostructures (Choi *et al.*, 1990; Choi *et al.*, 1996) albeit at scales where dimensional confinement is not expected to lead to modifications in the phonon modes. In particular, Choi *et al.* (1990) determined the current transfer ratio and its derivative r,

$$\alpha = I_C/I_E, \tag{10.31}$$

$$r = d\alpha/dV_E, \tag{10.32}$$

at 4.2 K. Here I_C is the collector current, I_E is the emitter current, and V_E is the bias voltage for the heterostructure shown in Figure 10.15.

The structure of Figure 10.15 consists of a 1000-ångstrom-wide GaAs emitter with n-type doping of 2×10^{17} cm^{-3}, an undoped 120-ångstrom-wide $Al_{0.35}Ga_{0.65}As$ injection barrier, a 1520-ångstrom-wide GaAs base doped at 2×10^{17} cm^{-3}, an undoped 165-ångstrom-wide $Al_{0.25}Ga_{0.75}As$ analyzer, and a 1000-ångstrom-wide gallium arsenide collector doped at 2×10^{17} cm^{-3}. Figure 10.16 presents the experimental and theoretical values of the transfer ratio and its derivative as functions of V_E.

The dominant peaks are at V_1 and V_2 and the peaks associated with phonon emission are the subsidiary peaks at V_{P1} and V_{P2}. These subsidiary peaks are separated by the LO phonon energy in GaAs – $\hbar\omega_0 = 36$ meV – as expected. In separate experiments, Choi *et al.* (1996) determined the hot-electron distribution $\rho(E)$, in a regime where both phonon emission and plasmon emission are important. The heterostructure used in these experiments consisted of a GaAs emitter layer, an AlGaAs injection barrier, a 1500-ångstrom-wide GaAs base doped at 2×10^{17} cm^{-3}, a double-barrier filter with a bandpass at 280 meV, and a GaAs collector. This structure and the theoretical and experimental electron energy distributions $\rho(E)$ are shown in Figure 10.17. The initial injected electron distribution is taken to be Gaussian and to have a width of 22 meV. The injection energy is taken to be $e|V_e|$.

The experimental photocurrent transfer ratio α_p, as a function of emitter voltage V_e, for the heterostructure shown in Figure 10.17 is presented in Figure 10.18 as a solid line. The theoretical contributions for plasmon emission, phonon emission, and both phonon and plasmon emission are designated by triangles, crosses, and circles respectively. The difference between the theoretical and experimental contributions at high V_e is due to the non-linear relation between the injection energy and V_e.

Based on these experimental and theoretical results, Choi *et al.* (1996) concluded that phonon emission dominates when the base doping density is less than 3.3×10^{17} cm^{-3}. For higher doping densities, plasmon emission becomes more important.

10.4 Phonon-enhanced population inversion in asymmetric double-barrier quantum-well lasers

Phonon-assisted interwell electronic transitions have been explored as a means of rapidly depopulating the lower laser level in intersubband semiconductor lasers

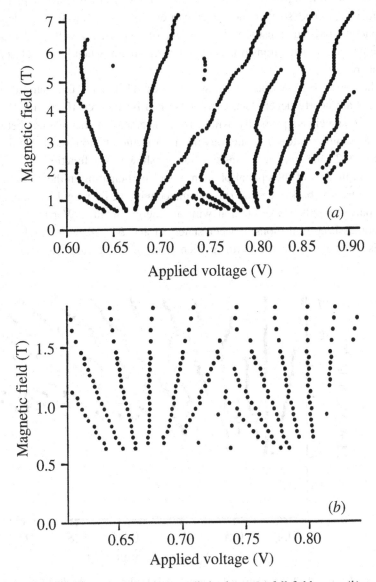

Figure 10.13. Magnetic field versus applied voltage: (*a*) full-field range; (*b*) low-field range. The heterostructure is a double-barrier structure with an 80-ångstrom-wide GaAs quantum well having a 33-ångstrom-wide AlAs emitter barrier and a 45-ångstrom-wide AlAs collector barrier. From Turley *et al.* (1993), American Physical Society, with permission.

(Faist *et al.*, 1996a, b; Gmachl *et al.*, 1998; Stroscio, 1996; Stroscio *et al.*, 1999). This fast depopulation promotes the lasing in intersubband lasers. Phonon-assisted depopulation rates can be enhanced in asymmetric double-quantum-well heterostructures if the subband being depopulated is degenerate with a second subband in the wider quantum well, as shown in Figure 10.19 (Stroscio *et al.*, 1999). The double-well active region of the AlAs/GaAs heterostructure depicted in this figure is designed so that the lower carrier state $A1$ of well A is depopulated by phonon-assisted transitions to state $B1$ of well B. As will be demonstrated, these phonon-assisted transitions are enhanced greatly when levels $A1$ and $B1$ are degenerate.

The interwell transition rate between states $A1$ and $B1$ is enhanced when $A1$ and $B2$ are nearly degenerate, as is illustrated by Figures 10.20 and 10.21. This effect is associated primarily with an enhancement in the overlap between the final and initial state wavefunctions involved in the phonon-assisted transition. In Figure 10.20 the interwell transition rate is shown as a function of a_1 for three different models of the optical phonon modes participating in the phonon-assisted transitions. The broken curve (a) is calculated with the Fermi golden rule, assuming that only the bulk GaAs phonon with an energy of 36 meV participates in the phonon-assisted transitions between states $A1$ and $B1$. Likewise, the broken curve (b) is calculated with the Fermi golden rule, assuming that only the bulk AlAs

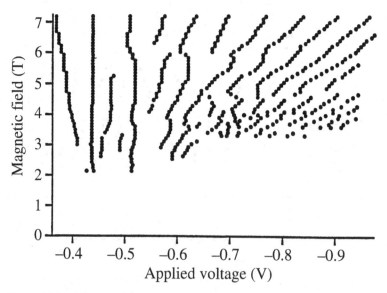

Figure 10.14. Magnetic field versus applied voltage: reverse bias diagram. The heterostructure is a double-barrier structure with an 80-ångstrom-wide GaAs quantum well having a 33-ångstrom-wide AlAs emitter barrier and a 45-ångstrom-wide AlAs collector barrier. From Turley *et al.* (1993), American Physical Society, with permission.

phonon with an energy of 51 meV participates in the phonon-assisted transitions between states $A1$ and $B1$. The remaining curve (solid line) is calculated with the Fermi golden rule, using the full spectrum of optical phonon modes as discussed in Chapter 7 and in Sections 10.1 and 10.2. The features on the solid curve between the thresholds for curves (a) and (b) are due to interface modes in the nominal reststrahlen band.

The peak values of the interwell phonon-assisted transition rates for conditions of near degeneracy of energy levels $A1$ and $B2$ are shown in Figure 10.21 for a series of values of the width a_3 of the wider well. For $a_3 = 18$ nm it is clear that the interwell transition rate is enhanced by about an order of magnitude as a result of the increased overlap of the final and initial electronic wavefunctions near conditions of degeneracy between levels $A1$ and $B2$.

In summary, the phonon-assisted transition rate τ_{out} for the intersubband laser structure of Figure 10.19 is enhanced significantly when the heterostructure is engineered so that there is a near degeneracy between states $A1$ and $B2$. In this case, the combination of band structure engineering and phonon engineering leads

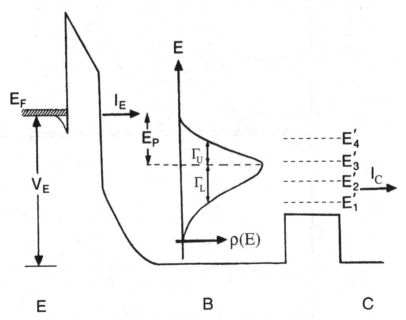

Figure 10.15. Band structure of the device studied by Choi *et al.* (1990) under a bias V_E. The energy distribution of the current, the so-called hot-electron distribution $\rho(E)$, is shown in the barrier region B. The current emerging from the emitter E is I_E and the current at the collector C is indicated by I_C. The states E'_n ($n = 1, 2, 3, 4$) are resonant states created in the barrier. E_F is the Fermi energy and E_P is the energy of the emitted phonon. Γ_U and Γ_L are the effective 'upper' and 'lower' contributions to the half-width of the energy distribution after phonon emission. From Choi *et al.* (1990), American Physical Society, with permission.

to a significant favorable enhancement in the expected population inversion for the double-well intersubband laser.

10.5   Confined-phonon effects in thin film superconductors

It has been recognized for a number of years that the dimensional confinement of carriers in thin film supeconductors leads to modification of the superconducting energy gap Δ and the critical temperature T_c (Blatt and Thompson, 1963; Thompson and Blatt, 1963; Yu *et al.*, 1976). In particular, it has been understood that size-

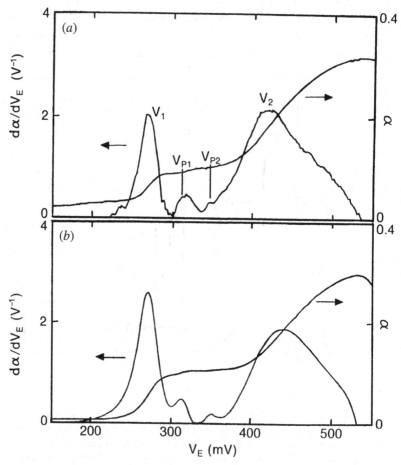

Figure 10.16. Transfer ratio α (smooth curve) and its derivative (peaked curve) as functions of V_E: (*a*) experimental values and (*b*) theoretical values. The subsidiary peaks at V_{P1} and V_{P2} are due to phonon emission. The arrows show the current directions. From Choi *et al.* (1990), American Physical Society, with permission.

quantization effects are manifest, as a function of the film thickness d, each time one of the two-dimensional subband energy levels $E_n(d)$ passes through the Fermi level as the film thickness is varied. Moreover, the Bardeen–Cooper–Schrieffer (BCS) theory of superconductivity is based on the formation of Cooper pairs whose binding energy is determined by the phonon-mediated pairing of electrons. Thus, it is reasonable to expect that the effects of phonon confinement play a role in determining the characteristics of thin film superconductors. Hwang *et al.* (2000) have recently investigated this possibility within the context of the BCS theory for thin films, by replacing the bulk phonon by the spectrum of confined phonons for the thin film. These results of Hwang *et al.* are expected to reveal selective qualitative features associated with phonon confinement in thin film superconductors but are not complete enough to give rigorous qualitative predictions. Only selected qualitative features are expected, since the calculations assume infinitely high electronic barriers at the interfaces of the thin film and since the interface phonon modes are not included in the spectrum of phonon modes. Figure 10.22 shows the

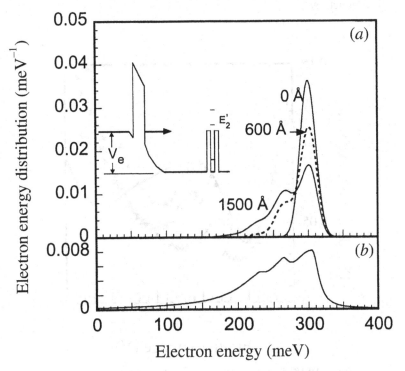

Figure 10.17. (*a*) Theoretical and (*b*) experimental electron energy distribution $\rho(E)$ for the heterostructure described previously and shown in the inset. The arrow in the inset shows the current direction. The theoretical electron energy distributions $\rho(E)$ are shown at 0 Å, 600 Å, and 1500 Å from the point of injection. The experimental curve for $\rho(E)$ was obtained at $V_e = -0.3$ V. From Choi *et al.* (1996), American Institute of Physics, with permission.

superconducting transition temperature T_c for a thin film with an electron density $n = 2 \times 10^{22}$ cm^{-3} as a function of the film thickness, d. The Debye energy, $\hbar\omega_D$, is taken to be 100 K. In this figure the broken line represents the results of Blatt and Thompson (1963) and Thompson and Blatt (1963); these results were obtained for the case where the electrons are treated as dimensionally confined but only bulk phonons are considered. The solid line represents the results of Hwang *et al.* (2000) where the BCS prediction for the superconducting transition temperature T_c is modified to take into account both electron confinement and phonon confinement.

The results of Hwang *et al.* (2000) demonstrate that phonon confinement effects play a role in the superconducting properties of thin film superconductors. The conclusion was expected since thin film superconductors may be as thin as several tens of ångstroms and since the binding energy of the phonon-mediated Cooper pairs is determined by the electron–phonon interaction. For films with thicknesses in the range shown in Figure 10.22 it is reasonable to expect that the interface phonons will also play an important role in determining selective properties of thin film superconductors.

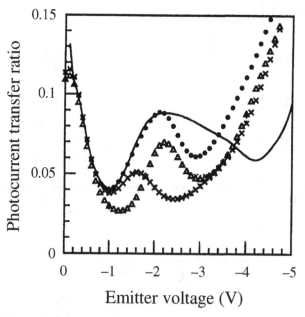

Figure 10.18. Experimental results for the photocurrent transfer ratio α are shown as a solid curve. The theoretical results for plasmon emission, phonon emission, and both phonon and plasmon emission are designated by triangles, crosses, and circles respectively. The difference between the theoretical and experimental contributions at high V_e is due to the non-linear relation between the injection energy and V_e. From Choi *et al.* (1996), American Institute of Physics, with permission.

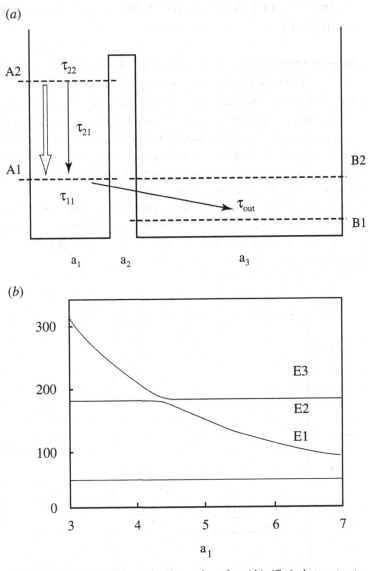

Figure 10.19. (a) Double-well active region of an AlAs/GaAs heterostructure designed so that the lower quasi-bound carrier state $A1$ of well A is depopulated by phonon-assisted transitions to state $B1$ of well B. The lasing transition is highlighted by the outlined arrow. Diagram (b) depicts the energies in meV, $E3$, $E2$, and $E1$ of the A1, B2, and B1 subbands respectively as functions of the width in nm, a_1, of the narrow well. The barrier thickness a_2 and the width of the wider well a_3 are 2 nm and 10 nm respectively. The lifetimes τ_{11}, τ_{22}, τ_{21}, and τ_{out} are respectively the intrasubband transition rates for the first and second subbands in well A, the intersubband transition rate for the second subband in well A, and the escape rate from state $A1$ to state $B1$. The lifetime τ_{out} should be short compared with τ_{21} in order to maintain a population inversion between $A2$ and $A1$. From Stroscio *et al.* (1999), American Institute of Physics, with permission.

10.6 Generation of acoustic phonons in quantum-well structures

Komirenko *et al.* (2000b) have calculated the rate of generation of high-frequency confined acoustic phonons via the deformation-potential interaction between drifting electrons and the confined acoustic phonon modes in a quantum well. These authors predict that the drifting electrons cause strong Cerenkov-like generation of confined acoustic phonons and that they determine the gain coefficient for this process as a function of the phonon frequency and parameters describing the quantum-well structure. A gain coefficient of several hundred cm^{-1} is predicted for a p-doped, 10-nm-wide Si/SiGe/Si quantum well.

To determine the rate of confined acoustic phonon generation, it is necessary to calculate the electron–confined-acoustic-phonon scattering rates in the Si/SiGe/Si quantum well. The carriers interact with acoustic phonons via the deformation-potential interaction, $H_{def}^{c,v}$, of Section 5.3, equation (5.39):

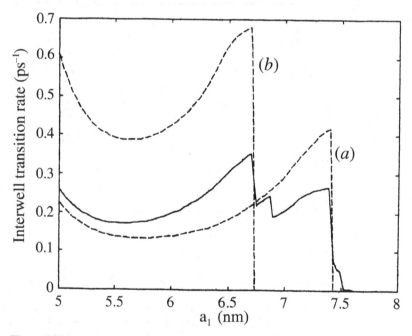

Figure 10.20. Interwell transition rates as a function of a_1 for phonon-assisted transitions between $A1$ and $B1$, as calculated by the Fermi golden rule for three different models of the optical phonons participating in the phonon-assisted process: the broken curve (*a*) assumes that only bulk GaAs phonons participate; the broken curve (*b*) assumes that only bulk AlAs phonons participate; and the solid curve includes the effects of the full spectrum of optical phonon modes for the heterostructure. As in Figure 10.19, $a_2 = 2$ nm and $a_3 = 10$ nm. From Stroscio *et al.* (1999), American Institute of Physics, with permission.

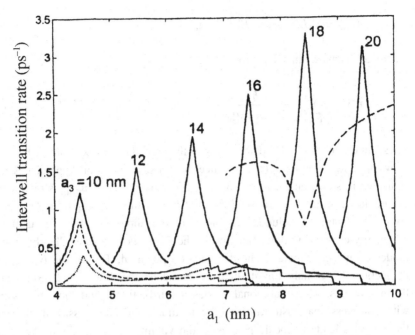

Figure 10.21. Peak values of the interwell phonon-assisted transition rates for conditions of near-degeneracy of energy levels $A1$ and $B2$ for a series of values of the width a_3 of the wider well. The curves corresponding to $a_3 = 10$ nm show the relative contributions of the confined phonons (broken curve) and interface phonons (dotted curve). The bold broken line with a minimum at about 8.5 nm is the rate of non-radiative intrawell intersubband transitions, $1/\tau_{12}$, for the case where $a_3 = 18$ nm. From Stroscio *et al.* (1999), American Institute of Physics, with permission.

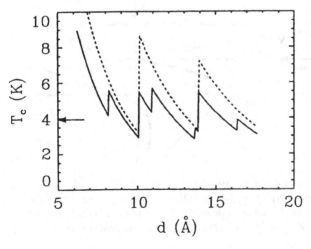

Figure 10.22. Superconducting critical temperature T_c for a thin film with an electron density $n = 2 \times 10^{22}$ cm^{-3}, as a function of the film thickness d. The solid line denotes the result for confined slab phonons and the broken line represents the corresponding result for bulk phonons. The Debye temperature, $\hbar\omega_D$, is taken to be 100 K. From Hwang *et al.* (2000), American Physical Society, with permission.

$$H_{\text{def}}^{c,v} = \Delta E_{c,v}(a) = E_1^{c,v} \nabla \cdot \mathbf{u}, \tag{10.33}$$

where, as discussed in subsection 7.6.1,

$$\mathbf{u}(\mathbf{r}) = \sum_{\mathbf{q}_{\parallel},n} \sqrt{\frac{\hbar}{2L^2 \rho \omega_n(\mathbf{q}_{\parallel})}} (a_{n,q} + a_{n,-q}^{\dagger}) \mathbf{w}_n(\mathbf{q}_{\parallel}, z) e^{i\mathbf{q}_{\parallel} \cdot \mathbf{r}_{\parallel}}. \tag{10.34}$$

These modes are normalized according to the procedures of Section 5.1 but in terms of $\mathbf{w}_n(\mathbf{q}_{\parallel}, z)$ instead of $\mathbf{u}_n(\mathbf{q}_{\parallel}, z)$, since the considerations of Appendix A make it clear that it is convenient to use $\mathbf{w}_n = \sqrt{\rho} \mathbf{u}_n$; $\mathbf{u}_n(\mathbf{q}_{\parallel}, z) \equiv \mathbf{u}_{n,\mathbf{q}_{\parallel}}$ is obtained from the mode amplitudes of subsection 7.6.2 by dividing by $\exp[i(\mathbf{q}_{\parallel} \cdot \mathbf{r}_{\parallel} - \omega_{\mathbf{q}_{\parallel}} t)]$, since this factor is included separately in the above equation for $\mathbf{u}(\mathbf{r})$ and in the energy-conserving delta function in the Fermi golden rule. Clearly, transverse modes do not contribute to the deformation potential. Two of the types of mode in Section 7.2 contain longitudinal components: the dilatational modes and the flexural modes. The dilatational modes are irrotational and they are associated with compressional distortions of the medium; the compressional character of these modes leads to local changes in the volume of the medium. As discussed in subsection 7.6.2, Wendler and Grigoryan (1988) derived the localized acoustic modes for an embedded quantum well. For a symmetric quantum well, the electrons couple via the deformation potential to the symmetric shear vertical (SSV) confined acoustic modes. The quantum-well heterostructure and the lowest-order SSV mode are depicted in Figures 10.23(a), (b).

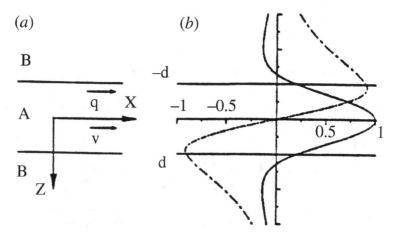

Figure 10.23. (a) Quantum-well heterostructure and (b) distribution of displacement components u_1 and u_3 for the SSV mode, for conditions corresponding to those of maximum amplification, as defined in Figure 10.24. v is the drift velocity of the electron. From Komirenko *et al.* (2000b), American Institute of Physics, with permission.

The electronic wavefunction for the lowest two-dimensional subband of the layer is given by

$$\Psi_{\mathbf{k}_{\parallel}}(\rho, z) = \frac{1}{\sqrt{S}} \exp(i\mathbf{k}_{\parallel} \cdot \rho)\chi(z), \tag{10.35}$$

where $\mathbf{k}_{\parallel}$ is the two-dimensional wavevector and, as defined previously, $\rho \equiv (x, y)$. Only this lowest electronic subband is taken to be occupied. Then, the probability of transitions between electronic states $\mathbf{k}_{\parallel}$ and $\mathbf{k}_{\parallel}'$ due to the emission or absorption of a confined phonon of band n and wavevector $\mathbf{q}_{\parallel}$ is given by

$$P^{(\pm)}(\mathbf{k}_{\parallel}, \mathbf{k}_{\parallel}' \,|\, n, \mathbf{q}_{\parallel}) = \frac{2\pi}{\hbar} |M(q)|^2 \left(n_{n,\mathbf{q}_{\parallel}} + \tfrac{1}{2} \pm \tfrac{1}{2} \right) \delta_{k_x \mp q, k_x'} \delta_{k_y, k_y'}$$

$$\times \delta \left(E(\mathbf{k}_{\parallel}) - E(\mathbf{k}_{\parallel}') \mp \hbar\omega_{n,q} \right)$$

$$\times F(\mathbf{k}_{\parallel}) \left[1 - F(\mathbf{k}_{\parallel}') \right], \tag{10.36}$$

where

$$M(q) = \frac{E_1^c}{\kappa^{el}(q)} \int_{-\infty}^{\infty} \nabla \cdot \mathbf{u}_{n,q} \chi^2(z) dz, \tag{10.37}$$

$n_{n,\mathbf{q}_{\parallel}}$ is the phonon occupation number of mode $\{n, \mathbf{q}_{\parallel}\}$, $\kappa^{el}(q)$ is the electron permittivity described in Appendix F (Bastard, 1988; Komirenko et al., 2000b), $F(\mathbf{k}_{\parallel}) = F(k_x, k_y)$ is the electron distribution function, and $q = |\mathbf{q}_{\parallel}|$. Appendix F provides a derivation of $\kappa^{el}(q)$ based on the Lindhart method as applied to a two-dimensional electron gas (Bastard, 1988).

To determine the net rate of generation of localized acoustic phonons, Komirenko et al. (2000b) considered

$$\frac{dn_{n,\mathbf{q}_{\parallel}}}{dt} = \gamma_{n,\mathbf{q}_{\parallel}}^{(+)} (1 + n_{n,\mathbf{q}_{\parallel}}) - \gamma_{n,\mathbf{q}_{\parallel}}^{(-)} n_{n,\mathbf{q}_{\parallel}} - \beta_{n,\mathbf{q}_{\parallel}} n_{n,\mathbf{q}_{\parallel}}, \tag{10.38}$$

where the $\gamma_{n,\mathbf{q}_{\parallel}}^{(\pm)}$ are determined by calculating the total rates of absorption and emission of phonons for mode $\{n, \mathbf{q}_{\parallel}\}$ and $\beta_{n,\mathbf{q}_{\parallel}}$ represents phonon losses such as non-electronic phonon absorption or anharmonic phonon decay. Defining the phonon increment by

$$\gamma_{n,\mathbf{q}_{\parallel}} \equiv \gamma_{n,\mathbf{q}_{\parallel}}^{(+)} - \gamma_{n,\mathbf{q}_{\parallel}}^{(-)}, \tag{10.39}$$

it follows that

$$\gamma_{n,\mathbf{q}_{\parallel}} = \frac{m^*}{\pi \hbar^3 q} |M(q)|^2 \left[\mathcal{I}^{(+)}(q) - \mathcal{I}^{(-)}(q) \right], \tag{10.40}$$

where m^* is the effective mass and

$$\mathcal{I}^{(\pm)}(q) = \int_{-\infty}^{\infty} dk_y \, F \left((\text{sign } q) \frac{m^* \omega_{n,\mathbf{q}_{\parallel}}}{\hbar |q|} \pm \frac{1}{2} q, k_y \right). \tag{10.41}$$

For the case where an external electric field causes the electrons to drift with a velocity V_{dr}, taking the electron distribution function to be a drifted Fermi distribution F_F, we have

$$F(k_x, k_y) = F_F \left(k_x - \frac{m^*}{\hbar} V_{dr}, k_y \right). \tag{10.42}$$

It then follows that $\alpha_{n,\mathbf{q}_\parallel} = \gamma_{n,\mathbf{q}_\parallel}/V_g > 0$ if the electron drift velocity exceeds the confined phonon phase velocity, $V_{dr} > \omega_{n,q}/|q|$, where the phonon group velocity $V_g = d\omega_{n,q}/dq$. This condition is the same as the criterion for Cerenkov emission.

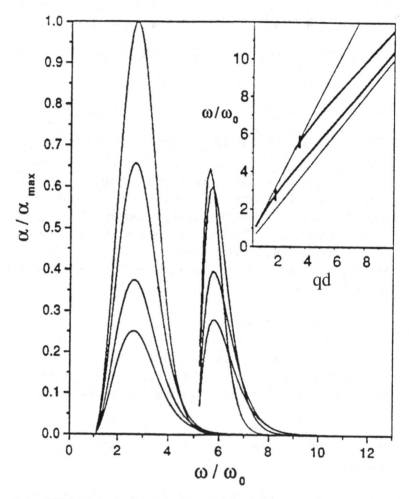

Figure 10.24. Dimensionless phonon amplification coefficient α/α_{max} versus phonon frequency ω/ω_0 for the two lowest SSV phonon branches. $T = 50$, 100, 150, and 200 K; the larger the value of T, the lower the value of α.
$\alpha_{max} = 290$ cm^{-1}, $\omega_0 = 110$ GHz. The values of ω/ω_0 and qd for the maxima of α are depicted in the inset for $T = 50$ K. From Komirenko *et al.* (2000b), American Institute of Physics, with permission.

Komirenko *et al.* (2000b) evaluated $\alpha_{n,\mathbf{q}_\parallel}$ for a p-doped, Si/SiGe/Si quantum well 10 nm thick, for the two lowest SSV phonon branches. The amplification coefficients for these phonon branches are shown in Figure 10.24 for temperatures T of 50, 100, 150, and 200 K. In these numerical results, the hole density is taken as 10^{12} cm^{-2}, the drift velocity is given by $V_{\mathrm{dr}} = 2.5\, c_{t1}$ and $c_{t1} = 3.4 \times 10^5$ cm s^{-1} for the SiGe layer.

The amplification coefficient for this p-doped Si/SiGe/Si quantum well is seen to be of the order of tens to hundreds of cm^{-1} for confined modes in the sub-THz frequency range.

Chapter 11

Concluding considerations

Now there is one outstandingly important fact regarding
Spaceship Earth, and that is that no instruction book came with
it.

R. Buckminster Fuller, *Operating Manual for Spaceship Earth*, 1969

11.1   Pervasive role of phonons in modern solid-state devices

As illustrated throughout this book, phonon effects are pervasive in modern solid-state devices. As is illustrated by the many examples for Chapters 7–10, the importance of these effects is usually at least as great for dimensionally confined structures as for bulk structures. Indeed, in Chapter 7 the effects of dimensional confinement were seen to be important even for biological structures! In this case, a cylindrical shell immersed in a fluid (Sirenko *et al.*, 1996b) was used to model the vibrational behavior of microtubules (MTs) immersed in water. In addition, the examples of Chapters 7 and 9 illustrate that the elastic continuum model provides an accurate description of acoustic phonons in dimensionally confined structures of many geometries including thin films, nanowires with rectangular and circular cross sections, and a variety of dot-like structures. These structures will inevitably be pervasive as elements of nanoscale structures mimicking the well known and larger microelectromechanical structures. Indeed, Cleland and Roukes (1996) reported a technique for fabricating nanometer-scale mechanical structures from bulk, single-crystal Si substrates. As another example of acoustic phonon effects in dimensionally confined structures, it was recently predicted theoretically that Cerenkov-like effects lead to the generation of high-frequency confined acoustic phonons in quantum wells (Komirenko *et al.*, 2000b); see Section 10.6.

In Chapter 8, values of carrier–optical-phonon scattering rates calculated for a variety of dimensionally confined semiconductor structures were found to exceed 10^{13} s^{-1}. Clearly, such rates are among the largest encountered in such structures and in many cases dominate over all other scattering mechanisms. It is well known that these relatively large carrier–phonon interactions play a major role in determining carrier mobilities (Ferry, 1991; Mitin *et al.*, 1999). In Chapter 10, phonons in dimensionally confined structures were seen to make a dominant contribution to the valley current in specific double-barrier quantum-well structures (Turley *et al.*, 1993) and the properties of thin film superconductors were shown to depend on the spectrum of confined phonons for sufficiently thin films (Hwang *et al.*, 2000).

Moreover, it was shown in Chapter 10 that dimensional confinement of phonons in intersubband semiconductor lasers changes the laser gain (Kisin *et al.*, 1997, 1998a, b) and leads to enhanced population inversions in some asymmetric double-barrier quantum-well lasers (Stroscio *et al.*, 1999). Indeed, Educato *et al.* (1993), Julien *et al.* (1995), Wang *et al.* (1996a, b), and Gauthier-Lafaye *et al.* (1997) have examined optically pumped intersubband scattering in coupled quantum-well lasers. These same authors have shown that interface–phonon-assisted transitions are important in such structures. Such results illustrate the importance of phonon confinement effects in intersubband lasers. As discussed in Chapter 10, the proper treatment of optical phonon confinement in optical systems such as intersubband semiconductor lasers depends critically on the detailed energy spectrum of the phonons. This is also true for a number of novel current-injection semiconductor intersubband lasers (Sun *et al.*, 1993; Zhang *et al.*, 1996; Sung *et al.*, 1996; Faist *et al.*, 1996a, b), as is illustrated by the calculations of Stroscio (1996) and Kisin *et al.* (1997, 1998a, b).

11.2 Future trends: phonon effects in nanostructures and phonon engineering

In this section, some speculative observations concerning phonons in nanostructures are made as a guide to potentially fruitful avenues of research on dimensionally confined phonons. In predicting future developments in the fields of phonon effects in nanostructures and phonon engineering, it is instructive to consider emerging international efforts for both nanostructures and bulk structures. Indeed, novel phonon effects in bulk materials are likely to have counterparts in nanostructures. Progress in femtosecond lasers and ultrafast spectroscopy and the continued development of novel techniques for fabricating nanostructures such as quantum dots (Empedocles, 1996) have been the basis for experimental observations of coherent oscillations of acoustic phonons in superlattices (Sun *et al.*, 1999), damped spherical acoustic breathing modes in quantum dots (Krauss and Wise, 1997, see the supporting analysis of Stroscio and Dutta, 1999), optical phonons near the surface of bulk GaAs (Cho *et al.*, 1990), optical phonons in Ge (Pfeifer *et al.*, 1992), and the excitation of coherent phonons in Sb, Bi, Te, and Ti_2O_3 (Cheng *et al.*, 1991). Coherent phonon effects are likely to represent one of the future trends in in nanostructures as well as in phonon engineering. Pötz and Shroeder (1999) pointed to the potential for achieving coherent control in atoms, molecules, and semiconductors.

Phonon engineering in nanostructures is likely to be a future avenue for device-related research and development. Indeed, the ability to model the phonon modes in dimensionally confined structures has been the basis for efforts to design nanostructures such that the resulting carrier and phonon states are tailored to yield dissipative and scattering mechanisms different from those of the corresponding bulk structures. Sakaki (1989) analyzed theoretically the electronic structure of quantum-wire superlattices and coupled quantum-box arrays with the purpose of engineering the electron energy subbands in such a way that carrier–optical-phonon scattering is suppressed. As another example, the use of metal–semiconductor heterointerfaces to reduce carrier–interface–phonon scattering rates (Stroscio *et al.*, 1992) represented one of the first approaches to phonon engineering in nanostructures. Indeed, the first calculations of carrier–LO-phonon scattering rates in polar-semiconductor quantum wires (Stroscio *et al.*, 1990; Kim *et al.*, 1991) revealed that the scattering of carriers with interface optical phonon modes dominates over the scattering associated with confined and half-space phonons. As a means of reducing the unwanted inelastic scattering caused by the interface LO phonons in semiconductor quantum wires, it was suggested (Stroscio *et al.*, 1992) that the large carrier–LO-phonon scattering caused by interface phonons could be eliminated by the encapsulation of the quantum wire in a metal. In a related example of phonon engineering, Leburton (1997) modeled dissipation- and scattering-time engineering in heterostructure-based quantum devices. Indeed, Leburton and his collaborators – Educato *et al.* (1993), Julien *et al.* (1995), Wang *et al.* (1996a, b), and Gauthier-Lafaye *et al.* (1997) – have exploited such effects in optically pumped intersubband scattering in coupled quantum-well lasers. Clearly, efforts in phonon engineering will be one of the future research trends based on the theories of confined phonons presented in this book.

Appendices

Appendix A: Huang–Born theory

Huang (1951) and Born and Huang (1954) took the most general form of the microscopic theory of diatomic polar crystals to be described by a pair of equations relating $\mathbf{w}$, $\mathbf{E}$, and $\mathbf{P}$,

$$\ddot{\mathbf{w}} = a\mathbf{w} + b\mathbf{E} \quad \text{and} \quad \mathbf{P} = d\mathbf{w} + c\mathbf{E}. \tag{A.1}$$

Born and Huang showed that d equals b as a result of energy conservation. In (A.1), $\mathbf{w} = \sqrt{\mu N/V}\,\mathbf{u}$; $\mathbf{u}$, μ, N, and V are the relative displacement of the two ions, the reduced mass of the ion pair, the number of unit cells in the crystal, and the volume of the crystal respectively. Assuming a time dependence $e^{-i\omega t}$ and eliminating $\mathbf{w}$ from these equations yields the relation

$$\mathbf{P} = \left(c + \frac{b^2}{-a - \omega^2}\right)\mathbf{E}. \tag{A.2}$$

Since $\mathbf{D} = \mathbf{E} + 4\pi\mathbf{P} = \epsilon(\omega)\mathbf{E}$, (A.2) may be rewritten as

$$\epsilon(\omega) = 1 + 4\pi c + \frac{4\pi b^2}{-a - \omega^2}. \tag{A.3}$$

From the generalized Lyddane–Sachs–Teller relation given in subsection 2.3.3, (2.32), it is clear that

$$\epsilon(\omega) = \epsilon(\infty)\frac{\omega_{LO}^2 - \omega^2}{\omega_{TO}^2 - \omega^2} = \epsilon(\infty)\frac{[(\omega_{TO}^2 - \omega^2) + (\omega_{LO}^2 - \omega_{TO}^2)]}{\omega_{TO}^2 - \omega^2}$$

$$= \epsilon(\infty) + \epsilon(\infty)\frac{\omega_{LO}^2/\omega_{TO}^2 - 1}{1 - \omega^2/\omega_{TO}^2}$$

$$= \epsilon(\infty) + \epsilon(\infty)\frac{\epsilon(0) - \epsilon(\infty)}{1 - \omega^2/\omega_{TO}^2}. \tag{A.4}$$

Comparing the relation (A.3) of Born and Huang with that based in the Lyddane–Sachs–Teller equation, it is evident that

$$c = \frac{\epsilon(\infty) - 1}{4\pi}, \quad a = -\omega_{TO}^2, \quad \text{and} \quad b = \left[\frac{\epsilon(0) - \epsilon(\infty)}{4\pi}\right]^{1/2}\omega_{TO}. \tag{A.5}$$

Thus, the pair of equations in (A.1) and (A.2) put forth by Huang and Born may be written as

$$(\omega_{TO}^2 - \omega^2)\mathbf{w} = \left[\frac{\epsilon(0) - \epsilon(\infty)}{4\pi}\right]^{1/2}\omega_{TO}\mathbf{E} \tag{A.6}$$

and

$$\mathbf{P} = \left[\frac{\epsilon(0) - \epsilon(\infty)}{4\pi}\right]^{1/2}\omega_{TO}\mathbf{w} + \frac{\epsilon(\infty) - 1}{4\pi}\mathbf{E}. \tag{A.7}$$

Re-introducing the relative displacement $\mathbf{u}$ these equations become

$$(\omega_{TO}^2 - \omega^2)\mathbf{u} = \left(\frac{V}{4\pi\mu N}\right)^{1/2}\sqrt{\epsilon(0) - \epsilon(\infty)}\,\omega_{TO}\mathbf{E}, \tag{A.8}$$

and

$$\mathbf{P} = \left(\frac{\mu N}{4\pi V}\right)^{1/2}\sqrt{\epsilon(0) - \epsilon(\infty)}\,\omega_{TO}\mathbf{u} + \frac{\epsilon(\infty) - 1}{4\pi}\mathbf{E}. \tag{A.9}$$

Appendix B: Wendler's theory

Wendler (1985) formulated a theory for the carrier–optical-phonon interactions in dielectric bilayer systems that includes the effect of the electronic polarizability on the phonon eigenmodes and their frequencies. Wendler's model predicts several relations that are useful in calculations of carrier–optical-phonon interactions in dimensionally confined systems. As in subsection 2.3.1, the displacement field in Wendler's model is related to the fields $\mathbf{E}(\mathbf{r})$ and $\mathbf{P}(\mathbf{r})$ through the driven oscillator equation and through the effective charge, e_n^* : for a binary medium labeled by subscript n,

$$-\mu_n\omega^2\mathbf{u}_n(\mathbf{r}) = -\mu_n\omega_{0n}^2\mathbf{u}_n(\mathbf{r}) + e_n^*\mathbf{E}^{\text{local}}(\mathbf{r}), \tag{B.1}$$

$$\mathbf{P}(\mathbf{r}) = n_n e_n^*\mathbf{u}_n(\mathbf{r}) + n_n\alpha_n\mathbf{E}^{\text{local}}(\mathbf{r}), \tag{B.2}$$

where $\mu_n = m_n M_n/(m_n + M_n)$ is the reduced mass, α_n is the electronic polarizability per unit cell and where by the Lorentz relation

$$\mathbf{E}^{\text{local}}(\mathbf{r}) = \mathbf{E}(\mathbf{r}) + \frac{4\pi}{3}\mathbf{P}(\mathbf{r}). \tag{B.3}$$

Here, $\mathbf{E}^{\text{local}}(\mathbf{r})$ is the local electric field associated with the optical modes acting on the electron shells of the ions. From Poisson's equation,

$$\Delta\Phi(\mathbf{r}, t) = -4\pi\varrho^{\text{total}}(\mathbf{r}, t), \tag{B.4}$$

where $\varrho^{\text{total}}(\mathbf{r}, t)$ includes both bulk and surface polarization charge density; using the Green's function approach Wendler took a solution to Poisson's equation of the form

$$\Phi(\mathbf{r}, t) = -\sum_\beta \int d^3\mathbf{r}' \frac{\partial}{\partial r_\beta} \frac{1}{|\mathbf{r} - \mathbf{r}'|} P_\beta(\mathbf{r}', t). \tag{B.5}$$

Then, with $E_\alpha(\mathbf{r}, t) = -\partial\Phi(\mathbf{r}, t)/\partial r_\alpha$ and with the Lorentz equation (B.3), it follows that

$$\begin{aligned}
E_\alpha^{\text{local}}(\mathbf{r}, t) = {} & \frac{4\pi}{3}(\lambda_n - \lambda_{0n})P_\alpha(\mathbf{r}, t) \\
& + 4\pi\sum_\beta \int d^3\mathbf{r}' \, \Gamma^{\alpha\beta}(\mathbf{r} - \mathbf{r}')P_\beta(\mathbf{r}', t),
\end{aligned} \tag{B.6}$$

where the Green's tensor is given by

$$\Gamma^{\alpha\beta}(\mathbf{r} - \mathbf{r}') = \frac{1}{4\pi} \frac{\partial^2}{\partial r_\alpha \partial r_\beta} \frac{1}{|\mathbf{r} - \mathbf{r}'|}. \tag{B.7}$$

By assuming that $\mathbf{P}(\mathbf{r}, t)$ depends sinusoidally on time, Wendler showed that

$$\frac{(\lambda_n - \lambda_{0n})\left(1 - \frac{4}{3}\pi n_n\alpha_n\right) + \frac{4}{3}\pi}{1 - n_n\alpha_n(\lambda_n - \lambda_{0n})} P_\alpha(\mathbf{r}) = -4\pi\sum_\beta \int d^3\mathbf{r}' \, \Gamma^{\alpha\beta}(\mathbf{r} - \mathbf{r}')P_\beta(\mathbf{r}'), \tag{B.8}$$

where

$$\lambda_n = 4\pi\omega^2/\omega_{\text{plasma},n}^2, \qquad \lambda_{0n} = 4\pi\omega_{0n}^2/\omega_{\text{plasma},n}^2, \tag{B.9}$$

and the plasma frequency squared, $\omega_{\text{plasma},n}^2$, is given by

$$\omega^2_{\text{plasma},n,a(b)} = 4\pi n_n e^{*2}_{n,a(b)}/\mu_{n,a(b)}. \tag{B.10}$$

By defining a two-dimensional Fourier transform for the electrostatic fields of the form

$$\mathbf{P}(\mathbf{r}) = \left(\frac{1}{2\pi}\right)\int_{-\infty}^{\infty} d^2 q_{\|} e^{i\mathbf{q}_{\|}\cdot\mathbf{r}_{\|}}\mathbf{P}(\mathbf{q}_{\|}, r_3), \tag{B.11}$$

with $\mathbf{r}_{\|}$ in the plane of the heterointerface of the bilayer system, Wendler showed that

$$\mathbf{E}^{\text{local}}(\mathbf{q}_{\|}, r_3) = \Phi_n \mathbf{P}(\mathbf{q}_{\|}, r_3), \tag{B.12}$$

where

$$\Phi_n = \frac{\lambda_n - \lambda_{0n}}{1 - n_n\alpha_n(\lambda_n - \lambda_{0n})} \tag{B.13}$$

and

$$\mathbf{u}(\mathbf{q}_{\|}, r_3) = \frac{\theta_n^{1/2}}{n_n e_n^*}\mathbf{P}(\mathbf{q}_{\|}, r_3), \tag{B.14}$$

with

$$\theta_n^{1/2} = \frac{1}{1 - n_n\alpha_n(\lambda_n - \lambda_{0n})}. \tag{B.15}$$

Moreover, the Green's function approach of Wendler yields several useful conditions:

$$\epsilon_n(\infty) = 1 + 4\pi\frac{n_n\alpha_n}{1 - \frac{4}{3}\pi n_n\alpha_n},$$

$$\omega^2_{\text{LO},n} = \omega^2_{0,n} + \frac{2}{3}\omega^2_{\text{plasma},n}\frac{1}{1 + \frac{8}{3}\pi n_n\alpha_n},$$

$$\omega^2_{\text{TO},n} = \omega^2_{0,n} + \frac{1}{3}\omega^2_{\text{plasma},n}\frac{1}{1 - \frac{4}{3}\pi n_n\alpha_n},$$

$$\omega^2_{\text{LO},n} - \omega^2_{\text{TO},n} = \frac{2}{3}\omega^2_{\text{plasma},n}\frac{1}{1 + \frac{8}{3}\pi n_n\alpha_n} + \frac{1}{3}\omega^2_{\text{plasma},n}\frac{1}{1 - \frac{4}{3}\pi n_n\alpha_n}$$

$$= \frac{\omega^2_{\text{plasma},n}}{(1 + \frac{8}{3}\pi n_n\alpha_n)(1 - \frac{4}{3}\pi n_n\alpha_n)}. \tag{B.16}$$

where n represents either material 1 or material 2. As may be verified algebraically, the Lyddane–Sachs–Teller relations of subsection 2.3.3 are satisfied by the frequencies in (B.16) (Wendler, 1985). Wendler's treatment of the effect of dielectric polarizability provides the basis for many works on optical phonons in dimensionally confined systems. The utility of Wendler's solution is illustrated in Chapter 7.

Appendix C: Optical phonon modes in double-heterointerface structures

Nash (1992) made an extremely important contribution to the field of phonons in nanostructures by showing that intrasubband and intersubband electron–phonon scattering rates are independent of the basis set used to describe the confined polar-optical phonons in a semiconductor slab. Comas *et al.* (1997) also formulated a model for treating both the mechanical and electrostatic fields associated with phonons in heterogeneous semiconductor structures.

In particular, Nash showed that the so-called slab modes derived with electrostatic boundary conditions, the guided modes derived with mechanical boundary conditions, and the reformulated slab modes of Huang and Zhu (1988) all predict the same scattering rates as long as each set of modes is orthogonal and complete. Moreover, Nash's analysis resolved the long-standing controversy over whether the normal modes satisfy mechanical or electromagnetic boundary conditions; specifically, Nash shows that the mechanical boundary conditions apply to $\mathbf{w} = \sqrt{\rho}\mathbf{u}$ (where $\mathbf{u}$ is the relative displacement of the ion pairs and ρ is the reduced mass per unit volume as discussed in Appendix A) and its derivatives, and the electromagnetic boundary conditions apply to Φ and $\mathbf{D}$. Furthermore, Nash showed that both types of boundary condition are necessary to obtain the normal modes of the heterostructure. Nash considered a Lagrangian density, for a polar material, of the form (Born and Huang, 1954),

$$\mathcal{L} = \frac{1}{2}\frac{\partial \mathbf{w}}{\partial t} \cdot \frac{\partial \mathbf{w}}{\partial t} - \frac{1}{2}\omega_{TO}^2 \mathbf{w} \cdot \mathbf{w}$$
$$+ \frac{1}{8\pi}\epsilon(\infty)\nabla\Phi \cdot \nabla\Phi - \gamma\mathbf{w} \cdot \nabla\Phi + \frac{1}{2}\sum_{ijkl} Z_{ijkl}\frac{\partial w_k}{\partial r_j}\frac{\partial w_l}{\partial r_i}, \tag{C.1}$$

where

$$\gamma = \left[\frac{1}{4\pi}\epsilon(\infty)(\omega_{LO}^2 - \omega_{TO}^2)\right]^{1/2}$$
$$= \left\{\frac{1}{4\pi}[\epsilon(0) - \epsilon(\infty)]\right\}^{1/2}\omega_{TO}$$
$$= \frac{\epsilon(\infty)}{4\pi}\omega_{LO}\left\{4\pi\left[\frac{1}{\epsilon(\infty)} - \frac{1}{\epsilon(0)}\right]\right\}^{1/2} \tag{C.2}$$

and where the displacement is described in terms of $\mathbf{w} = \sqrt{\rho}\mathbf{u}$ for the same reasons as in Appendix A. As will become apparent, it is convenient to define

$$\frac{1}{\beta} = \frac{4\pi}{\epsilon(\infty)}\gamma = \omega_{LO}\left\{4\pi\left[\frac{1}{\epsilon(\infty)} - \frac{1}{\epsilon(0)}\right]\right\}^{1/2}. \tag{C.3}$$

In the Lagrangian density (C.1), the first term is the kinetic energy, the second term is the potential energy of the lattice due to short-range forces, the third term is due to the potential energy of the macroscopic electric field in the absence of ionic motion, the fourth term is the potential energy associated with the coupling of the lattice to the macroscopic electric field, and the fifth term represents the quadratic dispersion associated with the short-range forces between ions, for the case of isotropic dispersion, $Z_{ijkl} = A\delta_{ij}\delta_{kl} + B\delta_{ik}\delta_{jl} + C\delta_{il}\delta_{jk}$.

The Euler–Lagrange equations derived by Nash from the variations of $\mathbf{w}$ and Φ to describe the optical phonon modes in a slab situated between $z = 0$ and $z = d$ are

$$\frac{\partial^2 \mathbf{w}}{\partial t^2} + \omega_{TO}^2 \mathbf{w} + \gamma \nabla \Phi + A \nabla^2 \mathbf{w} + (\mathbf{B} + \mathbf{C}) \nabla (\nabla \cdot \mathbf{w}) = 0 \tag{C.4}$$

and

$$\nabla \cdot [\epsilon(\infty) \nabla \Phi - 4\pi \gamma f(z) \mathbf{w}] = 0 \tag{C.5}$$

respectively. In these equations, γ is as given in (C.2) and $f(z) = \Theta(z) - \Theta(d - z)$, Θ being the Heaviside step function. Following Nash, $\mathbf{w}$ is written as the gradient of a mechanical potential $\bar{\chi}$, $\mathbf{w} = -\nabla \bar{\chi}$. The second of these equations is Poisson's equation,

$$\nabla \cdot [\epsilon(\infty) \nabla \Phi] = 4\pi \gamma \nabla [f(z) \mathbf{w}] = -4\pi \gamma \nabla^2 \bar{\chi} - 4\pi \gamma \nabla f(z) \nabla \bar{\chi}, \tag{C.6}$$

or

$$-\frac{1}{\epsilon(\infty)} \nabla \cdot [\epsilon(\infty) \nabla \Phi]$$
$$= \frac{4\pi \gamma}{\epsilon(\infty)} \left[\nabla^2 \bar{\chi} - \left.\frac{\partial \bar{\chi}}{\partial z}\right|_{z=d} \delta(z - d + \eta) + \left.\frac{\partial \bar{\chi}}{\partial z}\right|_{z=0} \delta(z - \eta) \right], \tag{C.7}$$

where the last two terms on the right-hand side of (C.7) arise from the surface polarization charge due to the discontinuity of the z-component of $\mathbf{w}$. In these results, $\mathbf{w}$ is taken to be normalized as described previously. That is, these modes are normalized (according to the procedures of Section 5.1) in terms of $\mathbf{w}_m(\mathbf{q}_\parallel, z)$ instead of $\mathbf{u}_m(\mathbf{q}_\parallel, z)$, since the considerations of Appendix A make it clear that it is convenient to use $\mathbf{w}_m = \sqrt{\rho} \mathbf{u}_m$:

$$\mathbf{u}_m(\mathbf{r}) = \sum_{\mathbf{q}_\parallel, m} \sqrt{\frac{\hbar}{2L^2 \rho \omega_m(\mathbf{q}_\parallel)}} (a_{m,q} + a_{m,-q}^\dagger) \mathbf{w}_m(\mathbf{q}_\parallel, z) e^{i\mathbf{q}_\parallel \cdot \mathbf{r}_\parallel}. \tag{C.8}$$

Here the subscript m is retained to remind us that this result holds for a general medium m. This last result is, of course, consistent with the normalization condition of subsection 7.3.1, (7.44),

$$\int L^2 [\sqrt{\mu n} \mathbf{u}_m(\mathbf{q}, z)]^* \cdot [\sqrt{\mu n} \mathbf{u}_m(\mathbf{q}, z)] \, dz = \frac{\hbar}{2\omega_m(q)}; \tag{C.9}$$

this equivalence follows straightforwardly by noting that $q_{\parallel} = q$. As discussed in Chapters 5 and 7, the interaction potential for these normalized modes is

$$\Phi_m(\mathbf{r}) = \sum_q \Phi_m(\mathbf{q}, z) \, e^{-i\mathbf{q} \cdot \boldsymbol{\rho}}. \tag{C.10}$$

Solutions of the form $\mathbf{w} = -\nabla \bar{\chi}$ satisfying these Euler–Lagrange equations are

$$(\omega_{LO}^2 - \omega^2)\nabla^2 \bar{\chi} + \mu \nabla^2 \nabla^2 \bar{\chi} = 0 \tag{C.11}$$

and

$$\Phi = \frac{1}{\gamma}(\omega_{TO}^2 - \omega^2)\bar{\chi} + \mu \nabla^2 \bar{\chi} - B_0, \tag{C.12}$$

for $0 < z < d$, with $\mu = A + B + C$ and B_0 a constant. For solutions outside the slab, Φ satisfies Poisson's equation. As discussed previously, the fields for a slab exhibit translational invariance in the direction normal to the surface of the slab; accordingly, $\bar{\chi}(\mathbf{r}) = \chi(z) \, e^{i\mathbf{q} \cdot \boldsymbol{\rho}}$ and $\Phi(\mathbf{r}) = \phi(z) \, e^{i\mathbf{q} \cdot \boldsymbol{\rho}}$, where $\mathbf{q}$ and $\boldsymbol{\rho}$ are as defined previously. Thus, for $q \neq 0$ it follows that $B_0 = 0$. Then the last two equations may be written as

$$\mu \left(\frac{d^2}{dz^2} + \frac{\omega_{LO}^2 - \omega^2}{\mu} - q^2 \right) \left(\frac{d^2}{dz^2} - q^2 \right) \chi(z) = 0 \tag{C.13}$$

and

$$\phi(z) = \frac{1}{\gamma}(\omega_{TO}^2 - \omega^2 - \mu q^2)\chi(z) + \mu \frac{d^2 \chi(z)}{dz^2}, \tag{C.14}$$

respectively. Nash has applied the methods used in analyzing the Sturm–Liouville equation to show that the normal-mode solutions for the slab must satisfy the following boundary conditions:

$$\Phi \Phi' = 0 \qquad \text{at } z = \pm\infty \tag{C.15}$$

and

$$w_x = w_y = w_z' = 0 \qquad \text{or} \qquad w_x w_x' + w_y w_y' + w_z = 0 \tag{C.16}$$

at $z = 0$ and d.

Again taking $\mathbf{w} = -\nabla \bar{\chi}$, the equations (C.16) reduce to

$$\chi = \chi'' = 0 \qquad \text{and} \qquad \chi' = 0 \tag{C.17}$$

respectively.

At this point it is instructive to consider each of the commonly used sets of normal modes and to compare the boundary conditions they satisfy with those obtained within the context of the Sturm–Liouville equation. Nash took $k_{nq}^2 = (\omega_{LO}^2 - \omega^2)/(\mu - q^2)$ in the fourth-order equation for $\chi(z)$ to define the fourth-order eigenvalue equation

$$\left(\frac{d^2}{dz^2} + k_{nq}^2\right)\left(\frac{d^2}{dz^2} - q^2\right)\chi_{nq}(z) = 0, \tag{C.18}$$

where k_{nq}^2 is the eigenvalue of $\chi_{nq}(z)$ and where the subscripts nq identify different members of the basis set. As usual, q is the wavevector in the xy-plane and n is here an integer used to distinguish different modes having the same value of q. It is this equation that Nash uses to generate complete sets of orthonormal functions representing the phonon modes. As discussed in Chapter 7, the orthogonality relation for the normal modes is

$$\int d^3r\, \mathbf{w}_\alpha(\mathbf{r}) \cdot \mathbf{w}_\beta(\mathbf{r}) = 0 \qquad \text{for } \alpha \neq \beta, \tag{C.19}$$

where α and β each denote the set of quantum numbers nq. Recalling that $\mathbf{w} = -\nabla\bar{\chi}$ and that $\bar{\chi}(\mathbf{r}) = \chi(z)\,e^{i\mathbf{q}\cdot\boldsymbol{\rho}}$ this orthogonality relation may be written as

$$\chi_{nq}^* \cdot \chi_{n'q'} = \frac{1}{2d}\int_0^d dz\, q^2\chi_{nq}^*(z) \cdot \chi_{n'q'}(z) + \frac{d\chi_{nq}^*}{dz}\frac{d\chi_{n'q'}}{dz}. \tag{C.20}$$

The various sets of confined and interface optical phonon modes, which have been discussed in the literature include the so-called slab modes that satisfy the electrostatic boundary conditions, the Huang–Zhu modes, which are based on a reformulation of the slab modes, and the so-called guided modes, which satisfy mechanical boundary conditions. For the slab modes,

$$\chi_{nq}(z) = \sin\frac{(n+1)\pi z}{d}; \tag{C.21}$$

the boundary conditions on $\chi_{nq}(z)$ at $z = 0$ and d are $\chi_{nq} = \chi_{nq}'' = 0$ and it is necessary to include the interface modes to obtain a complete set of modes. The electrostatic potentials for these modes, Φ_{nq}, are given by

$$\Phi_{nq} = -\frac{4\pi\gamma}{\epsilon(\infty)}\chi_{nq}(z). \tag{C.22}$$

The normal modes satisfying $\Phi\Phi' = 0$ at $z = \pm\infty$ and $\chi = \chi'' = 0$ at $z = 0$ and d are the slab modes.

For the modified Huang–Zhu modes given by Nash,

$$\chi_{nq}(z) = \begin{cases} \cos\mu_{nq}\pi z'/d + D_{nq}\cosh qz' & n \text{ even}, \\ \sin\mu_{nq}\pi z'/d + D_{nq}\sinh qz' & n \text{ odd}; \end{cases} \tag{C.23}$$

the boundary conditions on $\chi_{nq}(z)$ at $z = 0$ and d are $\chi_{nq} = \chi'_{nq} = 0$ and it is necessary to include the interface modes to obtain a complete set of modes. The Huang–Zhu (H–Z) modes have been defined to satisfy these boundary conditions at the interfaces since the available microscopic calculations exhibit this behavior. In practice, the boundary conditions $\chi_{nq} = \chi'_{nq} = 0$ are satisfied by selecting the necessary values of D_{nq} and μ_{nq}. The H–Z modes are referred to frequently as the reformulated slab modes and also as the reformulated modes of the dielectric continuum model (with electromagnetic boundary conditions). Huang and Zhu (1988) showed that the modes of the dispersionless microscopic theory at small q are approximated well by these reformulated slab modes. The electrostatic potentials for these modes, Φ_{nq}, are given by

$$\Phi_{nq} = -\frac{4\pi\gamma}{\epsilon(\infty)}\chi_{nq}(z). \tag{C.24}$$

For the guided modes,

$$\chi_{nq}(z) = \cos\frac{n\pi z}{d}; \tag{C.25}$$

the boundary conditions on $\chi_{nq}(z)$ at $z = 0$ and d are again $\chi'_{nq} = \chi''_{nq} = 0$ and, as pointed out by Nash, it is not necessary to include the interface modes to obtain a complete set of modes. The electrostatic potentials for these modes, Φ_{nq}, are given by

$$\Phi_{nq} = -\frac{4\pi\gamma}{\epsilon(\infty)}\begin{cases} \cos n\pi z/d - e^{-qd/2}\cosh q(z - d/2) & n \text{ even,} \\ \cos n\pi z/d + e^{-qd/2}\sinh q(z - d/2) & n \text{ odd.} \end{cases} \tag{C.26}$$

While these guided modes satisfy $\chi'_{nq} = 0$ at the interfaces, in agreement with the $q = 0$ modes observed in Raman backscatter experiments (Sood et al., 1985), they do not satisfy the conditions that $\chi_{nq}(z) = 0$ at $z = 0$ and d. Unlike the slab and H–Z modes, the guided modes have non-zero exponentially decreasing values outside the slab: $\Phi_{nq}(z) = \Phi_{nq}(0)\,e^{qz}$ for $z < 0$ and $\Phi_{nq}(z) = \Phi_{nq}(0)\,e^{q(d-z)}$ for $z > d$.

Finally, the interface modes that satisfy $\nabla^2 \bar{\chi} = 0$ are given by

$$\chi_{\text{IF},1q}(z) = \frac{\sinh q(z - d/2)}{\sinh qd/2} \quad \text{and} \quad \chi_{\text{IF},2q}(z) = \frac{\cosh q(z - d/2)}{\cosh qd/2}. \tag{C.27}$$

These expressions are consistent with those derived in Chapter 7; as argued there, the IF optical phonon mode potentials decrease exponentially with distance from the slab; that is, as e^{qz} for $z < 0$ and as $e^{q(d-z)}$ for $z > d$.

Characterizing the interaction potential for the region m, $\Phi_m(\mathbf{q}, z)$, in terms of the quantum numbers n and q for that region it follows for each of the sets of complete, normalized modes that the interaction Hamiltonian is

$$H_{\text{Fr}} = -e\Phi(\mathbf{r}) = -e\sum_{nq}\Phi_{nq}(\mathbf{q}, z)\,e^{-i\mathbf{q}\cdot\boldsymbol{\rho}}. \tag{C.28}$$

Clearly, of the three commonly used sets of phonon modes, only the slab modes satisfy the conditions that $\Phi\Phi' = 0$ at $z = \pm\infty$ and $\chi_{nq} = \chi''_{nq} = 0$ at $z = 0$ and d. The alternative set of boundary conditions, $\Phi\Phi' = 0$ at $z = \pm\infty$ and $\chi'_{nq} = 0$ at $z = 0$ and d, is not satisfied by any of the commonly used sets of phonon modes. As argued by Nash (1992), the Euler–Lagrange equations, (C.4) and (C.5) are satisfied in each material layer and provide the basis for determining the so-called connection rules at each heterointerface. In the dispersionless limit the Lagrangian density does not contain terms with spatial derivatives of $\mathbf{w}$ and, accordingly, the significance of mechanical boundary conditions is unclear in this case.

For this reason, Nash examined the Euler–Lagrange equations at the heterointerfaces for the case where the modes have non-zero dispersion, as described by the Lagrangian density of (C.1). In this case, the connection rules may be seen to be that $\mathbf{w}/\rho^{1/2}$, Φ, and the z-components of the electric displacement vectors, $D_z = (1/4\pi)\epsilon(\infty)\Phi' - \gamma w_z$, are continuous at the heterointerfaces. Thus in the dispersive continuum model there is no contradiction in applying mechanical boundary conditions to $\mathbf{w}$ and $\mathbf{w}'$ and at the same time applying electromagnetic boundary conditions to Φ and $\mathbf{D}$. Indeed, both types of boundary condition must be invoked to derive the full set of normal modes.

As an illustration leading to an appreciation of the importance of Nash's contribution to the understanding of the boundary conditions for the system at hand, it is instructive to oversimplify the current analysis artificially by considering the equation

$$\epsilon(\infty)\nabla\Phi - 4\pi\gamma\mathbf{w} = 0. \tag{C.29}$$

The left-hand side of this equation appears in the Euler–Lagrange equation obtained from variation of Φ in the Lagrangian density: $\nabla \cdot [\epsilon(\infty)\nabla\Phi - 4\pi\gamma\mathbf{w}] = 0$. Clearly, this simplified equation is not sufficient to describe the system fully but it is in fact the equation assumed in the mechanical models to derive longitudinal modes for the construction of normal modes. Let us consider (C.29) and show that it in fact gives the correct Fröhlich interaction Hamiltonian for electron–polar-optical-phonon interactions, as derived in Section 5.1. Substituting the expression for γ in terms of ω_{LO}, $\epsilon(0)$, and $\epsilon(\infty)$, it follows immediately that

$$\nabla\Phi = \omega_{LO}\left\{4\pi\left[\frac{1}{\epsilon(\infty)} - \frac{1}{\epsilon(0)}\right]\right\}^{1/2}\mathbf{w}. \tag{C.30}$$

Then using the expression of Appendix A, $\mathbf{w} = \sqrt{\mu N/V}\mathbf{u}$, and the normalized displacement of Section 5.1 for the case where $\omega_q = \omega_{LO}$ and only the longitudinal polarization contributes, it follows immediately that

$$
\nabla \Phi = \omega_{LO} \left\{ 4\pi \left[\frac{1}{\epsilon(\infty)} - \frac{1}{\epsilon(0)} \right] \right\}^{1/2}
$$

$$
\times \sqrt{\frac{N}{V}} \sqrt{\frac{1}{N}} \sum_q \sqrt{\frac{\hbar}{2\omega_{LO}}} (a_q e^{i\mathbf{q}\cdot\mathbf{r}} + a_q^\dagger e^{-i\mathbf{q}\cdot\mathbf{r}})
$$

$$
= \omega_{LO} \left\{ \frac{2\pi\hbar\omega_{LO}}{V} 4\pi \left[\frac{1}{\epsilon(\infty)} - \frac{1}{\epsilon(0)} \right] \right\}^{1/2}
$$

$$
\times \sum_q (a_q e^{i\mathbf{q}\cdot\mathbf{r}} + a_q^\dagger e^{-i\mathbf{q}\cdot\mathbf{r}}). \tag{C.31}
$$

Then, using the result that $\nabla \Phi = -i\mathbf{q}\Phi$ and multiplying Φ by $-e$ to obtain the Hamiltonian, it follows that

$$
H_{Fr} = -i \sqrt{\frac{2\pi e^2 \hbar \omega_{LO}}{V} \left[\frac{1}{\epsilon(\infty)} - \frac{1}{\epsilon(0)} \right]} \sum_q \frac{1}{q} (a_q e^{i\mathbf{q}\cdot\mathbf{r}} - a_q^\dagger e^{-i\mathbf{q}\cdot\mathbf{r}}),
$$

$$
\tag{C.32}
$$

which is precisely the result obtained in Section 5.2. Thus, the simplified result, $\epsilon(\infty)\nabla\Phi - 4\pi\gamma\mathbf{w} = 0$, gives the correct three-dimensional Fröhlich interaction Hamiltonian. However, this successful derivation does not justify the use of $\epsilon(\infty)\nabla\Phi - 4\pi\gamma\mathbf{w} = 0$ instead of $\nabla \cdot [\epsilon(\infty)\nabla\Phi - 4\pi\gamma\mathbf{w}] = 0$ in analyzing the boundary conditions for the phonon modes in a slab. Indeed, as argued by Nash (1992) the longitudinal waves obtained from this simplified equation do not provide sufficient degrees of freedom for the mechanical and electromagnetic boundary conditions to be satisfied simultaneously.

The understanding gained through Nash's paper (1992) has been critical in sorting out the correct approaches for calculating carrier–optical-phonon scattering rates in dimensionally confined structures. Of course, the key feature of such calculations is selecting a complete, orthogonal set of phonon modes. Notwithstanding the situation that only slab modes satisfy the desired boundary conditions, Nash demonstrated by explicit calculations that any of the three sets of complete and orthogonal modes – slab modes plus IF modes, H–Z modes plus IF modes, and guided modes – may be used as a basis set for determining the intrasubband and intersubband electron–phonon scattering rates. Nash showed for a quantum well of width d that

$$
e^2 \sum_q |\langle f | \Phi_{nq}(z) | i \rangle|^2 \propto \omega_{nq}^2 \int d^2\mathbf{q} \frac{1}{2q} n_1(\mathbf{q}) n_1^*(\mathbf{q}) f_n(\mathbf{q}), \tag{C.33}
$$

where the form factor, $f_n(\mathbf{q})$, is given by

$$
f_n(\mathbf{q}) = \left| \int dz\, n_0(z) \beta \Phi_{nq}(z) \right|^2 \frac{q}{d\chi_{nq}^*(z)\chi_{n'q'}(z)} \frac{\omega_{LO}^2}{\omega_{nq}^2}, \tag{C.34}
$$

and where

$$|i\rangle = \psi_0(z, i)\psi_1(\rho, i), \qquad |f\rangle = \psi_0(z, f)\psi_1(\rho, f),$$

$$n_1(q) = \int d^2\rho \exp(-i\mathbf{q} \cdot \boldsymbol{\rho})n_1(\boldsymbol{\rho}), \qquad\qquad (C.35)$$

$$n_1(\boldsymbol{\rho}) = \psi_1^*(\rho, f)\psi_1(\rho, i), \qquad n_0(z) = \psi_0^*(z, f)\psi_0(z, i).$$

Thus, for a given in-plane momentum transfer $\mathbf{q}$ the form factor $f_n(q)$ describes the probability that an electron will scatter from the initial subband $|i\rangle$ to the final subband $|f\rangle$ as a result of interaction with phonons of branch, n. The explicit numerical calculations of Nash (1992) showed that the form factors $f_n(q)$ corresponding to the three sets of complete and orthogonal modes – slab modes plus IF modes, H–Z modes plus IF modes, and guided modes – are identical. This result is evident from Figure C.1 for the case of intrasubband transitions and Figure C.2 for the case of intersubband transitions.

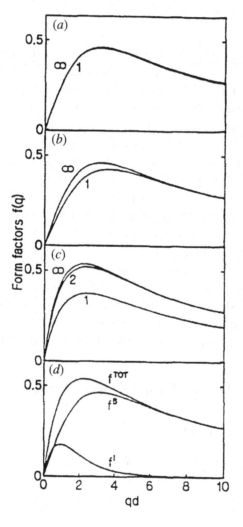

Figure C.1. Form factors $f_n(q)$ for intrasubband scattering for optical modes in a quantum well as functions of qd for three different cases: (a) slab modes, (b) reformulated slab modes, and (c) guided modes. The labels 1, 2, and ∞ designate the lowest-order mode, the second-order mode, and the infinite sum over all modes. (d) depicts the total form factor for bulk modes, f^B, as well as that for the interface modes, f^I; $f^{tot} = f^B + f^I$. From Nash (1992), American Physical Society, with permission.

These results indicate that any of the three sets of complete and orthogonal modes for a semiconductor layer mentioned above may be used to calculate electron–phonon scattering for the nearly degenerate longitudinal optical phonon modes in the layer. The key point is that the modes of each of the three sets must be complete and orthonormal. This result explains why the relatively simple modes of the so-called slab model may be used to perform such a scattering-rate calculation even though these modes are not the normal modes of a heterolayer.

As further proof of the validity of this approach, Nash showed that it is possible to perform unitary transformations between the slab modes plus IF modes, the H–Z modes plus IF modes, and the guided modes. These calculations demonstrate that it is essential to include the IF optical phonon modes in the set of modes for the

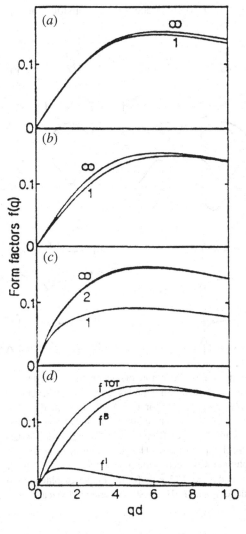

Figure C.2. Form factors $f_n(q)$ for intersubband scattering for optical modes in a quantum well as functions of qd for three different cases: (a) slab modes, (b) reformulated slab modes, and (c) guided modes. For a description of the labels, see the caption to the previous figure. From Nash (1992), American Physical Society, with permission.

slab model – also known as the dielectric continuum model with electromagnetic boundary conditions – and the Huang–Zhu models. Moreover, it is essential that the IF modes are not included in the set of guided modes; this set of guided modes is complete without the IF modes. Indeed, the guided modes explicitly include IF-like exponentially decaying components outside the slab, unlike the slab and H–Z modes. The dielectric continuum model with electromagnetic boundary conditions leads to a convenient set of modes – the slab modes plus the interface modes – for making relatively simple calculations of electron–phonon scattering rates in the case where it is not necessary to account for the dispersion of the confined modes, also referred to as the bulk-like modes in a slab. For GaAs and AlAs the total dispersion over the entire Brillouin zone is only a few meV and it follows that in typical carrier-transport calculations – including those of solid-state electronic devices with dimensionally confined structures – it is not necessary to consider this dispersion. As is evident in Chapter 10, it is necessary to consider energy differences

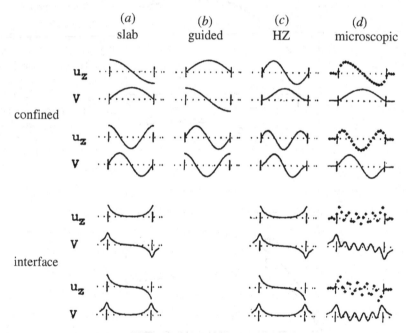

Figure C.3. The z-component of the atomic displacement, u_z, and the corresponding potential V for each of the modes reported by Rücker *et al.* (1991, 1992). The optical modes are presented for a 56 Å (001)-oriented GaAs quantum well surrounded by two AlAs layers. The confined modes of highest frequency and the two interface modes are displayed from top to bottom in order of decreasing phonon frequency for each type of mode. The heterointerfaces are designated by the vertical bars and the phonon wavevectors are $q_z = 0$ for the modes displayed in the upper half of the figure and $q_\parallel = 0.15$ Å^{-1} for the modes displayed in the lower half of the figure. From Rücker *et al.* (1991, 1992), American Physical Society, with permission.

as small as a few meV when modeling optoelectronic devices such as semiconductor quantum-well lasers.

In closing this appendix, it is extremely enlightening to consider a graphical comparison of the various macroscopic and microscopic optical phonon modes in two-dimensional heterostructures. Rücker *et al.* (1992) made direct comparisons of the slab, guided, and H–Z modes with the modes calculated with an *ab initio* microscopic model. Figure C.3 presents the z-component of the atomic displacement, u_z, as well as the corresponding potential, V, for each of these modes as reported by Rücker *et al.* (1991, 1992). The optical modes are presented for a 56 Å (001)-oriented GaAs quantum well surrounded by two AlAs layers. The heterointerfaces are designated by the vertical bars and the phonon wavevectors

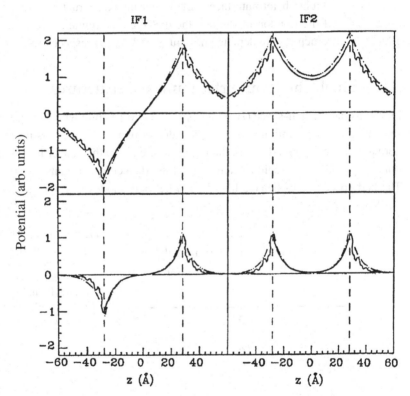

Figure C.4. Detailed comparison of potentials for the high-frequency symmetric (IF1) and antisymmetric (IF2) interface modes, as calculated via both macroscopic and microscopic models for $q = (q_x, 0, 0)$, where $q_x = 0.05$ Å^{-1} for the modes displayed in the upper half of the figure and $q_x = 0.15$ Å^{-1} for the modes displayed in the lower half of the figure. The broken-and-dotted and solid lines represent the potentials derived from the macroscopic model and from the microscopic model respectively. As expected, the agreement between the macroscopic and microscopic calculations is excellent. From Rücker *et al.* (1992), American Physical Society, with permission.

are taken to be $q_z = 0$ and $q_\parallel = 0.15$ Å^{-1}. Figure C.4 presents a detailed comparison of potentials for the interface modes as calculated via both macroscopic and microscopic models. As expected, the agreement between the macroscopic and microscopic calculations is excellent (Rücker *et al.*, 1992).

Appendix D: Optical phonon modes in single- and double-heterointerface würtzite structures

In subsection 7.3.1 the interface optical phonon modes for a single-heterointerface würtzite structure were derived on the basis of the dielectric continuum model and Loudon's model for uniaxial crystals. In this appendix, all the optical phonon modes in single- and double-heterointerface würtzite-like uniaxial structures are formulated on the basis of the Loudon model and the dielectric continuum model (Lee *et al.*, 1998). These structures are depicted in Figures D.1(*a*), (*b*) respectively.

D.1 Single-heterointerface uniaxial structures

First consider the half-space (HS) modes of a heterostructure with a single interface separating two semi-infinite polar-semiconductor regions. The c-axis is taken to be normal to the heterointerface. In the region where $z < 0$ the dielectric constants are $\epsilon(\omega)_{\perp(\parallel),2}$ and in the region where $z > 0$ the dielectric constants are $\epsilon(\omega)_{\perp(\parallel),1}$. The half-space modes behave like the normal bulk modes as $z \to \pm\infty$ and they

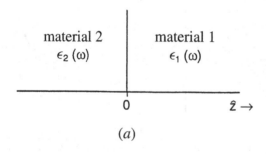

(*a*)

Figure D.1. (*a*) The basic single-heterointerface structure and (*b*) the basic double-heterointerface structure. From Lee *et al.* (1998), American Physical Society, with permission.

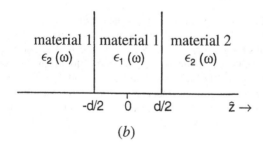

(*b*)

satisfy the electrostatic boundary conditions at $z = 0$. Consider the case in which an optical phonon mode propagates into region 2 from region 1 as depicted in Figure D1(a). The wavevector in region 1 is denoted by $(\mathbf{q}, q_{1,\parallel})$ and the angle between this wavevector and the z-axis is θ_1. When the allowed frequencies in region 1 do not overlap with the allowed frequencies of region 2, it follows that $\epsilon(\omega)_{\parallel,2}\epsilon(\omega)_{\perp,2} > 0$ and the modes in region 2 decay exponentially. Following an analysis similar to that of subsection 7.3.1,

$$\phi(r) = e^{i\mathbf{q}\cdot\rho} \times \begin{cases} A\cos q_{\parallel,1}z + B\sin q_{\parallel,1}z & z > 0, \\ Ce^{\kappa_2 z} & z < 0, \end{cases} \tag{D.1}$$

and in the absence of free charge, $\nabla \cdot \mathbf{D} = 0$ so that

$$\begin{aligned} \epsilon(\omega)_{\parallel,1}q_{\parallel,1}^2 + \epsilon(\omega)_{\perp,1}q^2 = 0 & \quad \epsilon(\omega)_{\parallel,1}\epsilon(\omega)_{\perp,1} > 0, \\ \epsilon(\omega)_{\parallel,2}\kappa_2^2 - \epsilon(\omega)_{\perp,2}q^2 = 0 & \quad \epsilon(\omega)_{\parallel,2}\epsilon(\omega)_{\perp,2} > 0. \end{aligned} \tag{D.2}$$

From the continuity of the tangential component of the electric field at $z = 0$ it then follows that $A = C$. From the continuity of the normal component of the electric displacement at $z = 0$, $\epsilon_{\parallel,1}q_{\parallel,1}B = \epsilon_{\parallel,2}\kappa_2 C$; thus, $C = \Delta_1 B$ with $\Delta_1 = \epsilon_{\parallel,1}k_{1,\parallel}/(\epsilon_{\parallel,2}\kappa_2)$. Thus

$$\phi(r) = \phi_0 e^{i\mathbf{q}\cdot\rho} \times \begin{cases} \Delta_1\cos q_{\parallel,1} + \sin q_{\parallel,1} & z > 0, \\ \Delta_1 e^{\kappa_2 z} & z < 0. \end{cases} \tag{D.3}$$

Now $\mathbf{E}_\perp$ and $\mathbf{E}_\parallel$ are given by the appropriate gradients of ϕ_0. The normalization condition of subsection 7.3.3,

$$\frac{1}{L^2}\frac{\hbar}{2\omega} = \int \left\{ \frac{1}{4\pi}\frac{1}{2\omega}\frac{\partial\epsilon(\omega)_{\perp,n}}{\partial\omega}\left|\mathbf{E}_{\perp,n}\right|^2 + \frac{1}{4\pi}\frac{1}{2\omega}\frac{\partial\epsilon(\omega)_{\parallel,n}}{\partial\omega}\left|\mathbf{E}_{\parallel,n}\right|^2 \right\} dz, \tag{D.4}$$

may be calculated in terms of $\left|\mathbf{E}_{\perp,n}\right|^2$ and $\left|\mathbf{E}_{\parallel,n}\right|^2$ through

$$\begin{aligned} \int_{-L/2}^{L/2}\left|\mathbf{E}_{\perp,1}\right|^2 dz &= \phi_0^2 q^2 \int_0^{L/2}\left|\sin q_{\parallel,1}z + \Delta_1\cos q_{\parallel,1}z\right|^2 dz \\ &= \phi_0^2 q^2 \frac{1}{2}(1+\Delta_1^2)\frac{L}{2}, \end{aligned} \tag{D.5}$$

$$\begin{aligned} \int_{-L/2}^{L/2}\left|\mathbf{E}_{\parallel,1}\right|^2 dz &= \phi_0^2 q_{\parallel,1}^2 \int_0^{L/2}\left|\sin q_{\parallel,1}z + \Delta_1\cos q_{\parallel,1}z\right|^2 dz \\ &= \phi_0^2 q_{\parallel,1}^2 \frac{1}{2}(1+\Delta_1^2)\frac{L}{2}, \end{aligned} \tag{D.6}$$

where the normalization length L in the z-direction is defined over the region $(-L/2, +L/2)$ and where the limit $L/2 \to \infty$ will be taken at a later stage of the calculation. Evidently, in the limit $L/2 \to \infty$ the exponentially decaying mode in region 2 makes a negligible contribution to the normalization integrals. Thus, it follows that

$$\phi_0^2 = \frac{4\pi\hbar}{L^3} \frac{4}{1+\Delta_1^2} \left(\frac{\partial \epsilon_{\perp,1}}{\partial \omega} q^2 + \frac{\partial \epsilon_{\parallel,1}}{\partial \omega} q_{1,\parallel}^2 \right)^{-1}, \tag{D.7}$$

and that

$$H_{HS} = \sum_q \sum_{q_{1,\parallel}} \sqrt{\frac{4\pi e^2 \hbar L^{-3}}{(\partial/\partial\omega)(\epsilon_{\perp,1}\sin^2\theta_1 + \epsilon_{\parallel,1}\cos^2\theta_1)}}$$

$$\times \frac{1}{\sqrt{q^2 + q_{\parallel,1}^2}} \frac{2}{\sqrt{\epsilon_{\parallel,1}^2 q_{\parallel,1}^2 + \epsilon_{\parallel,2}^2 \kappa_2^2}} e^{i\mathbf{q}\cdot\boldsymbol{\rho}}(a_\mathbf{q} + a_{-\mathbf{q}}^\dagger)$$

$$\times \begin{cases} \epsilon_{\parallel,1}q_{\parallel,1}\cos q_{\parallel,1}z + \epsilon_{\parallel,2}\kappa_2\sin q_{\parallel,1}z & z > 0, \\ \epsilon_{\parallel,1}q_{\parallel,1}e^{\kappa_2 z} & z < 0, \end{cases} \tag{D.8}$$

where the summation is taken only over values of $q_{\parallel,1}$ consistent with the condition $\epsilon(\omega)_{\parallel,2}\epsilon(\omega)_{\perp,2} > 0$. The case in which an optical phonon mode propagates into region 1 from region 2 may be treated in the same manner.

The angular dependence of the phonon frequencies in würtzite heterostructures leads to the situation – in contrast to the typical case for zincblende heterostructures – where there is an overlap in the allowed phonon frequencies in adjacent material regions. This situation leads to the occurence of modes that propagate across the heterointerface. Consider an optical phonon incident on region 2 from region 1. In this case, the phonon will propagate in region 2 if the phonon frequency in region 1, ω, is such that $\epsilon(\omega)_{\parallel,2}\epsilon(\omega)_{\perp,2} < 0$. As before, the allowed frequencies in region 1 are the solutions of $\epsilon(\omega)_{\parallel,1}q_{\parallel,1}^2 + \epsilon(\omega)_{\perp,1}q^2 = 0$ or $q_{\parallel,1} = \sqrt{-\epsilon(\omega)_{\perp,1}/\epsilon(\omega)_{\parallel,1}}q$. If $\epsilon(\omega)_{\parallel,2}\epsilon(\omega)_{\perp,2} < 0$, there exists a real solution for $q_{\parallel,2}$ which satisfies

$$\epsilon(\omega)_{\parallel,2}q_{\parallel,1}^2 + \epsilon(\omega)_{\perp,2}q^2 = 0, \tag{D.9}$$

or

$$q_{\parallel,2} = \sqrt{\frac{-\epsilon(\omega)_{\perp,2}}{\epsilon(\omega)_{\parallel,2}}}q = \sqrt{\frac{\epsilon(\omega)_{\parallel,1}\epsilon(\omega)_{\perp,2}}{\epsilon(\omega)_{\perp,1}\epsilon(\omega)_{\parallel,2}}}q_{\parallel,1}. \tag{D.10}$$

Writing the phonon potential as

$$\phi(r) = e^{i\mathbf{q}\cdot\boldsymbol{\rho}} \times \begin{cases} Ae^{-iq_{\parallel,1}z} + Be^{iq_{\parallel,1}z} & z > 0, \\ Ce^{-iq_{\parallel,2}z} & z < 0, \end{cases} \tag{D.11}$$

it follows that continuity of the tangential component of the electric field at the heterointerface leads to $A + B = C$ and continuity of the normal component of the electric displacement at the heterointerface results in the condition

$$(A - B)\Delta_2 = C \qquad \text{where} \quad \Delta_2 = \frac{\epsilon_{\|,1}k_{1,\|}}{\epsilon_{\|,2}k_{2,\|}}.$$

Eliminating B from these two conditions, it follows that $2A = (1 + \Delta_2^{-1})C$ so that $C = 2A\Delta_2/(1+\Delta_2)$ and $B = A(\Delta_2 - 1)/(1+\Delta_2)$. Then

$$\phi(r) = \phi_0 e^{i\mathbf{q}\cdot\boldsymbol{\rho}} \times \begin{cases} e^{-iq_{1,\|}z} + \dfrac{\Delta_2 - 1}{1 + \Delta_2} e^{iq_{1,\|}z} & z > 0, \\[2mm] \dfrac{2\Delta_2}{1 + \Delta_2} e^{-iq_{2,\|}z} & z < 0. \end{cases} \tag{D.12}$$

Once again, the normalization condition of subsection 7.3.1,

$$\frac{1}{L^2}\frac{\hbar}{2\omega} = \int \left\{ \frac{1}{4\pi}\frac{1}{2\omega}\frac{\partial\epsilon(\omega)_{\perp,n}}{\partial\omega}\left|\mathbf{E}_{\perp,n}\right|^2 + \frac{1}{4\pi}\frac{1}{2\omega}\frac{\partial\epsilon(\omega)_{\|,n}}{\partial\omega}\left|\mathbf{E}_{\|,n}\right|^2 \right\} dz, \tag{D.13}$$

may be calculated in terms of $\left|\mathbf{E}_{\perp,n}\right|^2$ and $\left|\mathbf{E}_{\|,n}\right|^2$ through

$$\int_0^{L/2} \left|\mathbf{E}_{\perp,1}\right|^2 dz = \phi_0^2 q^2 \left[1 + \left(\frac{\Delta_2 - 1}{1 + \Delta_2}\right)^2\right]\frac{L}{2},$$

$$\int_0^{L/2} \left|\mathbf{E}_{\|,1}\right|^2 dz = \phi_0^2 q_{1,\|}^2 \left[1 + \left(\frac{\Delta_2 - 1}{1 + \Delta_2}\right)^2\right]\frac{L}{2},$$

$$\int_{-L/2}^0 \left|\mathbf{E}_{\perp,2}\right|^2 dz = \phi_0^2 q^2 \left(\frac{2\Delta_2}{1 + \Delta_2}\right)^2 \frac{L}{2}, \tag{D.14}$$

$$\int_{-L/2}^0 \left|\mathbf{E}_{\|,2}\right|^2 dz = \phi_0^2 q_{1,\|}^2 \left(\frac{2\Delta_2}{1 + \Delta_2}\right)^2 \frac{L}{2}.$$

These normalization integrals determine the propagating (PR) optical phonon modes in a single-heterointerface würtzite structure. The corresponding electron–optical-phonon interaction Hamiltonians are:

$$H_{\text{PR}}^S = \sum_q \sum_{q_{1,\|}} \left(4\pi e^2 \hbar L^{-3}\right)^{1/2}$$

$$\times \left[\frac{\partial}{\partial\omega}\left(\left\{\epsilon_{\perp,1}(\omega)q^2 + \epsilon_{\|,1}(\omega)q_{1,\|}^2\right\} + \{1 \leftrightarrow 2\}\right)\right]^{-1/2}$$

$$\times e^{i\mathbf{q}\cdot\boldsymbol{\rho}}(a_{\mathbf{q}} + a_{-\mathbf{q}}^\dagger) \begin{cases} 2\cos q_{1,\|}z & z > 0, \\[2mm] 2\cos(q_{2,\|}z) & z < 0, \end{cases} \tag{D.15}$$

for the quasi-symmetric modes and

$$H_{PR}^A = \sum_q \sum_{q_{1,\parallel}} \left(4\pi e^2 \hbar L^{-3}\right)^{1/2}$$

$$\times \left[\frac{\partial}{\partial\omega}\{[\epsilon_{\perp,1}(\omega)q^2 + \epsilon_{\parallel,1}(\omega)q_{1,\parallel}^2]\epsilon_{\parallel,2}^2 q_{2,\parallel}^2\} + \{1 \leftrightarrow 2\}\right]^{-1/2}$$

$$\times e^{i\mathbf{q}\cdot\boldsymbol{\rho}}(a_\mathbf{q} + a_{-\mathbf{q}}^\dagger) \begin{cases} \epsilon_{\parallel,2}q_{2,\parallel}\sin q_{1,\parallel}z & z > 0, \\ \epsilon_{\parallel,1}q_{1,\parallel}\sin q_{2,\parallel}z & z < 0, \end{cases} \tag{D.16}$$

for the quasi-antisymmetric modes (Lee *et al.*, 1998).

D.2   Double-heterointerface uniaxial structures

First consider the interface modes of a uniaxial polar-semiconductor heterostructure consisting of a layer of material 1, of thickness d, bounded by two semi-infinite regions of material 2 occupying the regions $z < -d/2$ and $z > d/2$. The c-axis is taken to be normal to the heterointerfaces. In the region $|z| < d/2$ the dielectric constants are $\epsilon(\omega)_{\perp(\parallel),1}$ and in the semi-infinite regions $|z| > d/2$ the dielectric constants are $\epsilon(\omega)_{\perp(\parallel),2}$, as depicted in Figure D.1(b). As for the zincblende case discussed in Chapter 7, the potential $\phi(r)$ is taken to be of the form

$$\phi(r) = e^{i\mathbf{q}\cdot\boldsymbol{\rho}} \times \begin{cases} Ce^{\kappa_2(z+d/2)} & z < -d/2, \\ A\cosh\kappa_1 z + B\sinh\kappa_1 z & |z| < d/2, \\ De^{-\kappa_2(z-d/2)} & z > d/2. \end{cases} \tag{D.17}$$

In the absence of free charge, $\nabla \cdot \mathbf{D} = 0$ so that the dispersion relations in the two material regions are

$$\begin{aligned} \epsilon(\omega)_{\parallel,1}\kappa_1^2 - \epsilon(\omega)_{\perp,1}q^2 = 0 & \quad \epsilon(\omega)_{\parallel,1}\epsilon(\omega)_{\perp,1} > 0, \\ \epsilon(\omega)_{\parallel,2}\kappa_2^2 - \epsilon(\omega)_{\perp,2}q^2 = 0 & \quad \epsilon(\omega)_{\parallel,2}\epsilon(\omega)_{\perp,2} > 0, \end{aligned} \tag{D.18}$$

so it follows that $\kappa_1 = \sqrt{\epsilon(\omega)_{\perp,1}/\epsilon(\omega)_{\parallel,1}}\, q$ and $\kappa_2 = \sqrt{\epsilon(\omega)_{\perp,2}/\epsilon(\omega)_{\parallel,2}}\, q$. The continuity of the tangential component of the electric field at the heterointerfaces is satisfied if

$$C = A\cosh\kappa_1 d/2 - B\sinh\kappa_1 d/2 \quad z = -d/2, \tag{D.19}$$

and

$$D = A\cosh\kappa_1 d/2 + B\sinh\kappa_1 d/2 \quad z = d/2. \tag{D.20}$$

Likewise, the continuity of the normal component of electric displacement at the heterointerfaces is satisfied if

$$\epsilon(\omega)_{\parallel,2}\kappa_2 C = \epsilon(\omega)_{\parallel,1}\kappa_1(-A\sinh\kappa_1 d/2 + B\cosh\kappa_1 d/2) \quad z = -d/2, \tag{D.21}$$

and

$$-\epsilon(\omega)_{\parallel,2}\kappa_2 D = \epsilon(\omega)_{\parallel,1}\kappa_1 (A \sinh \kappa_1 d/2 + B \cosh \kappa_1 d/2) \qquad z = -d/2.$$

(D.22)

Eliminating C and D from these equations, it follows that

$$\epsilon(\omega)_{\parallel,2}\kappa_2 (A \cosh \kappa_1 d/2 - B \sinh \kappa_1 d/2)$$
$$= \epsilon(\omega)_{\parallel,1}\kappa_1 (-A \sinh \kappa_1 d/2 + B \cosh \kappa_1 d/2)$$

(D.23)

and

$$-\epsilon(\omega)_{\parallel,2}\kappa_2 (A \cosh \kappa_1 d/2 + B \sinh \kappa_1 d/2)$$
$$= \epsilon(\omega)_{\parallel,1}\kappa_1 (A \sinh \kappa_1 d/2 + B \cosh \kappa_1 d/2).$$

(D.24)

Accordingly, the condition for a non-trivial solution is

$$\frac{[\epsilon(\omega)_{\parallel,1}\kappa_1 \sinh \kappa_1 d/2 + \epsilon(\omega)_{\parallel,2}\kappa_2 \cosh \kappa_1 d/2]}{[\epsilon(\omega)_{\parallel,1}\kappa_1 \cosh \kappa_1 d/2 + \epsilon(\omega)_{\parallel,2}\kappa_2 \sinh \kappa_1 d/2]} = 0.$$

(D.25)

Clearly, two cases are possible. The case where

$$\epsilon(\omega)_{\parallel,1}\kappa_1 \tanh \kappa_1 d/2 + \epsilon(\omega)_{\parallel,2}\kappa_2 = 0$$

(D.26)

corresponds to $B = 0$, with the results that

$$C = D = \phi_0 \qquad \text{and} \qquad A = \frac{C}{\cosh \kappa_1 d/2} = \frac{\phi_0}{\cosh \kappa_1 d/2};$$

(D.27)

hence, the potential is symmetric and

$$\phi(r) = \phi_0 e^{i q \cdot \rho} \times \begin{cases} -e^{\kappa_2(z+d/2)} & z < -d/2, \\ (\cosh \kappa_1 z)/(\cosh \kappa_1 d/2) & |z| < d/2, \\ e^{-\kappa_2(z-d/2)} & z > d/2. \end{cases}$$

(D.28)

The case where

$$\epsilon(\omega)_{\parallel,1}\kappa_1 \coth \kappa_1 d/2 + \epsilon(\omega)_{\parallel,2}\kappa_2 = 0$$

(D.29)

corresponds to $A = 0$ with the results that

$$-C = D = \phi_0 \qquad \text{and} \qquad B = \frac{D}{\sinh \kappa_1 d/2} = \frac{\phi_0}{\sinh \kappa_1 d/2};$$

(D.30)

hence, the potential is antisymmetric and

$$\phi(r) = \phi_0 e^{i q \cdot \rho} \times \begin{cases} -e^{\kappa_2(z+d/2)} & z < -d/2, \\ (\sinh \kappa_1 z)/(\sinh \kappa_1 d/2) & |z| < d/2, \\ e^{-\kappa_2(z-d/2)} & z > d/2. \end{cases}$$

(D.31)

Once again, the normalization condition of subsection 7.3.1,

$$\frac{1}{L^2}\frac{\hbar}{2\omega} = \int \frac{1}{4\pi}\frac{1}{2\omega}\frac{\partial\epsilon(\omega)_{\perp,n}}{\partial\omega}\left|\mathbf{E}_{\perp,n}\right|^2 + \frac{1}{4\pi}\frac{1}{2\omega}\frac{\partial\epsilon(\omega)_{\parallel,n}}{\partial\omega}\left|\mathbf{E}_{\parallel,n}\right|^2 dz,$$

(D.32)

may be calculated in terms of the appropriate integrals of $\left|\mathbf{E}_{\perp,n}\right|^2 = q^2\left|\phi(z)\right|^2$ and $\left|\mathbf{E}_{\parallel,n}\right|^2 = \left|\partial\phi(z)/\partial z\right|^2$. For the symmetric case, these integrals are

$$\int_{-\infty}^{-d/2}\left|\mathbf{E}_{\perp,2}\right|^2 dz = \int_{d/2}^{\infty}\left|\mathbf{E}_{\perp,2}\right|^2 dz = \phi_0^2 q^2 \int_{d/2}^{\infty} e^{-2\kappa_2(z-d/2)}dz$$

$$= \phi_0^2\frac{q^2}{2\kappa_2},$$

$$\int_{-d/2}^{d/2}\left|\mathbf{E}_{\perp,1}\right|^2 dz = \phi_0^2 q^2\frac{1}{\cosh^2\kappa_1 d/2}\int_{-d/2}^{d/2}\cosh^2\kappa_1 z\, dz$$

$$= \frac{2\phi_0^2 q^2}{\cosh^2\kappa_1 d/2}\left(\frac{1}{2\kappa_1}\sinh\frac{\kappa_1 d}{2}\cosh\frac{\kappa_1 d}{2}+\frac{d}{4}\right),$$

$$\int_{-\infty}^{-d/2}\left|\mathbf{E}_{\parallel,2}\right|^2 dz = \int_{d/2}^{\infty}\left|\mathbf{E}_{\parallel,2}\right|^2 dz = \phi_0^2\kappa_2^2 \int_{d/2}^{\infty} e^{-2\kappa_2(z-d/2)}dz$$

$$= \phi_0^2\frac{\kappa_2}{2},$$

$$\int_{-d/2}^{d/2}\left|\mathbf{E}_{\parallel,1}\right|^2 dz = \phi_0^2\kappa_1^2\frac{1}{\cosh^2(\kappa_1 d/2)}\int_{-d/2}^{d/2}\sinh^2(\kappa_1 z)dz$$

$$= \frac{2\phi_0^2\kappa_1^2}{\cosh^2\kappa_1 d/2}\left(\frac{1}{2\kappa_1}\sinh\frac{\kappa_1 d}{2}\cosh\frac{\kappa_1 d}{2}-\frac{d}{4}\right).$$

(D.33)

For the antisymmetric case, these integrals are

$$\int_{-\infty}^{-d/2}\left|\mathbf{E}_{\perp,2}\right|^2 dz = \int_{d/2}^{\infty}\left|\mathbf{E}_{\perp,2}\right|^2 dz = \phi_0^2 q^2 \int_{d/2}^{\infty} e^{-2\kappa_2(z-d/2)}dz$$

$$= \phi_0^2\frac{q^2}{2\kappa_2},$$

$$\int_{-d/2}^{d/2}\left|\mathbf{E}_{\perp,1}\right|^2 dz = \phi_0^2 q^2\frac{1}{\sinh^2(\kappa_1 d/2)}\int_{-d/2}^{d/2}\sinh^2(\kappa_1 z)dz$$

$$= \frac{2\phi_0^2 q^2}{\sinh^2(\kappa_1 d/2)}\left[\frac{1}{2\kappa_1}\sinh\left(\frac{\kappa_1 d}{2}\right)\cosh\frac{\kappa_1 d}{2}-\frac{d}{4}\right],$$

$$\int_{-\infty}^{-d/2} |\mathbf{E}_{\parallel,2}|^2 \, dz = \int_{d/2}^{\infty} |\mathbf{E}_{\parallel,2}|^2 \, dz = \phi_0^2 \kappa_2^2 \int_{d/2}^{\infty} e^{-2\kappa_2(z-d/2)} dz$$

$$= \phi_0^2 \frac{\kappa_2}{2},$$

$$\int_{-d/2}^{d/2} |\mathbf{E}_{\parallel,1}|^2 \, dz = \phi_0^2 \kappa_1^2 \frac{1}{\sinh^2(\kappa_1 d/2)} \int_{-d/2}^{d/2} \cosh^2 \kappa_1 z \, dz$$

$$= \frac{2\phi_0^2 \kappa_1^2}{\sinh^2 \kappa_1 d/2} \left(\frac{1}{2\kappa_1} \sinh \frac{\kappa_1 d}{2} \cosh \frac{\kappa_1 d}{2} + \frac{d}{4} \right).$$

$$\text{(D.34)}$$

It then follows that the carrier–optical-phonon interaction Hamiltonian for the symmetric modes is

$$H_{\text{IF}}^S = \sum_q 4\pi e^2 \hbar L^{-2} \left\{ \frac{\partial}{\partial \omega} [\sqrt{\epsilon_{\perp,1}\epsilon_{\parallel,1}} \tanh(\sqrt{\epsilon_{\perp,1}/\epsilon_{\parallel,1}} q d/2) - \sqrt{\epsilon_{\perp,2}\epsilon_{\parallel,2}}] \right\}^{-1/2}$$

$$\times \frac{1}{\sqrt{2q}} e^{i\mathbf{q}\cdot\boldsymbol{\rho}} (a_{\mathbf{q}} + a_{-\mathbf{q}}^{\dagger}) \begin{cases} \dfrac{\cosh(\sqrt{\epsilon_{\perp,1}/\epsilon_{\parallel,1}} q z)}{\cosh(\sqrt{\epsilon_{\perp,1}/\epsilon_{\parallel,1}} q d/2)} & |z| < d/2, \\ e^{-\sqrt{\epsilon_{\perp,2}/\epsilon_{\parallel,2}} q(|z|-d/2)} & |z| > d/2, \end{cases} \quad \text{(D.35)}$$

where the frequency is the solution of the transcendental equation

$$\sqrt{\epsilon_{\perp,1}\epsilon_{\parallel,1}} \tanh(\sqrt{\epsilon_{\perp,1}/\epsilon_{\parallel,1}} q d/2) - \sqrt{\epsilon_{\perp,2}\epsilon_{\parallel,2}} = 0 \qquad \text{(D.36)}$$

with the range of frequencies determined by $\epsilon_{\parallel,1}\epsilon_{\parallel,2} < 0$ and, as discussed previously, $\epsilon_{\perp,1}\epsilon_{\parallel,1} > 0$ and $\epsilon_{\perp,2}\epsilon_{\parallel,2} > 0$. For the antisymmetric mode,

$$H_{\text{IF}}^A = \sum_q 4\pi e^2 \hbar L^{-2} \left\{ \frac{\partial}{\partial \omega} [\sqrt{\epsilon_{\perp,1}\epsilon_{\parallel,1}} \coth(\sqrt{\epsilon_{\perp,1}/\epsilon_{\parallel,1}} q d/2) - \sqrt{\epsilon_{\perp,2}\epsilon_{\parallel,2}}] \right\}^{-1/2}$$

$$\times \frac{1}{\sqrt{2q}} e^{i\mathbf{q}\cdot\boldsymbol{\rho}} (a_{\mathbf{q}} + a_{-\mathbf{q}}^{\dagger})$$

$$\times \begin{cases} \dfrac{\sinh(\sqrt{\epsilon_{\perp,1}/\epsilon_{\parallel,1}} q z)}{\sinh(\sqrt{\epsilon_{\perp,1}/\epsilon_{\parallel,1}} q d/2)} & |z| < d/2, \\ \text{sgn}(z)\, e^{-\sqrt{\epsilon_{\perp,2}/\epsilon_{\parallel,2}} q(|z|-d/2)} & |z| > d/2, \end{cases} \quad \text{(D.37)}$$

where the frequency is determined by

$$\sqrt{\epsilon_{\perp,1}\epsilon_{\parallel,1}} \coth(\sqrt{\epsilon_{\perp,1}/\epsilon_{\parallel,1}} q d/2) - \sqrt{\epsilon_{\perp,2}\epsilon_{\parallel,2}} = 0, \quad \text{with} \quad \epsilon_{\parallel,1}\epsilon_{\parallel,2} < 0.$$

The dispersion relations for these interface modes are displayed in Figure D.2 for a würtzite AlN/GaN/AlN heterostructure having a quantum well of thickness d.

For the symmetric confined modes in the double-heterointerface structure the potential, $\phi(r)$, is taken to be of the form

$$\phi(r) = \phi_0 e^{i\mathbf{q}\cdot\mathbf{\rho}} \times \begin{cases} \cos q_{\parallel,1} d/2 \, e^{\kappa_2(z+d/2)} & z < -d/2, \\ \cos q_{\parallel,1} z & |z| < d/2, \\ \cos q_{\parallel,1} d/2 \, e^{-\kappa_2(z-d/2)} & z > d/2. \end{cases} \tag{D.38}$$

In the absence of free charge, $\nabla \cdot D = 0$ so that the dispersion relations in the two material regions are

$$\begin{aligned} \epsilon(\omega)_{\parallel,1} q_{\parallel,1}^2 - \epsilon(\omega)_{\perp,1} q^2 = 0 \qquad & \epsilon(\omega)_{\parallel,1}\epsilon(\omega)_{\perp,1} < 0 \quad \text{for } |z| < d/2, \\ \epsilon(\omega)_{\parallel,2} q_{\parallel,2}^2 - \epsilon(\omega)_{\perp,2} q^2 = 0 \qquad & \epsilon(\omega)_{\parallel,2}\epsilon(\omega)_{\perp,2} > 0 \quad \text{for } |z| > d/2. \end{aligned}$$
$$\tag{D.39}$$

Clearly, the continuity of the tangential component of the electric field at the heterointerfaces is ensured by the choice of $\phi(r)$. The continuity of the normal component of the electric displacement at the heterointerfaces leads to the requirement that

$$\epsilon(\omega)_{\parallel,1} q_{\parallel,1} \sin q_{1,\parallel} d/2 - \epsilon(\omega)_{\parallel,2}\kappa_2 \cos q_{\parallel,1} d/2 = 0. \tag{D.40}$$

Once again, the normalization condition of subsection 7.3.1,

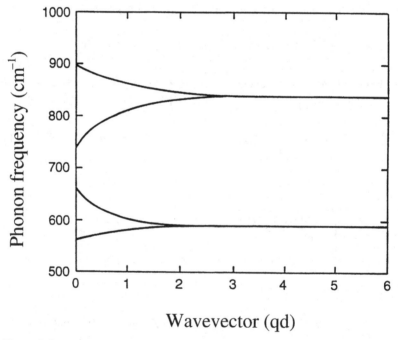

Figure D.2. Interface phonon dispersion relations for an AlN/GaN/AlN double-interface heterostructure with interfaces separated by a distance d. From Lee *et al.* (1998), American Physical Society, with permission.

$$
\frac{1}{L^2}\frac{\hbar}{2\omega} = \int \frac{1}{4\pi}\frac{1}{2\omega}\frac{\partial\epsilon(\omega)_{\perp,n}}{\partial\omega}\left|\mathbf{E}_{\perp,n}\right|^2 + \frac{1}{4\pi}\frac{1}{2\omega}\frac{\partial\epsilon(\omega)_{\parallel,n}}{\partial\omega}\left|\mathbf{E}_{\parallel,n}\right|^2 dz,
$$

$$(D.41)$$

may be calculated in terms of the appropriate integrals of $\left|\mathbf{E}_{\perp,n}\right|^2 = q^2\left|\phi(z)\right|^2$ and $\left|\mathbf{E}_{\parallel,n}\right|^2 = \left|\partial\phi(z)/\partial z\right|^2$. For the symmetric case, these integrals are

$$
\sum_{n=1}^{2}\int_{-\infty}^{+\infty}\frac{\partial\epsilon(\omega)_{\perp,n}}{\partial\omega}\left|\mathbf{E}_{\perp,n}\right|^2
$$

$$
= q^2\phi_0^2\left[\frac{\partial\epsilon(\omega)_{\perp,1}}{\partial\omega}\left(\frac{d}{2} + \frac{1}{q_{\parallel,1}}\sin\frac{q_{\parallel,1}d}{2}\cos\frac{q_{\parallel,1}d}{2}\right)\right]
$$

$$
+ q^2\phi_0^2\left(\frac{1}{\kappa_2}\frac{\partial\epsilon(\omega)_{\perp,2}}{\partial\omega}\cos^2\frac{q_{\parallel,1}d}{2}\right) \tag{D.42}
$$

and

$$
\sum_{n=1}^{2}\int_{-\infty}^{+\infty}\frac{\partial\epsilon(\omega)_{\parallel,n}}{\partial\omega}\left|\mathbf{E}_{\perp,n}\right|^2
$$

$$
= \phi_0^2\left[\frac{\partial\epsilon(\omega)_{\parallel,1}}{\partial\omega}\left(\frac{d}{2} - \frac{1}{q_{\parallel,1}}\sin\frac{q_{\parallel,1}d}{2}\cos\frac{q_{\parallel,1}d}{2}\right)\right]
$$

$$
+ \phi_0^2\left(\frac{\partial\epsilon(\omega)_{\parallel,2}}{\partial\omega}\kappa_2\cos^2\frac{q_{\parallel,1}d}{2}\right); \tag{D.43}
$$

thus

$$
\frac{1}{L^2}\frac{\hbar}{2\omega} = \frac{1}{4\pi}\frac{1}{2\omega}\phi_0^2\left\{\left[\frac{\partial\epsilon(\omega)_{\perp,1}}{\partial\omega}q^2 + \frac{\partial\epsilon(\omega)_{\parallel,1}}{\partial\omega}q_{\parallel,1}^2\right]\frac{d}{2}\right.
$$

$$
+ \left[\frac{\partial\epsilon(\omega)_{\perp,1}}{\partial\omega}\frac{q^2}{q_{\parallel,1}} - \frac{\partial\epsilon(\omega)_{\parallel,1}}{\partial\omega}q_{\parallel,1}\right]\sin\frac{q_{\parallel,1}d}{2}\cos\frac{q_{\parallel,1}d}{2}
$$

$$
\left. + \left[\frac{\partial\epsilon(\omega)_{\perp,2}}{\partial\omega}\frac{q^2}{\kappa_2} + \frac{\partial\epsilon(\omega)_{\parallel,2}}{\partial\omega}\kappa_2\right]\cos^2\frac{q_{1,\parallel}d}{2}\right\}
$$

$$
= \frac{1}{4\pi}\frac{1}{2\omega}\phi_0^2\left\{\left[\frac{\partial\epsilon(\omega)_{\perp,1}}{\partial\omega}q^2 + \frac{\partial\epsilon(\omega)_{\parallel,1}}{\partial\omega}q_{\parallel,1}^2\right]\frac{d}{2} - 2q\frac{\partial f_s}{\partial\omega}\cos\frac{q_{\parallel,1}d}{2}\right\}
$$

$$
= \frac{1}{4\pi}\frac{1}{2\omega}\phi_0^2\left\{\frac{\partial}{\partial\omega}\left[\epsilon(\omega)_{\perp,1}q^2 + \epsilon(\omega)_{\parallel,1}q_{\parallel,1}^2\right]\frac{d}{2} - 2q\frac{\partial f_s}{\partial\omega}\cos\frac{q_{\parallel,1}d}{2}\right\},
$$

$$(D.44)$$

where $f_s(\omega)$ is defined by

$$
f_s(\omega) = \text{sgn}[\epsilon(\omega)_{\parallel,1}]\sqrt{-\epsilon(\omega)_{\perp,1}\epsilon(\omega)_{\parallel,1}}\sin\frac{q_{\parallel,1}d}{2}
$$

$$
- \text{sgn}[\epsilon(\omega)_{\parallel,2}]\sqrt{\epsilon(\omega)_{\perp,2}\epsilon(\omega)_{\parallel,2}}\cos\frac{q_{1,\parallel}d}{2}. \tag{D.45}
$$

Thus,

$$\phi_0^2 = 4\pi\hbar L^{-2}\left\{\frac{\partial}{\partial\omega}\left[\epsilon(\omega)_{\perp,1}q^2 + \epsilon(\omega)_{\parallel,1}q_{\parallel,1}^2\right]\frac{d}{2} - 2q\frac{\partial f_s}{\partial\omega}\cos\frac{q_{\parallel,1}d}{2}\right\}^{-1}$$
(D.46)

and

$$H_C^S = \sum_q\sum_m e^{i\mathbf{q}\cdot\boldsymbol{\rho}}(a_{\mathbf{q}} + a_{-\mathbf{q}}^\dagger)4\pi\hbar L^{-2}\left\{\frac{\partial}{\partial\omega}\left[\epsilon(\omega)_{\perp,1}q^2 + \epsilon(\omega)_{\parallel,1}q_{m,1}^2\right]\frac{d}{2}\right.$$
$$\times\left. -2q\frac{\partial}{\partial\omega}f_s(\omega)\cos\frac{q_{m,1}d}{2}\right\}^{-1/2}$$
$$\times\begin{cases}\cos q_{m,1}z & |z| < d/2, \\ \cos q_{m,1}d/2\, e^{-\kappa_2(z-d/2)} & z > d/2,\end{cases}$$
(D.47)

where the discrete confined wavevectors $q_{1,m}$ are determined from

$$\epsilon(\omega)_{\parallel,1}q_{m,1}\sin q_{m,1}d/2 - \epsilon(\omega)_{\parallel,2}\kappa_2\cos q_{m,1}d/2 = 0,$$

$f_s(\omega)$ is evaluated at the discrete values of $q_{\parallel,1}$, $q_{m,1}$, and

$$2m\pi/d < q_{m,1} < 2(m+1)\pi/d.$$
(D.48)

As defined previously $\kappa_2 = \sqrt{\epsilon(\omega)_{\perp,2}/\epsilon(\omega)_{\parallel,2}}\, q$. Through an analogous derivation, it follows for the antisymmetric modes that

$$H_C^A = \sum_q\sum_m e^{i\mathbf{q}\cdot\boldsymbol{\rho}}(a_{\mathbf{q}} + a_{-\mathbf{q}}^\dagger)$$
$$\times 4\pi\hbar L^{-2}\left\{\frac{\partial}{\partial\omega}\left[\epsilon(\omega)_{\perp,1}q^2 + \epsilon(\omega)_{\parallel,1}q_{m,1}^2\right]\frac{d}{2}\right.$$
$$\left. -2q\frac{\partial}{\partial\omega}f_a(\omega)\sin\frac{q_{m,1}d}{2}\right\}^{-1/2}$$
$$\times\begin{cases}\sin q_{1,m}z & |z| < d/2, \\ \text{sgn}(z)\sin q_{m,1}d/2\, e^{-\kappa_2(z-d/2)} & z > d/2,\end{cases}$$
(D.49)

where

$$f_a(\omega) = \text{sgn}[\epsilon(\omega)_{\parallel,1}]\sqrt{-\epsilon(\omega)_{\perp,1}\epsilon(\omega)_{\parallel,1}}\cos\frac{q_{\parallel,1}d}{2}$$
$$+ \text{sgn}[\epsilon(\omega)_{\parallel,2}]\sqrt{\epsilon(\omega)_{\perp,2}\epsilon(\omega)_{\parallel,2}}\sin\frac{q_{\parallel,1}d}{2}$$
(D.50)

and where the discrete confined wavevectors $q_{1,m}$ are determined from

$$\epsilon(\omega)_{\parallel,1}q_{m,1}\cos q_{m,1}d/2 + \epsilon(\omega)_{\parallel,2}\kappa_2\sin q_{m,1}d/2 = 0$$
(D.51)

with $2(m-1)\pi/d < q_{m,1} < 2(m+1)\pi/d$ and $\kappa_2 = \sqrt{\epsilon(\omega)_{\perp,2}/\epsilon(\omega)_{\parallel,2}}q$.

The half-space modes in a double-heterointerface structure may be considered usefully in terms of symmetric and antisymmetric modes since such a heterostructure is symmetric about $z = 0$. For the symmetric modes take

$$\phi(r) = e^{i\mathbf{q}\cdot\boldsymbol{\rho}}$$

$$\times \begin{cases} A\cos[q_{\|,1}(z+d/2)] - B\sin[q_{\|,1}(z+d/2)] & z < -d/2, \\ C\cosh\kappa_1 z & |z| < d/2, \quad \text{(D.52)} \\ A\cos[q_{\|,1}(z-d/2)] + B\sin[q_{\|,1}(z-d/2)] & z > d/2. \end{cases}$$

Then, continuity of the tangential component of the electric field at the heterointerfaces yields $A = C\cosh\kappa_1 d/2$ and continuity of the normal component of the electric displacement at the heterointerfaces requires that $\epsilon(\omega)_{\|,2} q_{\|,2} B = \epsilon(\omega)_{\|,1}\kappa_2 C\sinh\kappa_1 d/2$. Since $\nabla\cdot\mathbf{D} = 0$ it follows that $\epsilon(\omega)_{\|,2} q_{\|,2}^2 + \epsilon(\omega)_{\perp,2} q^2 = 0$. Defining $\Delta_3 = \epsilon(\omega)_{\|,2} q_{\|,2}/\epsilon(\omega)_{\|,1}\kappa_1$ it follows that $A = B\Delta_3 \coth\kappa_1 d/2$. Then, with $B \equiv \phi_0$,

$$\phi(r) = \phi_0 e^{i\mathbf{q}\cdot\boldsymbol{\rho}}$$

$$\times \begin{cases} -\sin[q_{\|,1}(z+d/2)] + \Delta_3 \coth\kappa_1 d/2\cos[q_{\|,1}(z+d/2)] & z < d/2, \\ \Delta_3(\cosh\kappa_1 z)/(\sinh\kappa_1 d/2) & |z| < d/2, \\ \sin[q_{\|,1}(z-d/2)] + \Delta_3 \coth\kappa_1 d/2\cos[q_{\|,1}(z-d/2)] & z > d/2, \end{cases}$$

$$\text{(D.53)}$$

or

$$\phi(r) = \phi_0 e^{i\mathbf{q}\cdot\boldsymbol{\rho}} \times \begin{cases} \{\sin[q_{\|,1}(|z|-d/2)] \\ \quad + \Delta_3 \coth\kappa_1 d/2\cos[q_{\|,1}(|z|-d/2)]\} & |z| > -d/2, \\ \Delta_3(\cosh\kappa_1 z)/(\sinh\kappa_1 d/2) & |z| < d/2. \end{cases}$$

$$\text{(D.54)}$$

For the antisymmetric modes take

$$\phi(r) = e^{i\mathbf{q}\cdot\boldsymbol{\rho}} \times \begin{cases} -A\cos[q_{\|,1}(z+d/2)] + B\sin[q_{\|,1}(z+d/2)] & z < -d/2, \\ C\sinh\kappa_1 z & |z| < d/2, \\ A\cos[q_{\|,1}(z-d/2)] + B\sin[q_{\|,1}(z-d/2)] & z > d/2. \end{cases}$$

$$\text{(D.55)}$$

Then, continuity of the tangential component of the electric field at the heterointerfaces yields $A = C\sinh\kappa_1 d/2$ and continuity of the normal component of the electric displacement at the heterointerfaces requires that $\epsilon(\omega)_{\|,2} q_{\|,2} B = \epsilon(\omega)_{\|,1}\kappa_1 C\cosh\kappa_1 d/2$. Thus, it follows that $A = B\Delta_3 \tanh\kappa_1 d/2$. Then, with $B \equiv \phi_0$,

$$\phi(r) = \phi_0 e^{i\mathbf{q}\cdot\boldsymbol{\rho}}$$

$$\times \begin{cases} \sin[q_{\|,2}(z+d/2)] - \Delta_3 \tanh\kappa_1 d/2\cos[q_{\|,2}(z+d/2)] & z < -d/2, \\ \Delta_3 \sinh\kappa_1 z/(\cosh\kappa_1 d/2) & |z| < d/2, \\ \sin[q_{\|,2}(z-d/2)] + \Delta_3 \tanh\kappa_1 d/2\cos[q_{\|,2}(z-d/2)] & z > d/2, \end{cases}$$

$$\text{(D.56)}$$

or

$$\phi(r) = \phi_0 e^{i\mathbf{q}\cdot\rho}$$

$$\times \begin{cases} \mathrm{sgn}(z)\{\sin[q_{\|,2}(|z| - d/2)] \\ \quad + \Delta_3 \tanh \kappa_1 d/2 \cos[q_{\|,2}(|z| - d/2)]\} & |z| > -d/2, \\ \Delta_3(\sinh \kappa_1 z)/(\cosh \kappa_1 d/2) & |z| < d/2. \end{cases} \quad (D.57)$$

Once again, the normalization condition of subsection 7.3.1,

$$\frac{1}{L^2}\frac{\hbar}{2\omega} = \int \frac{1}{4\pi}\frac{1}{2\omega}\frac{\partial\epsilon(\omega)_{\perp,n}}{\partial\omega}\left|\mathbf{E}_{\perp,n}\right|^2 + \frac{1}{4\pi}\frac{1}{2\omega}\frac{\partial\epsilon(\omega)_{\|,n}}{\partial\omega}\left|\mathbf{E}_{\|,n}\right|^2 dz, \quad (D.58)$$

may be calculated in terms of the appropriate integrals of $\left|\mathbf{E}_{\perp,n}\right|^2 = q^2 |\phi(z)|^2$ and $\left|\mathbf{E}_{\|,n}\right|^2 = |\partial\phi(z)/\partial z|^2$. For the symmetric case, the integrals that make contributions in the limit $L \longrightarrow \infty$ are

$$\int_{d/2}^{L/2} \left|\mathbf{E}_{\perp,2}\right|^2 dz = \phi_0^2 q^2 \frac{1}{2}\left(1 + \Delta_3^2 \coth^2 \kappa_1 d/2\right)\frac{L}{2},$$

$$\int_{d/2}^{L/2} \left|\mathbf{E}_{\|,2}\right|^2 dz = \phi_0^2 q_{\|,2}^2 \frac{1}{2}\left[1 + \Delta_3^2 \coth^2(\kappa_1 d/2)\right]\frac{L}{2}. \quad (D.59)$$

For the antisymmetric case, the integrals that make contributions in the limit $L \to \infty$ are

$$\int_{-L/2}^{-d/2} \left|\mathbf{E}_{\perp,2}\right|^2 dz = \int_{d/2}^{L/2} \left|\mathbf{E}_{\perp,2}\right|^2 dz = \phi_0^2 q^2 \frac{1}{2}\left(1 + \Delta_3^2 \tanh^2 \kappa_1 d/2\right)\frac{L}{2},$$

$$\int_{-L/2}^{-d/2} \left|\mathbf{E}_{\|,2}\right|^2 dz = \int_{d/2}^{L/2} \left|\mathbf{E}_{\|,2}\right|^2 dz = \phi_0^2 q_{\|,2}^2 \frac{1}{2}\left(1 + \Delta_3^2 \tanh^2 \kappa_1 d/2\right)\frac{L}{2}. \quad (D.60)$$

Thus

$$\phi_0^2 = \frac{4\pi\hbar}{L^3} \times 2 \left[\frac{\partial\epsilon(\omega)_{\perp,2}}{\partial\omega}q^2 + \frac{\partial\epsilon(\omega)_{\perp,2}}{\partial\omega}q_{\|,2}^2\right]^{-1}$$

$$\times \begin{cases} (1 + \Delta_3^2 \coth^2 \kappa_1 d/2)^{-1} & \text{symmetric case,} \\ (1 + \Delta_3^2 \tanh^2 \kappa_1 d/2)^{-1} & \text{antisymmetric case.} \end{cases} \quad (D.61)$$

The interaction Hamiltonian for the symmetric HS modes is then

$$H_{\mathrm{HS}}^S = \sum_q \sum_{q_{\|,2}} 4\pi e^2 \hbar L^{-3} \left[\frac{\partial}{\partial\omega}(\epsilon_{\perp,2}\sin^2\theta_2 + \epsilon_{\|,2}\cos^2\theta_2)\right]^{-1/2} e^{i\mathbf{q}\cdot\rho}(a_{\mathbf{q}} + a_{-\mathbf{q}}^\dagger)$$

$$\times \sqrt{2}(q^2 + q_{\|,2}^2)^{-1/2}(\epsilon_{\|,1}^2\kappa_1^2 \sinh^2 \kappa_1 d/2 + \epsilon_{\|,2}^2 q_{\|,2}^2 \cosh^2 \kappa_1 d/2)^{-1/2}$$

$$\times \begin{cases} \{\epsilon_{\|,1}\kappa_1 \sinh \kappa_1 d/2 \sin[q_{\|,2}(|z| - d/2)] \\ \quad + \epsilon_{\|,2}q_{\|,2} \cosh \kappa_1 d/2 \cos[q_{\|,2}(|z| - d/2)]\} & |z| > d/2, \\ \epsilon_{\|,2}q_{\|,2} \cosh \kappa_1 z & |z| < d/2, \end{cases} \quad (D.62)$$

where the sum is over $q_{\parallel,2}$ and is taken only over those values (D.62) such that $\epsilon(\omega)_{\parallel,1}\epsilon(\omega)_{\perp,1} > 0$. Likewise, the interaction Hamiltonian for the antisymmetric HS modes is

$$H_{HS}^A = \sum_q \sum_{q_{\parallel,2}} 4\pi e^2 \hbar L^{-3} \left[\frac{\partial}{\partial\omega}(\epsilon_{\perp,2}\sin^2\theta_2 + \epsilon_{\parallel,2}\cos^2\theta_2) \right]^{-1/2} e^{i\mathbf{q}\cdot\mathbf{\rho}}(a_\mathbf{q} + a_{-\mathbf{q}}^\dagger)$$

$$\times \sqrt{2}(q^2 + q_{\parallel,2}^2)^{-1/2}(\epsilon_{\parallel,1}^2\kappa_1^2\cosh^2\kappa_1 d/2 + \epsilon_{\parallel,2}^2 q_{\parallel,2}^2\sinh^2\kappa_1 d/2)^{-1/2}$$

$$\times \begin{cases} \text{sgn}(z)\{\epsilon_{\parallel,1}\kappa_1\cosh\kappa_1 d/2\sin[q_{\parallel,2}(|z|-d/2)] \\ \qquad + \epsilon_{\parallel,2}q_{\parallel,2}\sinh\kappa_1 d/2\cos[q_{\parallel,2}(|z|-d/2)]\} & |z| > d/2, \\ \epsilon_{\parallel,2}q_{\parallel,2}\sinh\kappa_1 z & |z| < d/2. \end{cases} \quad (D.63)$$

The interaction Hamiltonians for the propagating (PR) modes may be obtained straightforwardly from those of the HS modes by making the substitution $\kappa_1 \to iq_{\parallel,1}$. Indeed, this substitution results in $\Delta_3 = -i\Delta_3'$ with

$$\Delta_3' = \frac{\epsilon(\omega)_{\parallel,2}q_{\parallel,2}}{\epsilon(\omega)_{\parallel,1}q_{\parallel,1}},$$

$$\cosh\kappa_1 z \to \cos q_{\parallel,1} z,$$

$$\sinh\kappa_1 z \to i\sin q_{\parallel,1} z, \quad (D.64)$$

$$\coth\kappa_1 d/2 \to -i\cot q_{\parallel,1} d/2,$$

$$\tanh\kappa_1 d/2 \to -i\tan q_{\parallel,1} d/2.$$

The interaction Hamiltonian for the symmetric PR modes is then

$$H_{PR}^S = \sum_q \sum_{q_{\parallel,2}} 4\pi e^2 \hbar L^{-3} \left[\frac{\partial}{\partial\omega}(\epsilon_{\perp,2}q^2 + \epsilon_{\parallel,2}q_{\parallel,2}^2) \right]^{-1/2} e^{i\mathbf{q}\cdot\mathbf{\rho}}(a_\mathbf{q} + a_{-\mathbf{q}}^\dagger)$$

$$\times \sqrt{2}(\epsilon_{\parallel,1}^2 q_{\parallel,1}^2\sin^2 q_{\parallel,1} d/2 + \epsilon_{\parallel,2}^2 q_{\parallel,2}^2\cos^2 q_{\parallel,1} d/2)^{-1/2}$$

$$\times \begin{cases} \{\epsilon_{\parallel,2}q_{\parallel,2}\cos q_{\parallel,1} d/2\cos[q_{\parallel,2}(|z|-d/2)] \\ \qquad - \epsilon_{\parallel,1}q_{\parallel,1}\sin q_{\parallel,1} d/2\sin[q_{\parallel,2}(|z|-d/2)]\} & |z| > d/2, \\ \epsilon_{\parallel,2}q_{\parallel,2}\cos q_{\parallel,1} z & |z| < d/2, \end{cases} \quad (D.65)$$

where the sum is over $q_{\parallel,2}$ and is taken only over these values such that $\epsilon(\omega)_{\parallel,1}\epsilon(\omega)_{\perp,1} < 0$. Likewise, the interaction Hamiltonian for the antisymmetric PR modes is

$$H_{PR}^A = \sum_q \sum_{q_{\parallel,2}} 4\pi e^2 \hbar L^{-3} \left[\frac{\partial}{\partial\omega}(\epsilon_{\perp,2}q^2 + \epsilon_{\parallel,2}q_{\parallel,2}^2) \right]^{-1/2} e^{i\mathbf{q}\cdot\mathbf{\rho}}(a_\mathbf{q} + a_{-\mathbf{q}}^\dagger)$$

$$\times \sqrt{2}(\epsilon_{\parallel,1}^2 q_{\parallel,1}^2\cos^2 q_{\parallel,1} d/2 + \epsilon_{\parallel,2}^2 q_{\parallel,2}^2\sin^2 q_{\parallel,1} d/2)^{-1/2}$$

$$\times \begin{cases} \text{sgn}(z)\{\epsilon_{\parallel,2}q_{\parallel,2}\sin q_{\parallel,1} d/2\cos[q_{\parallel,2}(|z|-d/2)] \\ \qquad + \epsilon_{\parallel,1}q_{\parallel,1}\cos q_{\parallel,1} d/2\sin[q_{\parallel,2}(|z|-d/2)]\} & |z| > d/2, \\ \epsilon_{\parallel,2}q_{\parallel,2}\sin q_{\parallel,1} z & |z| < d/2. \end{cases} \quad (D.66)$$

Appendix E: Fermi golden rule

The Fermi golden rule provides a convenient perturbative result for estimating carrier–phonon scattering rates. In calculating these rates in solids it is in fact usually possible to treat the carrier–phonon interactions as perturbations to the Hamiltonian describing the system in the absence of phonon interactions. Taking the unperturbed system to be described by the Hamiltonian H and the carrier–phonon interaction Hamiltonian to be $H_{c-p}(t)$, the wavefunction of the system, $\psi(x, t)$, obeys

$$[H + H_{c-p}(t)]\psi(x, t) = -\frac{\hbar}{i}\frac{\partial \psi(x, t)}{\partial t}, \tag{E.1}$$

where a one-dimensional system is considered for the sake of simplicity. The final result for the Fermi golden rule will be independent of the assumption that the system is one dimensional. Taking the wavefunction to be of the form

$$\psi(x, t) = \sum_n c_n(t)\phi_n(x) e^{-i E_n t/\hbar}, \tag{E.2}$$

where $\phi_n(x)$ is the eigenstate of the unperturbed system with quantum number n, it follows that,

$$[H + H_{c-p}(t)]\psi(x, t) = \sum_n c_n(t) E_n \phi_n(x) e^{-i E_n t/\hbar}$$
$$+ \sum_n c_n(t) e^{-i E_n t/\hbar} H_{c-p}(t)\phi_n(x) \tag{E.3}$$

and

$$-\frac{\hbar}{i}\frac{\partial \psi(x, t)}{\partial t} = -\frac{\hbar}{i}\sum_n \frac{dc_n(t)}{dt}\phi_n(x) e^{-i E_n t/\hbar}$$
$$+ \sum_n c_n(t) E_n \phi_n(x) e^{-i E_n t/\hbar}. \tag{E.4}$$

Thus, without approximation,

$$-\frac{\hbar}{i}\sum_n \frac{dc_n(t)}{dt}\phi_n(x) e^{-i E_n t/\hbar} = \sum_n c_n(t) e^{-i E_n t/\hbar} H_{c-p}(t)\phi_n(x). \tag{E.5}$$

Since the spectrum of eigenstates ϕ_n for a quantum mechanical system is complete, $\langle \phi_m | \phi_n \rangle = \delta_{mn}$. Thus, upon multiplying both sides of the last equation by ϕ_m^* and integrating over x, it follows that

$$i\hbar \frac{dc_m(t)}{dt} = \sum_n \langle \phi_m | H_{c-p}(t) | \phi_n \rangle e^{-i(E_n - E_m)t/\hbar} c_n(t). \tag{E.6}$$

Now if it is assumed that the system is initially in state n, so that $c_n(0) = 1$ and all other $c_m(0)$ are zero, the time evolution of state m is approximated by

$$c_m(t) \simeq -\frac{i}{\hbar} \int_0^t dt \left\langle \phi_m \left| H_{c-p}(t) \right| \phi_n \right\rangle e^{-i(E_n-E_m)t/\hbar}. \tag{E.7}$$

In the case where $H_{c-p}(t)$ is a step function in time,

$$c_m(t) = \left\langle \phi_m \left| H_{c-p} \right| \phi_n \right\rangle \frac{e^{-i(E_n-E_m)t/\hbar} - 1}{E_n - E_m} \tag{E.8}$$

or

$$|c_m(t)|^2 = \left| \left\langle \phi_m \left| H_{c-p} \right| \phi_n \right\rangle \right|^2 \frac{\sin^2[(E_n - E_m)t/(2\hbar)]}{[(E_n - E_m)/2]^2}. \tag{E.9}$$

In the case of nearly degenerate states $E_n \simeq E_m$, and where the transition from state n is to a dense group of final states centered around E_m, the probability of a transition, $|c_m(t)|^2$, takes the form

$$
\begin{aligned}
P &= \int_{E_n-\Delta E/2}^{E_n+\Delta E/2} dE \rho(E) \left| \left\langle \phi_m \left| H_{c-p} \right| \phi_n \right\rangle \right|^2 \frac{\sin^2[(E_n - E)t/(2\hbar)]}{[(E_n - E)/2]^2} \\
&\simeq 4\rho(E) \left| \left\langle \phi_m \left| H_{c-p} \right| \phi_n \right\rangle \right|^2 \int_{E_n-\Delta E/2}^{E_n+\Delta E/2} dE \frac{\sin^2[(E_n - E)t/(2\hbar)]}{(E_n - E)^2}, \tag{E.10}
\end{aligned}
$$

where the last result follows since $\rho(E) \left| \left\langle \phi_m \left| H_{c-p} \right| \phi_n \right\rangle \right|^2$ varies slowly with energy relative to $[\sin^2(E_n - E)t/(2\hbar)]/(E_n - E)^2$. Taking $x = (E - E_n)t/(2\hbar)$,

$$P = \frac{2t}{\hbar} \rho(E) \left| \left\langle \phi_m \left| H_{c-p} \right| \phi_n \right\rangle \right|^2 \int_{-t\Delta E/4\hbar}^{t\Delta E/4\hbar} dx \frac{\sin^2 x}{x^2}. \tag{E.11}$$

In the limit of large t,

$$P \rightarrow \frac{2\pi}{\hbar} \rho(E) \left| \left\langle \phi_m \left| H_{c-p} \right| \phi_n \right\rangle \right|^2 t; \tag{E.12}$$

the transition rate is given by

$$\frac{dP}{dt} = \frac{2\pi}{\hbar} \rho(E) \left| \left\langle \phi_m \left| H_{c-p} \right| \phi_n \right\rangle \right|^2. \tag{E.13}$$

This result is one of the expressions known as the Fermi golden rule.

An alternative form of the Fermi golden rule is obtained by returning to the case of a transition between two specific states (n and m), by taking the limit of large t,

$$t \gg \frac{2}{\omega_{mn}} = \frac{2\hbar}{E_m - E_n} \tag{E.14}$$

and using the relation

$$\frac{\sin^2 \omega_{mn} t/2}{(\omega_{mn} t/2)^2} \rightarrow 2\pi \delta(\omega_{mn} t) = \frac{2\pi \hbar}{t} \delta(\hbar \omega_{mn}) \tag{E.15}$$

for $t \gg 2/\omega_{mn}$; it then follows that for $t \gg 2/\omega_{mn}$

$$\frac{|c_m(t)|^2}{t} = \left(\frac{2\pi}{\hbar}\right) \left|\langle \phi_m \left| H_{c-p} \right| \phi_n \rangle\right|^2 \delta(\hbar\omega_{mn})$$

$$= \left(\frac{2\pi}{\hbar}\right) \left|\langle \phi_m \left| H_{c-p} \right| \phi_n \rangle\right|^2 \delta(E_m - E_n). \tag{E.16}$$

In the case where H_{c-p} is associated with a phonon field and thus has a harmonic time dependence, $e^{\pm i\omega_{\text{phonon}}t}$, the derivation of the last result is modified by replacing ω_{mn} by $\omega_{mn} \mp \omega_{\text{phonon}}$, the upper sign corresponds to phonon emission and the lower sign is associated with phonon absorption; thus,

$$P = \frac{|c_m(t)|^2}{t} = \left(\frac{2\pi}{\hbar}\right) \left|\langle \phi_m \left| H_{c-p} \right| \phi_n \rangle\right|^2 \delta(E_m - E_n \mp \hbar\omega_{\text{phonon}}). \tag{E.17}$$

This form of the Fermi golden rule is used throughout this book. Although the derivation of the Fermi golden rule is based on the assumption that $t \gg 2/\omega_{mn}$, there are many instances for nanoscale devices where this condition is not met. Indeed, for nanoscale devices, transit times are frequently 0.1 ps or less and energy differences are a few to tens of meV so that $2/\omega_{mn} \simeq 0.1$ ps. Hence, $t \approx 2/\omega_{mn}$ and $t \gg 2/\omega_{mn}$ is not satisfied. Nevertheless, the Fermi golden rule is used routinely in such situations and there are few cases where its failure to yield a reasonable approximation has been demonstrated convincingly. Moreover, the alternative means of making non-perturbative calculations such as those based on the Feynman path integral (Register, *et al.*, 1988; Komirenko, *et al.*, 2000c) are so computationally intense that they are rarely used in practice.

Appendix F: Screening effects in a two-dimensional electron gas

In this appendix, the role of screening is examined by the Lindhart method for the case of a two-dimensional electron gas (Bastard, 1988). The phonon field is taken to give rise to a potential $\Phi^{\text{phonon}}(\mathbf{r})$. In general, this potential will induce a charge distribution $\rho^{\text{induced}}(\mathbf{r})$, which produces a potential, $\Phi^{\text{induced}}(\mathbf{r})$. The total electrostatic potential is then the sum of that produced by the phonon field, $\Phi^{\text{phonon}}(\mathbf{r})$, and that associated with the induced charge density, $\Phi^{\text{induced}}(\mathbf{r})$:

$$\Phi^{\text{total}}(\mathbf{r}) = \Phi^{\text{phonon}}(\mathbf{r}) + \Phi^{\text{induced}}(\mathbf{r}). \tag{F.1}$$

Thus, the total interaction potential energy, $\bar{V}(\mathbf{r})$, is the sum of that associated with the phonons in the absence of the Coulomb screening, $-e\Phi^{\text{phonon}}(\mathbf{r})$, and that induced by the Coulomb screening effects, $-e\Phi^{\text{induced}}(\mathbf{r})$:

$$\bar{V}(\mathbf{r}) = -e[\Phi^{\text{phonon}}(\mathbf{r}) + \Phi^{\text{induced}}(\mathbf{r})] = -e\Phi^{\text{total}}(\mathbf{r}). \tag{F.2}$$

Taking this interaction to be a perturbation to the Hamiltonian of the carrier–phonon system without phonon interactions or screening, H_0, it follows from Appendix E that

$$i\hbar \frac{dc_i(t)}{dt} = \sum_k \langle \phi_i | \bar{V} | \phi_k \rangle c_k(t) \, e^{-i(E_i - E_k - \hbar\omega)t/\hbar} e^{\gamma t}, \tag{F.3}$$

where the time-dependent carrier–phonon interaction with screening effects included is taken as

$$H_{c-p}(t) = \bar{V}(\mathbf{r}) \, e^{-i\omega t + \gamma t}. \tag{F.4}$$

As in Appendix E, the system is assumed to be initially in state $|\phi_k\rangle$; then

$$i\hbar \frac{dc_i(t)}{dt} = \langle \phi_i | \bar{V} | \phi_k \rangle e^{-i(E_i - E_k - \hbar\omega)t/\hbar} e^{\gamma t}, \tag{F.5}$$

and

$$\begin{aligned}
c_i(t) &\simeq -\frac{i}{\hbar} \langle \phi_i | \bar{V} | \phi_k \rangle \int_{-\infty}^{t} dt \, e^{-i(E_i - E_k - \hbar\omega)t/\hbar} e^{\gamma t} \\
&= -\frac{\langle \phi_i | \bar{V} | \phi_k \rangle}{E_i - E_k - \hbar\omega - i\gamma} e^{-i(E_i - E_k - \hbar\omega)t/\hbar}.
\end{aligned} \tag{F.6}$$

The unperturbed wavefunction is taken to be that of a two-dimensional electron gas in the lowest subband,

$$\phi_{\mathbf{k}}(\mathbf{r}, t) = \frac{1}{\sqrt{S}} \chi_1(z) \, e^{i\mathbf{k}\cdot\boldsymbol{\rho} - iE_k t/\hbar}, \tag{F.7}$$

where S is the area of the two-dimensional electron gas, $\chi_1(z)$ describes the dependence of the wavefunction normal to the xy-plane of translational invariance, and $e^{i\mathbf{k}\cdot\boldsymbol{\rho}}$ is the wavefunction for a plane wave in the xy-plane with $\boldsymbol{\rho} = (x, y)$ and $\mathbf{k} = (k_x, k_y)$. The perturbed wavefunction is then given by

$$\begin{aligned}
\Psi_{\mathbf{k}}(\mathbf{r}, t) = {} & \frac{1}{\sqrt{S}} \chi_1(z) \, e^{i\mathbf{k}\cdot\boldsymbol{\rho} - iE_k t/\hbar} \\
& + \frac{1}{\sqrt{S}} \chi_1(z) \sum_{\mathbf{k}'} \frac{\langle \phi_{k'} | V | \phi_k \rangle \, e^{i\mathbf{k}'\cdot\boldsymbol{\rho}}}{E_{k'} - E_k + \hbar\omega + i\gamma} e^{-i(E_k + \hbar\omega)t/\hbar} \\
& + \frac{1}{\sqrt{S}} \chi_1(z) \sum_{\mathbf{k}'} \frac{\langle \phi_{k'} | V^* | \phi_k \rangle \, e^{i\mathbf{k}'\cdot\boldsymbol{\rho}}}{E_{k'} - E_k - \hbar\omega + i\gamma} e^{-i(E_k - \hbar\omega)t/\hbar},
\end{aligned} \tag{F.8}$$

where

$$\bar{V} = V e^{-i\omega t + \gamma t} + V^* e^{i\omega t + \gamma t},$$

$$\langle \phi_{k'} | V | \phi_k \rangle \equiv \int d^2\rho \, e^{-i\mathbf{k}'\cdot\boldsymbol{\rho}} V e^{i\mathbf{k}\cdot\boldsymbol{\rho}}, \tag{F.9}$$

$$\langle \phi_{k'} | V^* | \phi_k \rangle^* \equiv \langle \phi_{k'} | V | \phi_k \rangle.$$

The induced charge density is then proportional to the difference of the probability densities of the perturbed and unperturbed wavefunctions, multiplied by the Fermi–Dirac distribution function $F_{\mathbf{k}}$:

$$\rho^{\text{induced}}(\mathbf{r}) = -2e \sum_{\mathbf{k}} \left\{ |\phi_{\mathbf{k}}(\mathbf{r}, t)|^2 - \frac{1}{S} |\chi_1(z)|^2 \right\} F_{\mathbf{k}}, \tag{F.10}$$

where $-e$ is the charge of a single carrier and the factor of two is introduced since there are two spin states for every electronic quantum state. Thus,

$$
\begin{aligned}
\rho^{\text{induced}}(\mathbf{r}) = -2e \sum_{\mathbf{k}} \frac{\chi_1^2(z)}{S} &\left\{ \sum_{\mathbf{k}'} \frac{\langle k'|V|k\rangle e^{i(\mathbf{k}'-\mathbf{k})\cdot\boldsymbol{\rho}} e^{-i\omega t}}{E_k - E_{k'} + \hbar\omega + i\gamma} \right. \\
&+ \sum_{\mathbf{k}'} \frac{\langle k'|V|k\rangle^* e^{-i(\mathbf{k}'-\mathbf{k})\cdot\boldsymbol{\rho}} e^{i\omega t}}{E_k - E_{k'} + \hbar\omega - i\gamma} \\
&+ \sum_{\mathbf{k}'} \frac{\langle k'|V^*|k\rangle e^{i(\mathbf{k}'-\mathbf{k})\cdot\boldsymbol{\rho}} e^{i\omega t}}{E_k - E_{k'} - \hbar\omega + i\gamma} \\
&\left. + \sum_{\mathbf{k}'} \frac{\langle k'|V^*|k\rangle^* e^{-i(\mathbf{k}'-\mathbf{k})\cdot\boldsymbol{\rho}} e^{-i\omega t}}{E_k - E_{k'} - \hbar\omega - i\gamma} \right\} F_{\mathbf{k}} \\
= -2e \frac{\chi_1^2(z)}{S} \sum_{\mathbf{k},\mathbf{k}'} &\left\{ \frac{\langle k'|V|k\rangle e^{i(\mathbf{k}'-\mathbf{k})\cdot\boldsymbol{\rho}}}{E_k - E_{k'} + \hbar\omega + i\gamma} \right. \\
&\left. + \frac{\langle k'|V^*|k\rangle^* e^{-i(\mathbf{k}'-\mathbf{k})\cdot\boldsymbol{\rho}}}{E_k - E_{k'} - \hbar\omega - i\gamma} \right\} e^{-i\omega t} F_{\mathbf{k}} + \text{c.c.} \\
= -2e \frac{\chi_1^2(z)}{S} \sum_{\mathbf{k},\mathbf{k}'} &\langle k'|V|k\rangle e^{i(\mathbf{k}'-\mathbf{k})\cdot\boldsymbol{\rho}} \frac{F_{\mathbf{k}'} - F_{\mathbf{k}}}{E_k - E_{k'} + \hbar\omega + i\gamma}. \tag{F.11}
\end{aligned}
$$

Defining $\mathbf{q} = \mathbf{k}' - \mathbf{k}$, noting that the matrix element $\langle k'|V|k\rangle$ depends on the modulus of $\mathbf{k}' - \mathbf{k}$, and defining $V(|\mathbf{k}' - \mathbf{k}|) \equiv \langle k'|V|k\rangle$, it follows that

$$\rho^{\text{induced}}(\mathbf{r}) = -2e \frac{\chi_1^2(z)}{S} \sum_{\mathbf{k},\mathbf{k}'} V(|\mathbf{k}' - \mathbf{k}|) e^{i(\mathbf{k}'-\mathbf{k})\cdot\boldsymbol{\rho}} \frac{F_{\mathbf{k}'} - F_{\mathbf{k}}}{E_k - E_{k'} + \hbar\omega + i\gamma} \tag{F.12}$$

or, alternatively,

$$
\begin{aligned}
\rho^{\text{induced}}(\mathbf{r}) &= -2e \frac{\chi_1^2(z)}{S} \sum_{\mathbf{k}} \sum_{\mathbf{q}} V(|\mathbf{q}|) e^{i\mathbf{q}\cdot\boldsymbol{\rho}} \frac{F_{\mathbf{k}+\mathbf{q}} - F_{\mathbf{k}}}{E_k - E_{k+q} + \hbar\omega_q + i\gamma} \\
&= e\chi_1^2(z) \sum_{\mathbf{q}} V(|\mathbf{q}|) e^{i\mathbf{q}\cdot\boldsymbol{\rho}} A(\mathbf{q}, \omega), \tag{F.13}
\end{aligned}
$$

where

$$A(\mathbf{q}, \omega) \equiv -\frac{2}{S} \sum_{\mathbf{k}} \frac{F_{\mathbf{k}+\mathbf{q}} - F_{\mathbf{k}}}{E_k - E_{k+q} + \hbar \omega_q + i\gamma}. \tag{F.14}$$

$A(\mathbf{q}, \omega)$ represents the polarization of the two-dimensional electron gas. Moreover, $\rho^{\text{induced}}(\mathbf{r})$ is equivalent to the two-dimensional Fourier expansion of $\rho^{\text{induced}}(\mathbf{q}, z) = e \chi_1^2(z) V(|\mathbf{q}|) e A(\mathbf{q}, \omega)$. For the potential we have

$$\begin{aligned}
\Phi^{\text{induced}}(\mathbf{r}) &= 4\pi \int \frac{d\boldsymbol{\rho}' dz\, \rho^{\text{induced}}(\mathbf{r}')}{\sqrt{(\boldsymbol{\rho} - \boldsymbol{\rho}')^2 + (z - z')^2}} \\
&= \sum_{\mathbf{q}} \Phi^{\text{induced}}(\mathbf{q}, z)\, e^{i\mathbf{q}\cdot\boldsymbol{\rho}}.
\end{aligned} \tag{F.15}$$

Then

$$\left\langle \Phi^{\text{induced}}(\mathbf{q}) \right\rangle = -A(\mathbf{q}, \omega) \left\langle \Phi^{\text{total}}(\mathbf{q}) \right\rangle C(\mathbf{q}), \tag{F.16}$$

where the expectation values are of the form

$$\left\langle \Phi^{\text{induced}}(\mathbf{q}) \right\rangle = \int dz\, \chi_1^2(z) \Phi^{\text{induced}}(\mathbf{q}, z), \tag{F.17}$$

and

$$C(\mathbf{q}) = 4\pi e \int dz dz'\, \chi_1^2(z) \chi_1^2(z') \int \frac{d\boldsymbol{\rho}\, e^{i\mathbf{q}\cdot\boldsymbol{\rho}'}}{\sqrt{\rho'^2 + (z - z')^2}}. \tag{F.18}$$

The identity

$$\frac{1}{|\mathbf{r} - \mathbf{r}'|} = \frac{2\pi}{S} \sum_{\mathbf{q}} \frac{1}{q} e^{i\mathbf{q}\cdot(\boldsymbol{\rho}-\boldsymbol{\rho}')} e^{-q|z-z'|}, \tag{F.19}$$

may be used to write $C(\mathbf{q})$ as follows:

$$C(\mathbf{q}) = 8\pi^2 e \frac{1}{q} G(\mathbf{q}), \tag{F.20}$$

where

$$G(\mathbf{q}) = \int dz\, dz'\, \chi_1^2(z) \chi_1^2(z')\, e^{-q|z-z'|}. \tag{F.21}$$

Noting that $\left\langle \Phi^{\text{induced}}(\mathbf{q}) \right\rangle = \left\langle \Phi^{\text{total}}(\mathbf{q}) \right\rangle - \left\langle \Phi^{\text{phonon}}(\mathbf{q}) \right\rangle$, it follows that

$$\left\langle \Phi^{\text{total}}(\mathbf{q}) \right\rangle - \left\langle \Phi^{\text{phonon}}(\mathbf{q}) \right\rangle = -A(\mathbf{q}, \omega) \left\langle \Phi^{\text{total}}(\mathbf{q}) \right\rangle C(\mathbf{q}), \tag{F.22}$$

so that

$$\begin{aligned}
\left\langle \Phi^{\text{total}}(\mathbf{q}) \right\rangle &= \left\langle \Phi^{\text{phonon}}(\mathbf{q}) \right\rangle / [1 - A(\mathbf{q}, \omega) C(\mathbf{q})] \\
&= \left\langle \Phi^{\text{phonon}}(\mathbf{q}) \right\rangle / \kappa_{\text{el}}(\mathbf{q}),
\end{aligned} \tag{F.23}$$

where the electron permittivity, $\kappa_{el}(\mathbf{q})$, is given by

$$\kappa_{el}(\mathbf{q}) = 1 - A(\mathbf{q}, \omega)C(\mathbf{q})$$
$$= 1 - \frac{8\pi^2 e}{q} G(\mathbf{q})A(\mathbf{q}, \omega). \tag{F.24}$$

In Section 10.6, this expression for the electron permittivity, $\kappa_{el}(\mathbf{q})$, is used to take into account the effect of electron screening.

References

Alexson, D., Bergman, L., Dutta, M., Kim, K.W., Komirenko, S., Nemanich, R.J., Lee, B.C., Stroscio, M.A., and Yu, S. (1999), Confined phonons and phonon-mode properties of III-V nitrides with würtzite crystal structure, *Physics*, **B263** and **B264**, 510–513.

Alexson, D., Bergman, L., Nemanich, R.J., Dutta, M., Stroscio, M.A., Parker, C.A., Bedair, S.M., El-Masry, N.A., and Adar, F. (2000), UV Raman study of A_1(LO) and E_2 phonons in $In_x Ga_{1-x}$N alloys, *Applied Physics Letters*, **89**, 798–800.

Auld, B.A. (1973), *Acoustic Fields and Waves in Solids*. Wiley-Interscience, John Wiley & Sons, New York.

Azuhata, T., Sota, T., Suzuki, K., and Nakamura, S. (1995), Polarized Raman spectra in GaN, *Journal of the Physics of Condensed Matter*, **7**, L129–L135.

Bannov, N., Mitin, V., and Stroscio, M. (1994a), Confined acoustic phonons in semiconductor slabs and their interactions with electrons, *Physica Status Solidi B*, **183**, 131–138.

Bannov, N., Mitin, V., Stroscio, M.A., and Aristov, V. (1994b), Confined acoustic phonons: density of states, in *Proceedings of the 185th Meeting of the Electrochemical Society*, Vol. 94-1, 521; San Francisco, California, May 22–27, 1994.

Bannov, N., Aristov, V., Mitin, V., and Stroscio, M.A. (1995), Electron relaxation times due to deformation-potential interaction of electrons with confined acoustic phonons in a free-standing quantum well, *Physical Review*, **B51**, 9930–9938.

Bastard, G. (1988), *Wave Mechanics Applied to Semiconductor Heterostructures*. Halsted Press, New York.

Behr, D., Niebuhr, R., Wagnar, J., Bachem, K.H., and Kaufmann, U. (1997), Resonant Raman scattering in GaN/(AlGa)N single quantum wells, *Applied Physics Letters*, **70**, 363–365.

Belenky, G., Dutta, M., Gorfinkel, V.B., Haddad, G.I., Iafrate, G.J., Kim, K.W., Kisin, M., Luryi, S., Stroscio, M.A., Sun, J.P., Teng, H.B., and Yu, S. (1999), Tailoring of optical modes in nanoscale semiconductor structures: role of interface-optical phonons in quantum-well lasers, *Physica B*, **263–264**, 462–465.

Beltzer, A.I. (1988), *Acoustics of Solids*. Springer-Verlag, Berlin, Heidelberg, New York, London, Paris, and Tokyo.

Bergman, L., Alexson, D., Murphy, P.L., Nemanich, R.J., Dutta, M., Stroscio, M.A., Balkas, C., Shih, H., and Davis, R.F. (1999a), Raman analysis of phonon lifetimes in AlN and GaN of würtzite structure, *Physical Review*, **B59**, 12 977–12 982.

Bergman, L., Dutta, M., Balkas, C., Davis, R.F., Christman, J.A., Alexon, D., and Nemanich, R.J. (1999b), Raman analysis of the E_1 and A_1 quasi-longitudinal optical and quasi-transverse optical modes in würtzite AlN, *Journal of Applied Physics*, **85**, 3535–3539.

Bergman, L., Dutta, M., Kim, K.W., Klemens, P.G., Komirenko, S., and Stroscio, M.A. (2000), Phonons, electron–phonon interactions, and phonon–phonon interactions in III-V nitrides, *Proceedings of the SPIE*, Vol. 3914, 13–22.

Bhatt, A.R., Kim, K.W., Stroscio, M.A., Iafrate, G.J., Dutta, M., Grubin, H.L., Haque, R., and Zhu, X.T. (1993a), Reduction of LO-phonon interface modes using metal–semiconductor heterostructures, *Journal of Applied Physics*, **73**, 2338–2342.

Bhatt, A.R., Kim, K.W., Stroscio, M.A., and Higman, J.M. (1993b), Simplified microscopic model for electron–optical-phonon interactions in quantum wells, *Physical Review*, **B48**, 14 671–14 674.

Bhatt, A.R., Kim, K.W., and Stroscio, M.A. (1994), Theoretical calculation of the longitudinal-optical lifetime in GaAs, *Journal of Applied Physics*, **76**, 3905–3907.

Blakemore, J.S. (1985), *Solid State Physics, 2nd edition*. Cambridge University Press, Cambridge.

Blatt, J.M. and Thompson, C.J. (1963), Shape resonances in superconducting films, *Physical Review Letters*, **10**, 332–334.

Borer, W.J., Mitra, S.S., and Namjoshi, K.V. (1971), Line shape and temperature dependence of the first order Raman spectrum of diamond, *Solid State Communications*, **9**, 1377–1381.

Born, M. and Huang, K. (1954), *Dynamical Theory of Crystal Lattices*. Oxford University Press, Oxford.

Burns, G., Dacol, F., Marinace, J.C., and Scott, B.A. (1973), Raman scattering in thin-film waveguides, *Applied Physics Letters*, **22**, 356–357.

Campos, V.B., Das Sarma, S., and Stroscio, M.A. (1992), Hot carrier relaxation in polar-semiconductor quantum wires: confined LO-phonon emission, *Physical Review*, **B46**, 3849–3853.

Cardona, M. (1975), ed. *Light Scattering in Solids*. Springer-Verlag, New York.

Cardona, M. and Güntherodt, G., eds. (1982a), *Light Scattering in Solids II*. Springer-Verlag, Berlin.

Cardona, M. and Güntherodt, G., eds. (1982b), *Light Scattering in Solids III*. Springer-Verlag, Berlin.

Cardona, M. and Güntherodt, G., eds. (1984), *Light Scattering in Solids IV*. Springer-Verlag, Berlin.

Cardona, M. and Güntherodt, G., eds. (1989), *Light Scattering in Solids V*. Springer-Verlag, Berlin.

Cardona, M. and Güntherodt, G., eds. (1991), *Light Scattering in Solids VI*. Springer-Verlag, New York.

Castro, G.R. and Cardona, M., eds. (1987), *Lectures on Surface Science, Proceedings of the Fourth Latin-American Symposium*. Springer-Verlag, Berlin.

Chang, I.F. and Mitra, S.S. (1968), Application of modified random-element-isodisplacement model to long-wavelength optic phonons in mixed crystals, *Physical Review*, **172**, 924–933.

Cheng, T.K., Vidal, J., Zeiger, H.J., Dresselhaus, G., Dresselhaus, M.S., and Ippen, E.P. (1991), Mechanism for displacive excitation of coherent phonons in Sb, Bi, Te, and Ti_2O_3, *Applied Physics Letters*, **59**, 1923–1925.

Cho, G.C., Kutt, W., and Kurz, H. (1990), Subpicosecond time-resolved coherent-phonon oscillations in GaAs, *Physical Review Letters*, **65**, 764–766.

Choi, K.-K., Newman, P.G., and Iafrate, G.J. (1990), Quantum transport and phonon emission of nonequilibrium hot electrons, *Physical Review*, **B41**, 10 250–10 253.

Choi, K.-K., Tidrow, M.Z., Chang, W.H. (1996), Electron energy relaxation in infrared hot-electron transistors, *Applied Physics Letters*, **68**, 358–360.

Cingolani, A., Ferrara, M., Lugara, M., and Scamarcio, G. (1986), First order Raman scattering in GaN, *Solid State Communications*, **58**, 823–824.

Cleland, A.N. and Roukes, M.L. (1996), Fabrication of high frequency nanometer scale mechanical resonators from bulk Si crystals, *Applied Physics Letters*, **69**, 2653–2655.

Colvard, C., Merlin, R., Klein, M.V., and Gossard, A.C. (1980), Observation of folded acoustic phonons in semiconductor superlattices, *Physical Review Letters*, **45**, 298–301.

Comas, F., Trallero-Giner, C., and Cardona, M. (1997), Continuum treatment of phonon polaritons in semiconductor heterogeneous structures, *Physical Review*, **B56**, 4115–4127.

Constantinou, N.C. (1993), Interface optical phonons near perfectly conducting boundaries and their coupling to electrons, *Physical Review*, **B48**, 11 931–11 935.

Cowley, R.A. (1963), *Advances in Physics*, **12**, 421–480.

Cros, A., Angerer, H., Ambacher, O., Stutzmann, M., Hopler, R., and Metzger, T. (1997), Raman study of optical phonons in $Al_xGa_{1-x}N$ alloys, *Solid State Communications*, **104**, 35–39.

Das Sarma, S., Stroscio, M.A., and Kim, K.W. (1992), Confined phonon modes and hot-electron relaxation in semiconductor microstructures, *Semiconductor Science and Technology*, **7**, B60–B66.

de la Cruz, R.M., Teitsworth, S.W., and Stroscio, M.A. (1993), Phonon bottleneck effects for confined longitudinal optical phonons in quantum boxes, *Superlattices and Microstructures*, **13**, 481–486.

de la Cruz, R.M., Teitsworth, S.W., and Stroscio, M.A. (1995), Interface phonons in spherical $GaAs/Al_xGa_{1-x}As$ quantum dots, *Physical Review*, **B52**, 1489–1492.

Debernardi, A. (1998), Phonon linewidth in III-V semiconductors from density-functional perturbation theory, *Physical Review*, **B57**, 12 847–12 858.

Demangeot, F., Frandon, J., Renucci, M.A., Meny, C., Briot, O., and Aulombard, R.L. (1997), Interplay of electrons and phonons in heavily doped GaN epilayers, *Journal of Applied Physics*, **82**, 1305–1309.

Demangeot, F., Groenen, J., Frandon, J., Renucci, M.A., Briot, O., Clur, S., and Aulombard, R.L. (1998), Coupling of GaN- and AlN-like longitudinal optic phonons in $Ga_{1-x}Al_xN$ solid solutions, *Applied Physics Letters*, **72**, 2674–2676.

Di Bartolo, B. (1969), *Optical Interactions in Solids*. John Wiley and Sons, New York.

Dutta, M. and Stroscio, M.A., eds. (1998), *Quantum-based Electronic Devices and Systems*. World Scientific, Singapore, New Jersey, London, Hong Kong.

Dutta, M. and Stroscio, M.A., eds. (2000), *Advances in Semiconductor Lasers and Applications to Optoelectronics*. World Scientific, Singapore, New Jersey, London, Hong Kong.

Dutta, M., Grubin, H.L., Iafrate, G.J., Kim, K.W., and Stroscio, M.A. (1993), Metal-encapsulated quantum wire for enhanced charge transport, US Patent Number 5 264 711 issued November 23, 1993.

Dutta, M., Stroscio, M.A., and Kim, K.W. (1998), Recent developments on electron–phonon interactions in structures for electronic and optoelectronic devices, in *Quantum-Based Electronic Devices and Systems, Selected Topics in Electronics and Systems 14*. World Scientific, Singapore, New Jersey, London, Hong Kong.

Dutta, M., Stroscio, M.A., Bergman, L., Alexson, D., Nemanich, R.L., Dupuis, R. (2000), Observation of interface optical phonons in würtzite GaN/AlN superlattices, to be published.

Edgar, J.H., ed. (1994), *Properties of Group III Nitrides*. INSPEC, London.

Educato, J.L., Leburton, J.-P., Boucaud, P., Vagos, P., and Julien, F.H. (1993), Influence of interface phonons on intersubband scattering in asymmetric coupled quantum wells, *Physical Review*, **B47**, 12 949–12 952.

Empedocles, S.A., Norris, D.J., and Bawendi, M.G. (1996), Photoluminescence spectroscopy of single CdSe nanocrystallite quantum dots, *Physical Review Letters*, **77**, 3873–3876.

Engleman, R. and Ruppin, R. (1968a), Optical lattice vibrations in finite ionic crystals: I, *Journal of Physics C, Series 2*, **1**, 614–629.

Engleman, R. and Ruppin, R. (1968b), Optical Lattice vibrations in finite ionic crystals: II, *Journal of Physics C, Series 2*, **1**, 630–643.

Engleman, R. and Ruppin, R. (1968c), Optical lattice vibrations in finite ionic crystals: III, *Journal of Physics C, Series 2*, **1**, 1515–1531.

Esaki, L. and Tsu, R. (1970), Superlattice and negative differential conductivity in semiconductors, *IBM Journal of Research and Development*, **14**, 61–65.

Faist, J., Capasso, F., Sirtori, C., Sivco, D.L., Baillargeon, J.N., Hutchinson, L.A., Chu, Sung-Nee G., and Cho, A. (1996a), High power mid-infrared quantum cascade lasers operating above room temperature, *Applied Physics Letters*, **68**, 3680–3682; also see Hu, Q., and Feng, S. (1991), Feasibility of far-infrared lasers using multiple semiconductor quantum wells, *Applied Physics Letters*, **59**, 2923–2925.

Faist, J., Capasso, F., Sirtori, C., Sivco, D.L., Hutchnison, A.L., Hybertson, M.S., and Cho, A.Y. (1996b), Quantum cascade lasers without intersubband population inversion, *Physical Review Letters*, **76**, 411–414.

Fasol, G., Tanaka, M., Sakaki, H., and Horikosh, Y. (1988), Interface roughness and dispersion of confined LO phonons in GaAs/AlAs quantum wells, *Physical Review*, **B38**, 6056–6065.

Ferry, D.K. (1991), *Semiconductors*. Macmillan Co., New York.

Fuchs, R. and Kliewer, K.L. (1965), Optical modes of vibration in an ionic crystal slab, *Physical Review*, **140**, A2076–A2088.

Fuchs, R., Kliewer, K.L., and Pardee, W.J. (1966), Optical properties of an ionic crystal slab, *Physical Review*, **150**, 589–596.

Gauthier-Lafaye, O., Sauvage, S., Boucaud, P., Julien, F.H., Prazeres, R., Glotin, F., Ortega, J.-M., Thierry-Mieg, V., Planel, R., Leburton, J.-P., and Berger, V. (1997), Intersubband stimulated emission in GaAs/AlGaAs quantum wells: pump-probe experiments using a two-color free electron laser, *Physical Review Letters*, **70**, 3197–3200.

Gelmont, B., Gorfinkel, S., and Luryi, S. (1996), Theory of the spectral line shape and gain in quantum wells with intersubband transitions and ionic crystals, *Applied Physics Letters*, **68**, 2171–2173.

Giehler, M., Ramsteiner, M., Brandt, O., Yarg, H., and Ploog, K.H. (1995), Optical phonons of hexagonal and cubic GaN studied by infrared transmission and Raman spectroscopy, *Applied Physics Letters*, **67**, 733–735.

Gleize, J., Renucci, M.A., Frandon, J., and Demangeot, F. (1999), Anisotropy effects on polar optical phonons in würtzite GaN/AlN superlattices, *Physical Review*, **B60**, 15 985–15 992.

Gleize J., Demangeot, F., Frandon J., Renucci, M.A., Kuball M., Grandjean N., and Massies, J. (2000), Resonant Raman scattering in (Al,Ga)N/GaN quantum well structures, *Thin Solid Films*, **364**, 156–160.

Gmachl, C., Capasso, F., Tredicucci, D.L., Sivco, D.L., Hutchinson, A.L., Chu, S.N. G., and Cho, A.Y. (1998), Noncascaded intersubband injection lasers at $\lambda = 7.7$ μm, *Applied Physics Letters*, **73**, 3830–3832.

Göppert M., Hetterich, M., Dinger A., Klingshorn C., and O'Donnell, K.P. (1998), Infrared spectroscopy of confined optical and folded acoustical phonons in strained CdSe/CdS superlattices, *Physical Review*, **B57**, 13 068–13 071.

Gorczyca, I., Christensen, N.E., Peltzer y Blanca, E.L., and Rodriguez, C.O. (1995), Optical phonon modes in GaN and AlN, *Physical Review*, **B51**, 11 936–11 939.

Gorfinkel, V., Luryi, S., and Gelmont, B. (1996), Theory of gain spectra for quantum cascade lasers and temperature dependence of their characteristics at low and moderate carrier concentrations, *IEEE Journal of Quantum Electronics*, **32**, 1995–2003.

Harima, H., Inoue, T., Nakashima, S., Okumura, H., Ishida, Y., Koiguari, T., Grille, H., and Bechstedt, F. (1999), Raman studies on phonon modes in cubic AlGaN alloys, *Applied Physics Letters*, **74**, 191–193.

Hayashi, K., Itoh, K., Sawaki, N., and Akasaki, I. (1991), Raman scattering in $Al_x Ga_{1-x} N$ alloys, *Solid State Communications*, **77**, 115–118.

Hayes, W. and Loudon, R. (1978), *Scattering of Light by Crystals*. John Wiley and Sons, New York.

Hess, Karl (1999), *Advanced Theory of Semiconductor Devices*. IEEE Press, New Jersey.

Hewat, A.W. (1970), Lattice dynamics of ZnO and BeO, *Solid State Communications*, **8**, 187–189.

Hoben, M.V. and Russell, J.P. (1964), The Raman spectrum of gallium phosphide, *Physics Letters*, **13**, 39–41.

Hon, D.T. and Faust, W.L. (1973), Dielectric parameterization of Raman lineshapes for GaP with a plasma of charge carriers, *Applied Physics Letters*, **1**, 241–256.

Hu, Q. and Feng, S. (1991), Feasibility of far-infrared lasers using multiple semiconductor quantum wells, *Applied Physics Letters*, **59**, 2923–2925.

Huang, K. (1951), On the interaction between the radiation field and ionic crystals, *Proceedings of the Royal Society A*, **208**, 352–365.

Huang, K. and Zhu, B. (1988), Dielectric continuum model and Fröhlich interaction in superlattices, *Physical Review*, **B38**, 13 377–13 386.

Hwang, E.H., Das Sarma, S., and Stroscio, M.A. (2000), Role of confined phonons in thin-film superconductivity, *Physical Review*, **B61**, 8659–8662.

Iafrate, Gerald J., and Stroscio, M.A. (1996), Application of quantum-based devices: trends and challenges, *IEEE Transactions on Electron Devices*, **43**, 1621–1625.

Iijima, S. (1991), Helical microtubules of graphitic carbon, *Nature*, **354**, 56–57.

Irmer, G., Toporov, V.V., Bairamov, B.H., and Monecke, J. (1983), Determination of the charge carrier concentration and mobility in n-GaP by Raman spectroscopy, *Physics Status Solidi*, **B119**, 595–603.

Julien, F.H., Sa'ar, A., Wang, J., and Leburton, J.-P. (1995), Optically pumped intersubband emission in quantum wells, *Electronics Letters*, **31**, 838–839.

Jusserand, B. and Cardona M. (1991), Raman spectroscopy of vibrations in supertlattices, in *Light Scattering in Solids V*, ed. M. Cardona, and G. Güntherodt. Springer-Verlag, Berlin.

Jusserand, B. and Paguet, D. (1986), Comment on 'Resonance Raman Scattering by confined LO and TO phonons in GaAs–AlAs superlattices', *Physical Review Letters*, **56**, 1752.

Kash, J.A. and Tsang, J.C. (1991), Light scattering and other secondary emission studies of dynamical processes in semiconductors, in *Light Scattering in Solids VI*, M. Cardona, ed. Springer-Verlag, Berlin.

Kash, J.A., Jha, S.S., and Tsang, J.C. (1987), Picosecond Raman studies of the Fröhlich interaction in semiconductor alloys, *Physical Review Letters*, **58**, 1869–1872; see also Nash, K.J., and Skolnick, M.S. (1988), Comment on 'Picosecond Raman studies of the Fröhlich interaction in semiconductor alloys', *Physical Review Letters*, **60**, 863.

Kash, J.A., Jha, S.S., and Tsang, J.C. (1988), Kash, Jha, and Tsang reply, *Physical Review Letters*, **60**, 864.

Kash, J.A., Tsang, J.C., and Hvam, J.M. (1985), Subpicosecond time-resolved Raman spectroscopy of LO phonons in GaAs, *Physical Review Letters*, **54**, 2151–2154.

Keating, P.N. (1966), Theory of the third-order elastic constants of diamond-like crystals, *Physical Review*, **149**, 674–678.

Kim, K.W. and Stroscio, M.A. (1990), Electron–optical-phonon interaction in binary/ternary heterostructures, *Journal of Applied Physics*, **68**, 6289–6292.

Kim, K.W., Stroscio, M.A., Bhatt, A., Mickevicius, R., and Mitin, V.V. (1991), Electron–optical-phonon scattering rates in a rectangular semiconductor quantum wire, *Journal of Applied Physics*, **70**, 319–327.

Kim, K.W., Bhatt, A.R., Stroscio, M.A., Turley, P.J., and Teitsworth, S.W. (1992), Effects of interface phonon scattering in multi-heterointerface structures, *Journal of Applied Physics*, **72**, 2282–2287.

Kirillov, D., Lee, H., and Harris, J.S. (1996), Raman study of GaN films, *Journal of Applied Physics*, **80**, 4058–4062.

Kisin, M.V., Gorfinkel, V.B., Stroscio, M.A., Belenky, G., and Luryi, S. (1997), Influence of complex phonon spectra on intersubband optical gain, *Journal of Applied Physics*, **82**, 2031–2038.

Kisin, M.V., Stroscio, M.A., Belenky, G., and Luryi, S. (1998a), Electron–plasmon relaxation in quantum wells with inverse subband occupation, *Applied Physics Letters*, **73**, 2075–2077.

Kisin, M.V., Stroscio, M.A., Belenky, G., Gorfinkel, V.B., and Luryi, S. (1998b), Effects of interface phonon scattering in three-interface heterostructures, *Journal of Applied Physics*, **83**, 4816–4822.

Kitaev, Yu. E., Limonov, M.F., Tronc, P., and Yushkin, G.N. (1998), Raman-active modes in würtzite $(GaN)_m(AlN)_n$ superlattices, *Physical Review*, **B57**, 14 209–14 212.

Kittel, C. (1976), *Introduction to Solid State Physics, 5th edition*. John Wiley & Sons, New York.

Klein, M.V. (1975), Electronic Raman scattering, pp. 147–204, in *Light Scattering in Solids*, M. Cardona, ed. Springer-Verlag, Heidelberg.

Klein, M.V. (1986), Phonons in semiconductor superlattices, *IEEE Journal of Quantum Electronics*, **QE-22**, 1760–1770.

Klein, M.V., Ganguly, B.N., and Colwell, P.J. (1972), Theoretical and experimental study of Raman scattering from coupled LO-phonon–plasmon modes in silicon carbide, *Physical Review*, **B6**, 2380–2388.

Klemens, P.G. (1958), Thermal conductivity and lattice vibrational modes, in *Solid State Physics: Advances in Research and Applications*, Vol. 7, eds. F. Seitz and D. Turnbull, pp. 1–98. Academic Press, New York.

Klemens, P.G. (1966), Anharmonic decay of optical phonons, *Physical Review*, **148**, 845–848.

Klemens, P.G. (1975), Anharmonic decay of optical phonons in diamond, *Physical Review*, **B11**, 3206–3207.

Kliewer, K.L. and Fuchs, R. (1966a), Optical modes of vibration in an ionic slab including retardation. I.: Non-radiative region, *Physical Review*, **144**, 495–503.

Kliewer, K.L. and Fuchs, R. (1966b), Optical modes of vibration in an ionic slab including retardation. II.: Radiative region, *Physical Review*, **150**, 573–588.

Knipp, P.A. and Reinecke, T.L. (1992), Interface phonons of quantum wires, *Physical Review*, **B45**, 9091–9102.

Komirenko, S.M., Kim, K.W., Stroscio, M.A., and Dutta, M. (1999), Dispersion of polar optical phonons in würtzite heterostructures, *Physical Review*, **B59**, 5013–5020.

Komirenko, S.M., Kim, K.W., Stroscio, M.A., and Dutta, M. (2000a), Energy-dependent electron scattering via interaction with optical phonons in würtzite crystals and quantum wells, *Physical Review*, **B61**, 2034–2040.

Komirenko, S.M., Kim, K.W., Demidenko, A.A., Kochelap, V.A., and Stroscio, M.A. (2000b), Cerenkov generation of high-frequency confined acoustic phonons in quantum wells, *Applied Physics Letters*, **76**, 1869–1871.

Komirenko, S.M., Kim, K.W., Stroscio, M.A., and Dutta, M. (2000c), Applicability of the Fermi golden rule, Fröhlich polaron in the Feynman model and new perspectives for possibility of quasiballistic transport in nitrides, to be published.

Kozawa, T., Kachi, T., Kano, H., Taga, Y., Hashimoto, M., Koide, N., and Manabe, K. (1998), Raman scattering from LO-phonon–plasmon coupled modes in gallium nitride, *Journal of Applied Physics*, **75**, 1098–1101.

Krauss, Todd D. and Wise, Frank W. (1997), Coherent acoustic phonons in a quantum dot, *Physical Review Letters*, **79**, 5102–5105.

Kroto, H.W., Heath, J.R., O'Brien, S.C., Curl, R.F., and Smalley, R.E. (1985), C_{60}: Buckminsterfullerine, *Nature*, 318, 162–163.

Kwon, H.J., Lee, Y.H., Miki, O., Yamano, H., and Yoshida, A. (1996), Raman scattering of indium nitride thin films grown by microwave-excited metalorganic vapor phase epitaxy on (0001) sapphire substrates, *Applied Physics Letters*, **69**, 937–939.

Leburton, J.-P. (1984), Size effects on polar optical phonon scattering of 1D and 2D electron gas in synthetic semiconductors, *Journal of Applied Physics*, **56**, 2850–2855.

Leburton, J.-P. (1997), Dissipation and scattering time engineering in quantum devices, in *Proceedings of the International Conference on Quantum Devices*, eds. K. Ismail, S. Bandyopadhyay, and J.-P. Leburton, pp. 242–252. World Scientific, Singapore.

Lee, B.C., Kim, K.W., Dutta, M., and Stroscio, M.A. (1997), Electron–optical-phonon scattering in würtzite crystals, *Physical Review*, **B56**, 997–1000.

Lee, B.C., Kim, K.W., Stroscio, M.A., and Dutta, M. (1998), Optical phonon confinement and scattering in würtzite heterostructures, *Physical Review*, **B58**, 4860–4865.

Lee, I., Goodnick, S.M., Gulia, M., Molinari, E., and Lugli, P. (1995), Microscopic calculations of the electron–optical-phonon interaction in ultrathin $GaAs/Al_xGa_{1-x}As$ alloy quantum-well systems, *Physical Review*, **B51**, 7046–7057.

Lemos, V., Arguello, C.A., and Leite, R.C.C. (1972), Resonant Raman scattering of $TO(A_1)$, $TO(E_1)$ and E_2 optical phonons in GaN, *Solid State Communications*, **11**, 1351–1353.

Licari, J.J. and Evrard R. (1977), Electron–phonon interaction in a dielectric slab: effect of the electronic polarizability, *Physical Review*, **B15**, 2254–2264.

Loudon, R. (1964), The Raman effect in crystals *Advances in Physics*, **13**, 423–482; erratum, *ibid*. (1965), **14**, 621.

Lucovsky, G. and Chen, M.F. (1970), Long wave optical phonons in the alloy systems: $Ga_{1-x}In_xAs$, $GaAs_{1-x}Sb_x$, and $InAs_{1-x}Sb_x$, *Solid State Communications*, **8**, 1397–1401.

Lyubomirsky, I. and Hu, Q. (1998), Energy level schemes for far-infrared quantum well lasers, *Applied Physics Letters*, **73**, 300–302; see also Dutta, M., and Stroscio, M.A. (1999), Comment on 'Energy level schemes for far-infrared quantum well lasers', *Applied Physics Letters*, **74**, 2555.

McNeil, L.E., Grimsditch, M., and French, R.H. (1993), Vibrational spectroscopy of aluminum nitride, *Journal of the American Ceramics Society*, **76**, 1132–1136.

Manchon, D.D., Barker, A.S., Dean, P.J., and Zetterstrom, R.B. (1970), Optical studies of the phonons and electrons in gallium nitride, *Solid State Communications*, **8**, 1227–1231.

Marcatili, E.J. (1969), Dielectric rectangular waveguide and directional coupler for integrated optics, *Bell Systems Technical Journal*, **48**, 2071–2102.

Markus, Štefan (1988), *The Mechanics of Vibrations of Cylindrical Shells*. Elsevier, Amsterdam.

Menéndez, J. and Cardona, M. (1984), Temperature dependence of the first-order Raman scattering by phonons in Si, Ge, and α-Sn: anharmonic effects, *Physical Review*, **B29**, 2051–2059.

Mitin, V.V., Kochelap, V.A., and Stroscio, M.A. (1999), *Quantum Heterostructures: Microelectronics and Optoelectronics*. Cambridge University Press, Cambridge.

Miwa, K. and Fukumoto, A. (1993), First-principles calculation of the structural, electronic, and vibrational properties of gallium nitride and aluminum nitride, *Physical Review*, **B48**, 7897–7902.

Molinari, E., Bungaro, C., Gulia, M., Lugli, P., and Rücker, H. (1992), Electron–phonon interactions in two-dimensional systems: a microscopic approach, *Semiconductor Science and Technology*, **7**, B67–B72.

Molinari, E., Bungaro, C., Rossi, F., Rota, L., and Lugli, P. (1993), Phonons in GaAs/AlAs nanostructures: from two-dimensional to one-dimensional systems, in *Phonons in Semiconductor Nanostructures*, eds. J.-P. Leburton, J. Pascual, and C. Sotomayer-Torres, NATO Advanced Study Institutes, Series E, Vol. 236. Kluwer Academic Publishers, Dordrecht, 39–48.

Mooradian, A., and Wright, G.B., eds. (1969), *Light Scattering Spectra in Solids*, Springer-Verlag, New York, p. 297.

Mori, N. and Ando, T. (1989), Electron–optical phonon interaction in single and double heterostructures, *Physical Review*, **B40**, 6175–6188.

Mori, N., Taniguchi, K., and Hamaguchi, C. (1992), Effects of electron–interface-phonon interaction on resonant tunneling in double-barrier heterostructures, *Semiconductor Science and Technology*, **7**, B83–B87.

Morse, R.W. (1948), Dispersion of compressional waves in isotropic rods of rectangular cross section, *Journal of the Acoustical Society of America*, **20**, 833–838.

Morse, R.W. (1949), The dispersion of compressional waves in isotropic rods of rectangular cross section, Ph.D.Thesis, Brown University, Providence, RI.

Morse, R.W. (1950), The velocity of compressional waves in rods of rectangular cross section, *Journal of the Acoustical Society of America*, **22**, 219–223.

Murugkar, S., Merlin, R., Botchkarev, A., Salvador, A., and Morkoç, H. (1995), Second order Raman spectroscopy of the würtzite form of GaN, *Journal of Applied Physics*, **77**, 6042–6043.

Nabity, J.C. and Wybourne, M.N. (1990a), Phonon trapping in thin metal films, *Physical Review*, **B42**, 9714–9716.

Nabity, J.C. and Wybourne, M.N. (1990b), Evidence of two-dimensional phonons in thin metal films, *Physical Review*, **B44**, 8990–8999.

Nash, K.J. (1992), Electron–phonon interactions and lattice dynamics of optic phonons in semiconductor heterostructures, *Physical Review*, **B46**, 7723–7744.

Nipko, J.C. and Loong, C.K. (1998), Phonon excitations and related thermal properties of aluminum nitride, *Physical Review*, **B57**, 10 055–10 554.

Nipko, J.C., Loong, C.K., Balkas, C.M., and Davis, R.F. (1998), Phonon density of states of bulk gallium nitride, *Applied Physics Letters*, **73**, 34–36.

Nusimovici, M.A. and Birman, J.L. (1967), Lattice dynamics of würtzite: CdS, *Physical Review*, **156**, 925–938.

Omar, M.A. (1975), *Elementary Solid State Physics: Principles and Applications.* Addison-Wesley, Reading, MA.

Patel, C.K.N. and Slusher, R.E. (1968), Light scattering by plasmons and Landau levels of electron gas in InAs, *Physical Review*, **167**, 413–415.

Perlin, P., Polian, A., and Suski, T. (1993), Raman scattering studies of aluminum nitride at high pressure, *Physical Review*, **B47**, 2874–2877.

Pfeifer, T., Kutt, W., Kurz, H., and Scholz, H. (1992), Generation and detection of coherent optical phonons in germanium, *Physical Review Letters*, **69**, 3248–3251.

Platzman, P.M. and Wolff, P.A. (1973), Waves and interactions in solid state plasmas, *Solid State Physics Series, Supplement 13*, eds. H. Ehrenreich, Seitz, F., and Turnbull, D. Academic Press, New York.

Ponce, F.A., Steeds, J.W., Dyer, C.D., and Pitt, G.D. (1996), Direct imaging of impurity-induced Raman scattering in GaN, *Applied Physics Letters*, **69**, 2650–2652.

Pötz, W. and Schroeder, W.A., eds. (1999), *Coherent Control in Atoms, Molecules, and Semiconductors*. Kluwer Academic Publishers, Dordrecht, The Netherlands.

Register, L.F., Stroscio, M.A., and Littlejohn, M.A. (1988), A highly efficient computer algorithm for evaluating Feynman path-integrals, *Superlattices and Microstructures*, **6**, 233–236.

Register, L.F., Stroscio, M.A., and Littlejohn, M.A. (1991), Conservation law for confined polar-optical phonon influence functionals, *Physical Review*, **B44**, 3850–3857.

Richter, E. and Strauch, D. (1987), Lattice dynamics of GaAs/AlAs superlattices, *Solid State Communications*, **64**, 867–870.

Ridley, B.K. (1996), The LO phonon lifetime in GaN, *Journal of Physics: Condensed Matter*, **8**, L511–L513.

Ridley, B.K. (1997), *Electrons and Phonons in Semiconductor Multilayers*. Cambridge University Press, Cambridge.

Ridley, B.K. and Gupta, R. (1991), Nonelectronic scattering of longitudinal-optical phonons in bulk polar semiconductors, *Physical Review*, **B43**, 4939–4944.

Rücker, H., Molinari, E., and Lugli, P. (1992), Microscopic calculation of the electron–phonon interactions in quantum wells, *Physical Review*, **B45**, 6747–6756.

Ruppin, R., and Engleman, R. (1970), Optical phonons in small crystals, *Reports on the Progress of Physics*, **33**, 149–196.

Sakaki, H. (1989), Quantum wire superlattices and coupled quantum box arrays: a novel method to suppress optical phonon scattering in semiconductors, *Japanese Journal of Applied Physics*, **28**, L314–L316.

Seyler, J. and Wybourne, M.N. (1992), Phonon subbands observed in electrically heated metal wires, *Physics of Condensed Matter*, **4**, L231–L236.

Scott, J.F., Leite, R.C.C., Damen, T.C., and Shah, J. (1969), Resonant Raman effects in semiconductors, *Physical Review*, **188**, 1285–1290.

Shah, J., Leite, R.C.C., and Scott, J.F. (1970), Photo-excited hot LO phonons in GaAs, *Solid State Communications*, **9**, 1089–1093.

Shapiro, S.M. and Axe, J.D. (1972), Raman scattering from polar phonons, *Physical Review*, **B6**, 2420–2427.

Siegle, H., Kaczmarczyk, G., Filippidis, L., Litvinchuk, A.P., Hoffmann, A., and Thomsen, C. (1997), Zone-boundary phonons in hexagonal and cubic GaN, *Physical Review*, **B55**, 7000–7004.

Singh, J. (1993), *Physics of Semiconductors and Their Heterostructures*. McGraw-Hill, New York.

Sirenko, Y.M., Kim, K.W., and Stroscio, M.A. (1995), Elastic vibrations of biological and artificial microtubules and filaments, *Electrochemical Society Proceedings*, **95-17**, 260–271.

Sirenko, Y.M., Stroscio, M.A., and Kim, K.W. (1996a), Dynamics of cytoskeleton filaments, *Physical Review*, **E54**, 1816–1823.

Sirenko, Y.M., Stroscio, M.A., and Kim, K.W. (1996b), Elastic vibrations of microtubules in a fluid, *Physical Review*, **E53**, 1003–1011.

Sood, A.K., Menéndez, J., Cardona, M., and Ploog, K. (1985), Resonance Raman scattering by confined LO and TO phonons, *Physical Review Letters*, **54**, 2111–2114.

Sood, A.K., Menéndez, J., Cardona, M., and Ploog, K. (1986), Sood *et al.* respond, *Physical Review Letters*, **56**, 1753.

Stroscio, M.A. (1989), Interaction between longitudinal-optical phonon modes of a rectangular quantum wire with charge carriers of a one-dimensional electron gas, *Physical Review*, **B40**, 6428–6431.

Stroscio, M.A. (1996), Interface-phonon-assisted transitions in quantum well lasers, *Journal of Applied Physics*, **80**, 6864–6867.

Stroscio, M.A. and Kim, K.W. (1993), Piezoelectric scattering of carriers in confined acoustic modes in cylindrical quantum wires, *Physical Review*, **B48**, 1936–1939.

Stroscio, M.A. and Dutta, M. (1999), Damping of nonequilibrium acoustic phonon modes in semiconductor quantum dot, *Physical Review*, **B60**, 7722–7724.

Stroscio, M.A., Kim, K.W., and Littlejohn, M.A. (1990), Theory of optical-phonon interactions in rectangular quantum wires, in *Proceedings of the Society of Photo-optical Instrumentation Engineers Conference on Physical Concepts of Materials for Novel Optoelectronic Device Applications II: Device Physics and Applications*, **1362** ed. M. Razeghi, pp. 556–579.

Stroscio, M.A., Iafrate, G.J., Kim, K.W., Littlejohn, M.A., Goronkin, H., and Maracas, G. (1991a), Transitions from LO-phonon to SO-phonon scattering in short-period AlAs–GaAs superlattices, *Applied Physics Letters*, **59**, 1093–1096.

Stroscio, M.A., Kim, K.W., and Rudin, S. (1991b), Boundary conditions for electron–LO-phonon interaction in polar semiconductor quantum wires, *Superlattices and Microstructures*, **10**, 55–58.

Stroscio, M.A., Kim, K.W., Iafrate, G.J., Dutta, M., and Grubin, H.L. (1991c), Reduction and control of inelastic longitudinal-optical phonon scattering in nanoscale and mesoscopic device structures, in *Proceedings of the 1991 International Device Research Symposium*, pp. 87–89, ISBN 1-880920-00-X, University of Virginia Engineering Academic Outreach Publication, Charlottesville.

Stroscio, M.A., Kim, K.W., Iafrate, G.J., Dutta, M., and Grubin H. (1992), Dramatic reduction of the longitudinal-optical phonon emission rate in polar-semiconductor quantum wires, *Philosophical Magazine Letters*, **65**, 173–176.

Stroscio, M.A., Iafrate, G.J., Kim, K.W., Yu, S., Mitin, V., and Bannov, N. (1993), Scattering of carriers from confined acoustic modes in nanostructures, *Proceedings of the 1993 International Device Research Symposium*, pp. 873–875, ISBN 1-880920-02-6, University of Virginia Engineering Academic Outreach Publication, Charlottesville.

Stroscio, M.A., Kim, K.W., Yu, S., and Ballato, A. (1994), Quantized acoustic phonon modes in quantum wires and quantum dots, *Journal of Applied Physics*, **76**, 4670–4673.

Stroscio, M.A., Sirenko, Yu. M., Yu, S., and Kim, K.W. (1996), Acoustic phonon quantization in buried waveguides and resonators, *Journal of Physics: Condensed Matter*, **8**, 2143–2151.

Stroscio, M.A., Kisin, M.V., Belenky, G., and Luryi, S. (1999), Phonon enhanced inverse population in asymmetric double quantum wells, *Applied Physics Letters*, **75**, 3258–3260.

Sun, C.-K., Liang, J.-C., Stanton, C.J., Abare, A., Coldren, L., and DenBaars, S. (1999), Large coherent acoustic-phonon oscillations observed in InGaN/GaN multiple-quantum wells, *Applied Physics Letters*, **75**, 1249–1251.

Sun, H.C., Davis, L., Sethi, S., Singh, J., and Bhattacharya, P. (1993), Properties of a tunneling injection quantum-well laser: recipe for a 'cold' device with a large modulation bandwidth, *IEEE Photonics Technology Letters* **5**, 870–872.

Sun, J.-P., Teng, H.B., Haddad, G.I., Stroscio, M.A., and Iafrate, G.J. (1997), Intersubband relaxation in step quantum well structures, *International Conference on Computational Electronics, University of Notre Dame, 28–30 May 1997*, private communication.

Sung, C.Y., Norris, T.B., Afzali-Kushaa, A., and Haddad, G.I. (1996), Femtosecond intersubband relaxation and population inversion in stepped quantum well, *Applied Physics Letters*, **68** , 435–437; also see the references in this paper.

Teng, H.B., Sun, J.P., Haddad, G.I., Stroscio, M.A., Yu, S., and Kim, K.W. (1998), Phonon assisted intersubband transitions in step quantum well structures, *Journal of Applied Physics*, **84**, 2155–2164.

Thompson, C.J. and Blatt, J.M. (1963), Shape resonances in superconductors II: simplified theory, *Physics Letters*, **5**, 6–9.

Tsen, K.T. (1992), Picosecond time-resolved Raman studies of electron–optical phonon interactions in ultrathin GaAs–AlAs multiple quantum well structures, *Semiconductor Science and Technology*, 7, B191–B194.

Tsen, K.T. and Morkoç, H. (1988a), Picosecond Raman studies of the optical phonons in the AlGaAs layers of GaAs–AlGaAs multiple-quantum-well structures, *Physical Review*, B37, 7137–7139.

Tsen, K.T. and Morkoç, H. (1988b), Subpicosecond time-resolved Raman spectroscopy of LO phonons in GaAs–$Al_x Ga_{1-x}$As multiple-quantum-well structures, *Physical Review*, **B38**, 5615–5616.

Tsen, K.T., Joshi, R.P., Ferry, D.K., and Morkoc, H. (1989), Time-resolved Raman scattering of nonequilibrium LO phonons in GaAs quantum wells, *Physical Review*, **B39**, 1446–1449.

Tsen, K.T., Joshi, R.P., Ferry, D.K., Botchkarev, A., Sverdlov, B., Salvador, A., and Morkoc, H. (1996), Nonequilibrium electron distributions and phonon dynamics in würtzite GaN, *Applied Physics Letters*, **68**, 2990–2992.

Tsen, K.T., Ferry, D.K., Botchkarev, A., Sverdlov, B., Salvador, A., and Morkoc, H. (1997), Direct measurement of electron-longitudinal optical phonon scattering rates in würtzite GaN, *Applied Physics Letters*, **71**, 1852–1853.

Tsen, K.T., Ferry, D.K., Botchkarev, A., Sverdlov, B., Salvador, A., and Morkoc, H. (1998), Time-resolved Raman studies of the decay of the longitudinal optical phonons in würtzite GaN, *Applied Physics Letters*, **72**, 2132–2136.

Tua, P.F. (1981), Lifetime of high-frequency longitudinal-acoustic phonons in CaF_2 at low crystal temperatures, Ph.D. thesis, Indiana University.

Tua, P.F. and Mahan, G.D. (1982), Lifetime of high-frequency longitudinal-acoustic phonons in CaF_2 at low crystal temperatures, *Physical Review*, **B26**, 2208–2215.

Turley, P.J., Wallis, C.R., and Teitworth, S.W. (1991a), Phonon-assisted tunneling due to localized modes in double-barrier structures, *Physical Review*, **B44**, 8181–8184.

Turley, P.J., Wallis, C.R., and Teitworth, S.W. (1991b), Electronic wave functions and electron-confined-phonon matrix elements in GaAs/Al$_x$Ga$_{1-x}$As double-barrier resonant tunneling structures, *Physical Review*, **B44**, 3199–3210.

Turley, P.J., Wallis, C.R., and Teitworth, S.W. (1991c), Effects of localized phonon modes on magnetotunneling spectra in double-barrier structures, *Physical Review*, **B44**, 12 959–12 963.

Turley, P.J., Wallis, C.R., and Teitworth, S.W. (1992), Theory of localized phonon modes and their effects on electron tunneling in double-barrier structures, *Journal of Applied Physics*, **76**, 2356–2366.

Turley, P.J., Wallis, C.R., Teitworth, S.W., Li, W., and Bhattacharya, P.K. (1993), Tunneling measurements of symmetric-interface phonons in GaAs/AlAs double-barrier structures, *Physical Review*, **B47**, 12 640–12 648.

Vogl, P. (1980), The electron–phonon interaction in semiconductors, in *Physics of Nonlinear Transport in Semiconductors, Proceedings of the NATO Advanced Study Institute Seminar*, eds. D.K. Ferry, J.R. Barker, and C. Jacoboni. Plenum, New York.

von der Linde, D., Kuhl, J., and Klingenberg, H. (1980), Raman scattering from nonequilibrium LO phonons with picosecond resolution, *Physical Review Letters*, **44**, 1505–1508.

Wang, Jin, Leburton, J.-P., Moussa, Z., Julien, F.H., and Sa'ar, A. (1996a), Simulation of optically pumped mid-infrared intersubband semiconductor laser structures, *Journal of Applied Physics*, **80**, 1970–1978.

Wang, Jin, Leburton, J.-P., Julien, F.H., and Sa'ar, A. (1996b), Design and performance optimization of optically-pumped mid-infrared intersubband semiconductor lasers, *IEEE Photonics Technology Letters*, **8**, 1001–1003.

Wang, X.F. and Lei, X.L. (1994), Polar-optic phonons and high-field electron transport in cylindrical GaAs/AlAs quantum wires, *Physical Review*, **B49**, 4780–4789.

Waugh, J.L.T. and Dolling, G. (1963), Crystal dynamics of gallium arsenide, *Physical Review*, **132**, 2410–2412.

Wendler, L. (1985), Electron–phonon interaction in dielectric bilayer systems: effects of the electronic polarizability, *Physica Status Solidi B*, **129**, 513–530.

Wendler, L. and Grigoryan, V.G. (1988), Acoustic interface waves in sandwich structures, *Surface Science*, **206**, 203–224.

Wetzel, C., Walukiewicz, W., Haller, E.E., Ager, III, J., Grzegory, I., Porowski, S., and Suski, T. (1996), Carrier localization of as-grown n-type gallium nitride under large hydrostatic pressure, *Physical Review*, **B53**, 1322–1326.

Wisniewski, P., Knap, W., Malzak, J.P., Camassel, J., Bremser, M.D., Davis, R.F., and Suski, T. (1998), Investigation of optically active E$_1$ transversal optic phonon modes in Al$_x$Ga$_{1-x}$N layers deposited on 6H-SiC substrates using infrared reflectance, *Applied Physics Letters*, **73**, 1760–1762.

Worlock, J.M. (1985), Phonons in superlattices, in *Proceedings of the Second International Conference on Phonon Physics*, J. Kollár, N. Kroó, N. Menyhárd, and T. Siklós., eds., pp. 506–520. World Scientific, Singapore.

Xu, B., Hu, Q., and Mellock, M.R. (1997), Electrically pumped tunable terahertz emitter based on intersubband transition, *Applied Physics Letters*, **71**, 440–442.

Yafet, Y. (1966), Raman scattering by carriers in Landau levels, *Physical Review*, **152**, 858–862.

Yu, P.Y. and Cardona, M. (1996), *Fundamentals of Semiconductors*. Springer, Berlin.

Yu, M., Strongin, M., and Paskin, A. (1976), Consistent calculation of boundary effects in thin superconducting films, *Physical Review*, **B14**, 996–1001.

Yu, S., Kim, K.W., Stroscio, M.A., Iafrate, G.J., and Ballato, A. (1994a), Electron–acoustic-phonon scattering rates in rectangular quantum wires, *Physical Review*, **B50**, 1733–1738.

Yu, S., Kim, K.W., Stroscio, M.A., and Iafrate, G.J. (1994b), Electron–acoustic-phonon scattering rates in cylindrical quantum wires, *Physical Review*, **B51**, 4695–4698.

Yu, S., Kim, K.W., Stroscio, M.A., Iafrate, G.J., and Ballato, A. (1996), Electron–confined-acoustic-phonon scattering rates in cylindrical quantum wires via deformation potential, *Journal of Applied Physics*, **80**, 2815–2822.

Yu, S., Kim, K.W., Stroscio, M.A., Iafrate, G.J., Sun, J.P., and Haddad, G.I. (1997), Transfer matrix method for interface optical-phonon modes in multiple-interface heterostructure systems, *Journal of Applied Physics*, **82**, 3363–3367.

Yu, SeGi, Kim, K.W., Bergman, L., Dutta, M., Stroscio, M.A., and Zavada, J.M. (1998), Long-wavelength optical phonons in ternary nitride-based crystals, *Physical Review*, **B58**, 15 283–15 287.

Zhang, J.M., Ruf, T., Cardona, M., Ambacher, O., Stutzmann, M., Wagner, J.M., and Bechstedt, F. (1997), Raman spectra of isotropic GaN, *Physical Review*, **B56**, 14 399–14 406.

Zhang, X., Gutierrez-Aitken, A.L., Klotzkin, D., Bhattacharya, P., Caneau, C., and Bhat, R. (1996), 0.98-micrometer multi-quantum well tunneling injection lasers with ultra-high modulation bandwidths, *Electronics Letters*, **32**, 1715–1716.

Index

A_1 mode, 17
acoustic modes, localized, 105
acoustic phonons
   in cylindrical quantum wires, 175
   in cylindrical structures, 112
   in cylindrical waveguide, 113
   in double-interface heterostructures, 100
   in free-standing layers, 97
   in nanostructures, 39, 56, 60
   in quantum dots, 124
   in quantum-well structures, 212
   in rectangular quantum wires, 105
   in superlattices, 95, 219
   in thin metallic foils, 60
   in unconstrained layers, 96
   in wires, 60
alloying, 28
anharmonic coupling, of phonons, 45
anharmonic effects, in würtzite structures, 50
anharmonic interactions, 8, 30, 46
anharmonic phonon decay, 215
anharmonic terms, 45
anharmonic third-order potential, 47
annihilation operators, 35
antisymmetric radial–axial modes, 115, 116
antisymmetric shear vertical modes, 102
antisymmetric torsional mode, 123

Bardeen–Cooper–Schrieffer theory of
      superconductivity, 3
binary polar semiconductor, 53
binary semiconductor layer, 66
Brillouin zone, 9
Boltzmann factor, 37
Bose–Einstein condensation, 35
Bose–Einstein distribution, 38
Bose–Einstein occupation number, 47
breathing mode, in spherical quantum dot, 127, 128
bulk dispersion relations, 102

canonical transformation, 40
carrier concentrations, 34
carrier mobilities, 219
carrier–optical-phonon interaction, 79
carrier–phonon scattering, 35
carrier relaxation, 48
carrier wavefunctions, 38
Cerenkov emission, 216
Cerenkov-like generation, 212
classical acoustics, 177
clamped-surface boundary condition, 177
coherent phonons, 219
compressional solutions, 60
conduction bands, 27
confined-phonon effects, in thin film
      superconductors, 208
continuum models of phonons, 52
corner modes, in rectangular quantum wire, 154
Coulomb interaction, 41
coupled radial–longitudinal modes, 120
creation operators, 35
critical points in semiconductors, 27
current transfer ratio, 204
cylindrical quantum wire, 175
cylindrical shell, 112
cylindrical shell, immersed in fluid, 122
cylindrical waveguide, 112
cytoskeletal filaments, 123
crystal symmetry, 95
cubic crystals, 6

damped spherical acoustic modes, 219
decay of phonons, 30, 49
deformation potential constant, 172
deformation potential interaction, 31, 43, 110
   in rectangular quantum wire, 181
deformation potential scattering, in bulk zincblende
      structures, 172
diatomic lattice, 7

dielectric continuum model, 52, 60, 93
dielectric function, 13
dielectric polarizability, 224
dilatation, of medium, 58
dilatational modes, 99
dilatational solutions, 60
dilatational waves, 98
dimensional confinement, 39
dispersion curves, in würtzites 32, 77, 78
dispersion relations, 8, 26
displacement, 8
displacement eigenmodes, 98
displacement field, quantized, 40
dissipative mechanisms, 220
distortional solutions, 60
double-barrier heterostructure, 88
drifted Fermi distribution, 216
driven-oscillator equation, 14, 63
dynamical screening, in polar quantum wires, 162, 165

E_1 mode, 17
effective charge, 53
elastic continuum model, 56, 176
elastic continuum theory, 60
elasticity, theory of, 47
electron–acoustic-phonon scattering
   in cylindrical quantum wire, 176
   in rectangular quantum wire, 112
electron permittivity, 215
electron–phonon interaction, 30
   for slab modes, 66
electronic polarizability, 68
electrostatic boundary conditions, 66, 79, 152
energy-conserving delta function, 142
energy loss rate, 48
energy–wavevector relationship, 97
equivoluminal solutions, 60
Euler–Lagrange equations, 226
excitonic states, 27
extraordinary waves, 19

face-centered cubic lattices, 7
femtosecond lasers, 219
Fermi golden rule, 39, 47
flexural modes, 100, 123
flexural thickness acoustic modes, in quantum dot, 128
flexural waves, 98
folded acoustic modes, in semiconductor
   superlattices, 60
force equations, 59, 118
Fourier decomposition, 38
free-standing cylindrical structure, 180
free-surface boundary condition, 177
form factor, 232
Fröhlich Hamiltonian, for two-dimensional slab, 69
Fröhlich interaction, 31, 40
Fröhlich interaction Hamiltonian, 41
   for polar uniaxial crystal, 136
   for quantum box, 168
   for two-dimensional slab, 140

gain in intersubband laser, 195, 219
Grüneisen constant, 47
guided modes, 225

half-space optical phonon modes, 152
Hamiltonian for harmonic oscillator, 35
harmonic interactions, 8
harmonic modes, 45
harmonic oscillator, 35
hexagonal würtzite structures, 53
high-temperature electronics, 54
Hooke's law, 8, 56
hot electron distribution, 204
hot-phonon-bottleneck effect, 164
hot phonon decay, 3
hot phonons, in polar quantum wires, 163
Huang–Born equations, 20, 54, 221
Huang–Zhu modes, 66, 228

impurity, 28
   in-gap, 27
inhomogeneous broadening effects, 31
infrared-active modes, 20
infrared-active phonons, 138
   LO-like, 138
   TO-like, 138
interface disorder, 28
interface modes
   for optical phonons, 66, 68, 80, 83, 152
   in slab, 61
   optical phonon interaction Hamiltonian, 65
intersubband lasers, 196, 219
intersubband scattering rates, 91
intrasubband transition rates, 187
interwell phonon-assisted transition, 207
intrasubband scattering rates, 91
inversion symmetry, 43
ionic bonding, 7
irrotational solutions, 60
isotopic mass, 33
isotropic medium, 43

Kane wavefunction, 196
Klemens' channel, 2, 46, 163

Lagrangian density, 225, 230
Lamé's constants, 58, 101, 118
Landau fans, 203
lifetimes, longitudinal optical phonons, 3, 49
linear-chain model, 7
localized acoustic modes, 214
longitudinal acoustic mode, 9, 17
longitudinal electromagnetic disturbance, 66
longitudinal electromagnetic wave, 14
longitudinal optical mode, 9, 17
longitudinal solutions, 60
longitudinal sound speed, 57, 59
   in thin plate, 119
Loudon model, 14, 15, 23, 54
Lyddane–Sachs–Teller relation, 13, 15, 52, 80, 221

macroscopic theory of polar modes, 14
magnetotunneling spectra, 202

metal–semiconductor heterointerfaces, 220
metal–semiconductor structures, 165
micro-Raman techniques, 26
microscopic models for phonons, 90, 91
microtubules, 122, 218
mode normalization condition, 61
modified random-element isodisplacement model, 33

nanometer-scale mechanical structures, 218
neutron scattering measurements, 96
non-equilibrium phonons, 31, 48, 162, 163
normal-mode phonon displacement, 38
normal vibrational modes, 16
normalization condition, 81, 226

occupation number representation, 35
one-mode behavior, 32
optical phonons
    in nanostructures, 39
    in würtzite structures, 18
ordinary waves, 19

periodic boundary conditions, 52
phonon-assisted electron intersubband transition
    rates, 186
phonon-assisted transition, 207
phonon-assisted tunneling, 202
phonon-assisted tunneling peak, 203
phonon bottleneck, 50
    in quantum-well lasers, 50
phonon decay, in nitride materials, 51
phonon decay time, 164
phonon eigenstates, 38
phonon emission, 204
phonon engineering, 219
phonon-enhanced population inversion, 205
phonon lifetimes, 30
phonon linewidths, 31
phonon matrix elements, 38
phonon occupation number, 37, 38, 215
phonon operators, 38
phonon potential, 82
photocurrent transfer ratio, 204
picosecond Raman spectroscopy, 31
piezoelectric coupling, 173
piezoelectric crystal, 44
piezoelectric interaction, 43
piezoelectric interaction potential, 179, 181
piezoelectric polarization, 44, 173
piezoelectric scattering, in bulk semiconductor
    structures, 173
plasma frequency, 82
plasmon emission, 204
plasmon–phonon modes, 33
plasmons, 33
Poisson ratio, 101, 118
Poisson's equation, 227
polar-optical phonons, 40
polar semiconductors, 7
polaritons, 11
polarization charge density, 68

polarization vector, 38
population inversion, 205
power loss, 164
pure torsional mode, 120

quantum-box arrays, 220
quantum-cascade laser, 196
quantum-well laser, 2
quantum-wire superlattices, 220
quantum wires, with circular and elliptical cross
    sections, 161

Raman-active modes, 95
Raman analysis, of anharmonic phonon decay, 51
Raman experiments, at ultraviolet wavelengths, 33
Raman scattering, 4, 26
    in bulk zincblende structures, 26
    in bulk würtzite structures, 26
Raman scattering measurements, 66, 93, 95
Raman techniques, 26
Raman tensor, 26
reduced-mass density, 51
reformulated dielectric continuum model, 61
reformulated modes, 66, 225
resonant Raman studies, 95
Ridley channel, 50
rotational vector-potential solutions, 60

saturation velocity, 2
scalar potential, 59, 113
scattering rates, electron–LO-phonon
    in quantum dots, 167
    in quantum wires, 146, 150
    in würtzite quantum-well structures, 146
    in würtzite semiconductors, 131
    in zincblende quantum-well structures, 141
    in zincblende semiconductors, 131
scattering rates, interface phonon, in quantum
    wires, 154
screening of Coulomb-like interaction, 134
second-order susceptibilities, 28
shear horizontal modes, 104
shear solutions, 60
shear waves, 98, 99
short-range forces, 226
short-wavelength optoelectronic devices, 54
simple harmonic oscillator, 35
slab modes, 61, 152, 225
    in confined würtzite structures, 71
space group, 29
space-group symmetry, 28
strain, 28, 56, 103
stress, 56, 103
stress–strain relation, 57
stress tensor, 97
Sturm–Liouville equation, 227
sub-picosecond Raman spectroscopy, 31
surface charge density, 68
symmetric shear vertical modes, 102

T_2 representation, 28
ternary alloys, 32
ternary polar semiconductor, 53

thermalization time, of carriers, 50
thickness modes, in rectangular quantum wire, 107, 109
thin film superconductors, 5, 210, 219
third-order elastic coefficients, 47
III-V nitrides, 16
III-V nitride materials, 29
three-phonon decay process, 30
time-resolved Raman scattering, 31
torsional mode, in spherical quantum dot, 127, 128
traction force, 97
transfer matrix model, for multi-heterointerface structures, 79, 85
transverse acoustic mode, 17
transverse optical phonons, 11
transverse solutions, 60
transverse sound speed, 59
tunneling currents, 202
tunneling injection lasers, 50
two-mode behavior, 32

Umklapp process, 45
uniaxial crystals, 18
uniaxial materials, 53

uniaxial polar materials, 54
uniaxial semiconductors, 71
unit cell, 16, 53

valence bands, 27
valley current, 202, 219
vector potential, 59
volume charge density, 68

wavevector, 8, 35
Wendler's conditions, for two-layer system, 69
Wendler's model, 222
width modes, in rectangular quantum wire, 109
würtzite nitride system, 95
würtzite nitrides, 32
würtzite structure, 6, 16
würtzite superlattice, 76

Young's modulus, 56, 101, 118

zincblende crystals, 6
zincblende structures, 16
zone-center phonon frequency, 133